T0260939

"Have you ever wondered why it's important for everyone to understand evolution? This masterful book presents a clear and approachable picture of evolution and what it tells us about our lives and interactions with other species. Compelling examples of the ever-present impact of evolution in many, often unexpected, facets of our lives provides a broad new perspective on its meaning and relevance. If you've ever doubted that evolution matters to your life, this book will convince you that it does."
—MARVALEE H. WAKE, University of California, Berkeley

"From human evolution to the evolution of antibiotic resistance, from cultural evolution to the cultural importance of evolutionary thought, this is a wonderful and essential volume."
—SIMON LEVIN, Princeton University

"Yet another book about evolution? Yes, and thank goodness. With engaging brief essays by leading experts, this book illustrates the importance of evolution for our lives and planet so vividly that I plan to create a course to share it with students—and to have an excuse to read it again myself."
—RANDOLPH M. NESSE, M.D., director of the Arizona State University Center for Evolution and Medicine

"Evolution is everywhere, it changes everything, and it is relevant to everyone. This volume shows us how, with chapters on topics ranging from language to medicine to biodiversity. Moving beyond superficial summaries, each of the essays is a thoughtful consideration of just how much evolution matters."
—MARLENE ZUK, author of *Paleofantasy: What Evolution Really Tells Us about Sex, Diet, and How We Live*

HOW EVOLUTION SHAPES OUR LIVES

HOW EVOLUTION SHAPES OUR LIVES

Essays on Biology and Society

Jonathan B. Losos and Richard E. Lenski, editors

Princeton University Press
Princeton and Oxford

Copyright © 2016 by Princeton University Press
Published by Princeton University Press, 41 William Street, Princeton, New Jersey 08540
In the United Kingdom: Princeton University Press, 6 Oxford Street, Woodstock,
Oxfordshire OX20 1TR

press.princeton.edu
Cover images courtesy of Shutterstock

ISBN 978-0-691-17187-6

ISBN (pbk.) 978-0-691-17039-8

Library of Congress Cataloging-in-Publication Data

Names: Losos, Jonathan B., editor. | Lenski, Richard, editor.
Title: How evolution shapes our lives : essays on biology and society / [13 contributors] ;
edited by Jonathan B. Losos, Richard E. Lenski.
Description: Princeton, NJ : Princeton University Press, 2016. | Includes bibliographical
references and index.
Identifiers: LCCN 2016016195| ISBN 9780691171876 (hardback) | ISBN 9780691170398
(paperback)
Subjects: LCSH: Human evolution. | Human population genetics.
Classification: LCC GN281.4 H69 2016 | DDC 599.93/8—dc23 LC record available at
https://lccn.loc.gov/2016016195

British Library Cataloging-in-Publication Data is available

This book has been composed in Minion Pro text with Futura Std display

Printed on acid-free paper. ∞

Printed in the United States of America

10 9 8 7 6 5 4 3 2 1

CONTENTS

PREFACE

For more than 150 years, since the publication of *On the Origin of Species*, biologists have focused on understanding the evolutionary chronicle of diversification and extinction, and the underlying evolutionary processes that have produced it. Although progress in evolutionary biology has been steady since Darwin's time, developments in the last 20 years have ushered in a golden era of evolutionary study in which biologists are on the brink of answering many of the fundamental questions in the field.

These advances have come from a confluence of technological and conceptual innovations. In the laboratory, the rapid and inexpensive sequencing of large amounts of DNA is producing a wealth of data on the genomes of many species; comparisons of these genomes are allowing scientists to pinpoint the specific genetic changes that have occurred over the course of evolution. In parallel, spectacular fossil discoveries have filled many of the most critical gaps in our documentation of the evolutionary pageant, detailing how whales evolved from land-living animals, snakes from their four-legged lizard forebears, and humans from our primate ancestors. In addition, providing the data that Darwin could only imagine, field biologists are now tracking populations, directly documenting natural selection as it occurs, and monitoring the resulting evolutionary changes that occur from one generation to the next.

This volume is an abridged and updated follow-up to *The Princeton Guide to Evolution*. Published in 2013, the guide's 107 chapters provide in-depth, yet accessible, coverage of the entirety of modern evolutionary biology. In this volume, we take a more focused approach, concentrating on the interplay of evolution and modern society.

The reason is simple: even as this is a golden time for evolutionary science, it is also one in which understanding evolution has never been more important; its impact is now being felt throughout human society. Many current issues—such as the rise of new diseases, the increased resistance of pests and microorganisms to efforts to control them, and the effect of changing environmental conditions on natural populations—revolve around aspects of natural selection and evolutionary change. Many disparate areas of modern life—medicine, the legal system, computing—increasingly employ evolutionary thinking and use methods developed in evolutionary biology. Paradoxically, even as our understanding of evolution and its importance to society has never been greater, substantial proportions of the population in a number of countries—most notably the United States and Turkey—dispute the scientific findings of evolutionary biologists and resist the teaching of evolution in schools.

For these reasons, we have selected for this volume 22 chapters from *The Princeton Guide to Evolution* that have particular relevance to humans and modern society. Most of these chapters have been revised, some substantially. In addition, we have included a new chapter on evolutionary resilience to climate change. The resulting volume should be of use to anyone—scientist, student, government planner, or physician, to name just a few—interested in the impact of evolutionary change on the world around us.

This volume could not have been possible without the efforts of the authors, whose contributions were essential to such a wide-ranging and authoritative work, and the editors of *The Princeton Guide to Evolution* who helped assemble the original volume: David Baum, Douglas Futuyma, Hopi Hoekstra, Allen Moore, Catherine Peichel, Dolph Schluter, and Michael Whitlock. In addition, the editorial staff at Princeton University Press was indispensable in putting together this volume. The entire project was skillfully overseen by senior editor Alison Kalett, and day-to-day management moved smoothly and efficiently under the watchful eyes of editorial assistant Betsy Blumenthal and production editor Karen Carter.

Jonathan B. Losos and Richard E. Lenski

CONTRIBUTORS

Dan I. Andersson, Professor, Department of Medical Biochemistry and Microbiology, Uppsala University, Sweden

Francisco J. Ayala, University Professor and Donald Bren Professor of Biological Sciences, University of California, Irvine

Amy Cavanaugh, Assistant Professor of Biology, Truckee Meadows Community College

Cameron R. Currie, Ira L. Baldwin Professor of Bacteriology, University of Wisconsin–Madison

Dieter Ebert, Professor of Zoology and Evolutionary Biology, University of Basel, Switzerland

Andrew D. Ellington, Wilson and Kathryn Fraser Research Professor in Biochemistry, The University of Texas at Austin

Elizabeth Hannon, Associate Director, Forum for European Philosophy, London School of Economics

John Hawks, Associate Professor of Anthropology, University of Wisconsin–Madison

Paul Keim, Regents' Professor and Cowden Chair, Department of Biological Sciences, Northern Arizona University

Richard E. Lenski, John Hannah Distinguished Professor of Microbial Ecology, Department of Microbiology and Molecular Genetics, Michigan State University

Tim Lewens, Professor of Philosophy of Science, University of Cambridge, United Kingdom

Jonathan B. Losos, Professor, Department of Organismic and Evolutionary Biology; Curator of Herpetology, Museum of Comparative Zoology, Harvard University

Virpi Lummaa, Reader in Evolutionary Biology, Department of Animal and Plant Sciences, University of Sheffield, United Kingdom

Jacob A. Moorad, Lecturer in Quantitative Genetics, The University of Edinburgh, United Kingdom

Craig Moritz, Professor, Research School of Biology, Australian National University, Canberra

Martha M. Muñoz, Postdoctoral Fellow, Department of Biology, Duke University

Mark Pagel, FRS, Professor of Evolutionary Biology, University of Reading, United Kingdom

Talima Pearson, Assistant Director at The Center for Microbial Genetics & Genomics; Research Assistant Professor of Biological Sciences, Northern Arizona University

Robert T. Pennock, Professor, Lyman Briggs College and Departments of Philosophy, Computer Science and Engineering, and Ecology, Evolutionary Biology and Behavior, Michigan State University

Daniel E. L. Promislow, Professor, Departments of Pathology and Biology, University of Washington

Erik M. Quandt, Graduate Student in Cell and Molecular Biology, The University of Texas at Austin

David C. Queller, Spencer T. Olin Professor, Department of Biology, Washington University in St. Louis

Robert C. Richardson, Charles Phelps Taft Professor of Philosophy, University Distinguished Research Professor, University of Cincinnati

Eugenie C. Scott, Founding Executive Director, National Center for Science Education

H. Bradley Shaffer, Distinguished Professor, Department of Ecology and Evolutionary Biology, University of California, Los Angeles

Joan E. Strassmann, Charles Rebstock Professor of Biology, Washington University in St. Louis

Alan R. Templeton, Charles Rebstock Professor of Biology Emeritus, Washington University in St. Louis

Paul E. Turner, Professor, Chair of Ecology and Evolutionary Biology, Yale University

Carl Zimmer, Columnist, *New York Times*

HOW EVOLUTION SHAPES OUR LIVES

CHAPTER 1

HOW EVOLUTION SHAPES OUR LIVES

Richard E. Lenski

OUTLINE

1. Biological foundations
2. Evolution in health and disease
3. Reshaping our world
4. Evolution in the public sphere
5. Nature and nurture

Many people think of evolution as a fascinating topic, but one with little relevance to our lives in the modern world. After all, most people first encounter the idea of evolution in museums, where they see the fossilized remnants of organisms that lived long ago. Later exposure to evolution may come in courses that present the basic theory along with evidence from the tree of life and the genetic code shared by all life on earth. For those enamored of wildlife, evolution might also be discussed in programs about exotic organisms in far-away lands, often showing nature "red in tooth and claw." So it is easy to overlook the fact that evolution is important for understanding who we are, how we live, and the challenges we face.

The comic strip shown here comes from Garry Trudeau's *Doonesbury* series, and it reminds us that evolution is highly relevant to our lives and to society. In fact, it touches on several themes in this volume. The conversation between the doctor and patient reminds us that despite our efforts to control nature, we remain targets for organisms that have evolved, and continue to evolve, to exploit our bodies for their own propagation. At the same time, the cartoon emphasizes that humans have acquired another mode of response—the use of technology—that allows us to combat diseases far more quickly (and with less suffering)

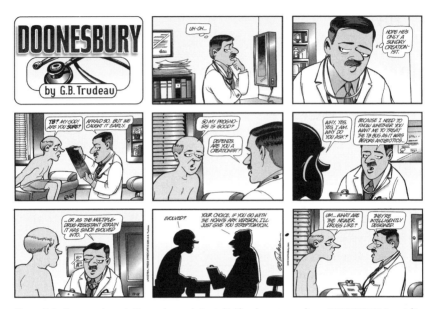

Figure 1-1. Cartoon strip reminding us that evolution is highly relevant to our lives. (DOONESBURY Copyright 2005 G. B. Trudeau. Reprinted with permission of Universal Uclick. All rights reserved.)

than if we had to rely on a genetically determined evolutionary response. More subtly, the technology, institutions, and language (including humor) that make human societies what they are today all reflect a process of cultural evolution that emerged from, and now often overwhelms, its natural counterpart by virtue of the speed and flexibility of cultural systems. Finally, Trudeau jabs us with the needle of the conflict between evolutionary science and religion that dominates many discussions of evolution in the public sphere, despite the overwhelming and continually growing body of evidence for evolution.

1. BIOLOGICAL FOUNDATIONS

To set the stage, we begin this volume with a question: What is evolution? From subtle shifts in the genetic makeup of a single population to the entire tree of life, evolution is the process by which life changes from one generation to the next and from one geological epoch to another.

The study of evolution encompasses both the historical pattern of evolution—who gave rise to whom, and when, in the tree of life—and the ecological and genetic mechanisms that underlie the evolutionary process (see chapter 2).

When Charles Darwin published *The Descent of Man* in 1871, he used the comparative method to make sense of the evolution of our species. That is, he looked to the similarities and differences in the appearance and behavior of humans and our living relatives to understand how we came into being. Some Neanderthal bones had been discovered, but otherwise there was no fossil record of hominids—the taxonomic family that includes humans and the great apes—in his time. It would be several decades before much older fossils were discovered that began to fill in some of the so-called "missing links," and the insights from DNA lay more than a century ahead. Today we have a bounty of fossil hominids, and DNA from both living organisms and fossils is providing new insights into what makes us human, where we came from, and even who mated with whom (see chapter 3).

One of the important attributes of humans is that we live in social groups that require a substantial level of cooperation—in hunting game, rearing a family, and dividing the tasks of labor seen in modern societies. But cooperation is not unique to humans, or even to our close relatives (see chapter 4). Social insects, for example, exhibit remarkable cooperation and division of labor. Within ant colonies and beehives, most individuals forgo reproduction while supporting reproduction by their queen, who is usually the sister of the nonreproductive workers. In other cases, unrelated individuals cooperate, as in mutualisms involving different species, such as the fungi and algae that together make lichens. Understanding the evolutionary forces that promote these different forms of cooperation sheds light on our own behaviors as humans, as well as on some of our commonalities with other organisms whose behaviors have been shaped by these same forces.

One approach to understanding the evolution of human behavior is to ask whether our actions are, in some sense, optimal. Do humans consume food in a manner that optimizes nutritional status? Are children born at intervals that maximize reproductive success? The recognition that there might be trade-offs—for example, between the number of children and the probability they will survive to adulthood—allows the possibility, at least, of defining optimal strategies in mathematical terms (see chapter 5). However, the conditions in which our species exists have changed greatly as a result of our ancestors' migrations across the planet, as well as technological innovations that affect food availability, life expectancy, and so on. As a consequence, there might be critical mismatches between the behaviors that were evolutionarily advantageous in the past—which continue to influence how we respond—and those that would be most beneficial at present (see chapter 6). By understanding how evolution helped shape our psychology, individuals and societies might make better decisions about how to respond to the challenges and choices we face today.

2. EVOLUTION IN HEALTH AND DISEASE

Diseases are usually studied with a focus on proximate causes. For example, what organ is having problems, and how can it be repaired? Did an infectious agent cause the problem, and if so, how can it be eliminated? But one can also ask questions about the evolutionary forces that shape disease, although this is rarely done, and that failure may leave important stones unturned (see chapter 7). For example, why might one group of people be more susceptible to a particular disease than another group? Why are some diseases more prevalent now than in the past, despite improved sanitation and increased access to food? Why do we senesce, losing our reproductive capacity as we age (see chapter 8)? And why, in particular, do human females lose the capacity to reproduce even while they are still healthy and vigorous?

Turning our attention to infectious diseases: why do some pathogens and parasites make us very sick, or even kill us, when closely related microbes are harmless? For many years the conventional wisdom was that evolution would favor those parasites and pathogens that were harmless to their hosts. If parasites killed their hosts (so the thinking went), then they would drive their hosts and themselves to extinction. From this perspective, a highly virulent parasite was seen as a transient aberration, perhaps indicative of a pathogen that had recently jumped to a new host—one that would, over time, evolve to become less virulent if it did not burn out first. But this view has been challenged by more rigorous analyses. Even lethal infections do not usually drive their hosts extinct, and the optimum virulence, from the parasite's perspective, likely depends on the balance between within-host growth and between-host transmission (see chapter 9).

The antibiotics that scientific researchers and pharmaceutical companies developed to treat bacterial infections were hailed as a triumph of technology over nature. Only a few decades ago, the most dangerous infections were largely conquered in developed countries. Schools of public health shifted their attention from infectious diseases to other threats, while the public looked forward to a cure for the common cold (along with personal jet packs). This benign outlook was shaken, however, in the 1980s by the AIDS epidemic and the discovery that it was caused by a virus. And it continues to be shaken by reports of emerging and reemerging diseases that threaten denizens of even the wealthiest nations, from the SARS virus and bird flu (the H5N1 influenza virus) to multidrug-resistant strains of dangerous bacteria including *Mycobacterium tuberculosis* and *Staphylococcus aureus*. The reemergence of these bacterial pathogens reflects the evolution of varieties resistant to some or all of the antibiotics that were previously used to treat them (see chapter 10). Thousands of tons of antibiotics are used each year, causing intense selection for bacteria that can survive and grow in their presence. As a consequence, pharmaceutical companies must spend vast sums to develop new antimicrobial compounds that will allow us, we hope, to keep up with fast-evolving microbes. Meanwhile, emerging diseases that are new to humankind typically derive from pathogens that infect other animals.

The toolbox of molecular evolution and phylogenetic methods is now widely used to determine the source of zoonotic infections and to track a pathogen's transmission through the host population based on mutations that arise as the pathogen continues to evolve during an outbreak. And if these challenges were not enough, some terrorists have deployed pathogens. Investigators must identify the precise source of the microbes deployed in an attack, using evolutionary approaches similar to those used to track natural outbreaks. The "Amerithrax" case, in which spores of *Bacillus anthracis* (the bacterium that causes anthrax) were spread via the US Postal Service, demonstrated the power of new genome sequencing methods to discover tiny genetic differences among samples that may identify relevant sources (see chapter 11).

3. RESHAPING OUR WORLD

Some 10,000 years ago, humans began to harness the power of evolution by selectively breeding various plants and animals for food, clothing materials, and transportation. However, humans were not the first species to invent agriculture. That distinction belongs to ants, some of which began cultivating fungi for food millions of years ago (see chapter 12). Some ants even tend other insects, such as aphids. Humans and other farming species change the environment of domesticated species by providing them with shelter, nutrients, and reproductive assistance. Selection in this protective environment reshapes the morphology, physiology, and behavior of the domesticated varieties. While we usually think of the farmer as controlling the domesticated species, their relationship is really a mutualism; the farmer, too, may evolve greater dependence on agriculture. For example, humans have evolved an unusual trait among mammals that enables many (but not all) adults to continue to produce lactase, an enzyme that allows milk sugar to be digested.

Agriculture is the most familiar way in which humans use evolution for practical purposes, but it is not the only way. Over the past few decades, molecular biologists have developed systems that allow populations of molecules to evolve even outside the confines of living cells (see chapter 13). For example, RNA molecules have been selected to perform new functions in vitro, such as binding to targets of interest. After a random library of sequences has been generated, sequences that have bound to the target are separated from those that have not. The former are then amplified (replicated) using biochemical methods that introduce new variants by mutation and recombination. This Darwinian process of replication, variation, and selection is repeated many times, allowing the opportunity for further improvement in binding to the target. Similar approaches allow the directed evolution of proteins, so that today RNA and protein molecules produced by directed evolution are used to treat certain diseases.

Perhaps even more remarkably, computer scientists and engineers are harnessing evolution to write code and solve complex problems (see chapter 14). They do so by implementing the processes of biological evolution—replication, variation, and selection—inside a computer. This approach has been used in biology to test hypotheses that are difficult to study in natural systems, and in engineering to facilitate the discovery of solutions to complex problems. For example, the design for an antenna used on some NASA satellites was generated not by a team of engineers but in a population of evolving programs (variant codes) that were selected based on the predicted functional properties of the objects encoded in their virtual genomes. Of course, it was necessary to build the physical objects and test them to see whether they would perform as intended, which they did.

With the success of agriculture and other technologies, the human population has grown tremendously in size and pushed into geographic areas that would otherwise be inhospitable to our species. As a result, humans have altered—by habitat destruction, introduction of nonnative species, and pollution—many environments to which other organisms have adapted and on which they depend, leading to the extinction of some species and threatening many others. Evolutionary biology contributes to conservation efforts in several ways (see chapter 15). For example, phylogenetic analyses are used to quantify branch lengths on the tree of life and determine, in effect, how much unique evolutionary history would be lost by the extinction of one species or another. Given limited resources for conservation efforts, this information can inform where those resources will have the biggest impact. Also, the mathematical framework of population genetics, which underpins evolutionary theory, is widely used in the management of endangered populations. In particular, captive-breeding programs and even the physical structure of wildlife reserves can be designed to maximize the preservation of genetic diversity and minimize the effects of inbreeding depression. However, the challenges are only growing. Through many of the activities we take for granted—the production of desirable products, our travel for work and pleasure, and the heating and air-conditioning that keep us comfortable—we are changing the earth's climate (see chapter 16). As we do so, we impose selection on many other organisms. Which species and lineages will survive, and which will go extinct because they cannot cope with the changes? And how will the survivors cope? Might they simply migrate to new habitats and locations that match their previously evolved requirements? Or will they evolve new preferences and traits that allow them to tolerate their altered world?

4. EVOLUTION IN THE PUBLIC SPHERE

Evolutionary biology attracts substantial public attention for two reasons. First, many people find it fascinating to understand how humans and other species came into being. Indeed, that question has interested people since the dawn of history, with different cultures and religions providing diverse narratives about the origins of the world and its inhabitants. Second, evolutionary biology often attracts attention because its findings are inconsistent with those narratives. The resulting tension is complex, with many different positions held by scientists and nonscientists alike (see chapter 17). Some people reject religions whose narratives are contradicted by established bodies of scientific evidence. Others emphasize the difference between evidence and faith, viewing them as separate domains of human understanding; these people may retain some religious beliefs and sensibilities while rejecting the literal interpretation of prescientific narratives. Yet other people have suggested that evolution may, in fact, illuminate theology by providing a deeper understanding of nature as it was created.

While some view evolution as an affront to their religion, evolutionary biologists often feel that their field of study is under attack, especially in the United States, where opposition to teaching evolution is a hot-button issue that generates loud and emotional responses. However, the opposition is far from unified; instead, it is a coalition of creationists expressing views that are inconsistent not only with the scientific evidence but also with the beliefs of other coalition members (see chapter 18). Efforts have been made to unify the creationist coalition and give it a veneer of credibility by hiding these differences and obscuring their religious basis under the gloss of "intelligent design." However, several US court decisions have recognized the religious nature of the opposition to evolution and they have disallowed such nonscientific ideas to be presented as scientific alternatives to evolution in public schools.

Fortunately, the exciting discoveries of evolutionary biology also receive considerable media attention (see chapter 19). The field has many gifted writers and communicators among its practitioners, and the excitement draws reporters and authors, some of whom have immersed themselves in the questions and evidence. Newspapers, books, and television once dominated the media coverage of evolution and usually exerted some quality control; but today websites, blogs, and tweets present a more complex and uneven terrain, even as they provide more opportunities than ever for the public to explore and examine the discoveries and implications of evolutionary biology.

5. NATURE AND NURTURE

How have humans become such an unusual and dominant species on earth? Agriculture, medicine, and other technological innovations are certainly key elements of this story. But the development of technologies depended on the prior emergence of other traits, including the language and culture that enable us to communicate among individuals and across generations, building on prior discoveries and allowing innovations to spread far more quickly than if they had to be hardwired into our genomes. In essence, culture provides a second, and extraordinarily powerful, way of evolving. Other organisms communicate with sounds, chemicals, or visual displays, but human language is unique in its compositional form, which allows an infinite variety of ways for ideas to be combined and expressed (see chapter 20). It is not known when our ancestors evolved language, but studies comparing the morphology and genomes of humans with our relatives (in some cases even extinct species) are providing clues as to how and when the potential for speech evolved. Of course, understanding the capacity for speech does not explain *why* it evolved, but it seems likely that the evolution of human language and social behaviors were tightly connected. In any case, once language emerged, it underwent rapid diversification, with the patterns and processes that govern linguistic evolution similar in some respects to biological evolution. Even so, there are important differences between the evolution of culture and that of biology (see chapter 21). Genes encode information about phenotypic solutions to problems that organisms encountered in the past, and that information is transmitted only from parents to offspring. By contrast, cultural information—knowledge, technology, ideas and preferences—can be disseminated broadly, and the information can accumulate within a single generation. We all obtain our genetic information in discrete bits from just two individuals (our parents), whereas we can obtain cultural information from many sources and blend that information in myriad ways. Moreover, cultural information offers the potential to plan for the future—for example, by anticipating and ameliorating changes caused by our actions—in ways that biological evolution does not.

Although technological innovation and other cultural influences dominate modern life, we are also the products of biological evolution. As a consequence, we differ from one another by virtue of our genealogical pedigrees as well as our cultural and biological environments, and these differences lead to debates about the contributions of "nature" and "nurture" to various attributes. The most obvious differences among individuals (in terms of being quickly perceived) are the varied colors of our skin and hair. Based on these superficial differences, people are then categorized into different races. Are these races biologically meaningful, or are they cultural constructs? The word *race* has a particular meaning in biology, corresponding to a genetically distinct lineage within a species (see chapter 22). Indeed, one can quantify the amount and distribution of individual variation within and among populations of the same species. For example, chimpanzees were split into four races, or subspecies, based on morphological differences, whereas genetic analyses indicate there are only three chimpanzee races. Even so, these racial differences account for about 30 percent of the genetic variation in chimpanzees. By contrast, studies of human genetic diversity do not support the existence of biologically defined races. Although differences in ancestry can be detected and associated with geography, "races" account for only a few percent of the variation among humans. Moreover, even those small differences indicate a history of genetic admixture as populations spread around the world. Still, humans have adapted to their local environments; variation in pigmentation, for example, likely reflects compromises between avoiding damage to cells caused by UV radiation and producing vitamin D, which requires UV radiation. However, local adaptation is not equivalent to biological races owing to the few genes involved and the history of population admixture.

And what does the future hold for human evolution (see chapter 23)? Has our biological evolution stopped now that cultural evolution, including agriculture and medicine, has become so powerful? In fact, the opposite may sometimes be true. As agriculture spread across the globe, so, too, did the standing water that mosquitoes need to breed; and with mosquitoes came malaria, leading to recent evolution in some human populations by favoring mutations that confer disease resistance. Medicine, too, may promote evolution by relaxing constraints. The large heads that hold the brains that give our species the capacity to communicate and innovate pose a severe risk during childbirth, one that has caused the deaths of countless mothers and infants. Will our species evolve even larger brains and greater intelligence as cesarean births become increasingly common? As Yogi Berra said, "It's tough to make predictions, especially about the future." But one prediction seems safe, namely, that the small genetic differences among populations will eventually disappear as a consequence of the increased migration our technologies allow—that is, unless the colonization of other planets, or some catastrophe here on earth, produces new barriers to migration and gene flow.

PART I

Biological Foundations

CHAPTER 2

WHAT IS EVOLUTION?

Jonathan B. Losos

OUTLINE

1. What is evolution?
2. Evolution: Pattern versus process
3. Evolution: More than changes in the gene pool
4. In the light of evolution
5. Critiques and the evidence for evolution
6. The pace of evolution
7. Evolution, humans, and society

Evolution refers to change through time as species become modified and diverge to produce multiple descendant species. *Evolution* and *natural selection* are often conflated, but evolution is the historical occurrence of change, and natural selection is one mechanism—in most cases the most important—that can cause it. Recent years have seen a flowering in the field of evolutionary biology, and much has been learned about the causes and consequences of evolution. The two main pillars of our knowledge of evolution come from knowledge of the historical record of evolutionary change, deduced directly from the fossil record and inferred from examination of phylogeny, and from study of the process of evolutionary change, particularly the effect of natural selection. It is now apparent that when selection is strong, evolution can proceed considerably more rapidly than was generally envisioned by Darwin. As a result, scientists are realizing that it is possible to conduct evolutionary experiments in real time. Recent developments in many areas, including molecular and developmental biology, have greatly expanded our knowledge and reaffirmed evolution's central place in the understanding of biological diversity.

GLOSSARY

Evolution. Descent with modification; transformation of species through time, including both changes that occur within species, as well as the origin of new species.

Natural Selection. The process in which individuals with a particular trait tend to leave more offspring in the next generation than do individuals with a different trait.

Approximately 375 million years ago, a large and vaguely salamander-like creature plodded from its aquatic home and began the vertebrate invasion of land, setting forth the chain of evolutionary events that led to the birds that fill our skies, the beasts that walk our soil, me writing this chapter, and you reading it. This was, of course, just one episode in life's saga: millions of years earlier, plants had come ashore, followed soon thereafter—or perhaps simultaneously—by arthropods. We could go back much earlier, 4 billion years or so, to that fateful day when the first molecule replicated itself, an important milestone in the origin of life and the beginning of the evolutionary pageant. Moving forward, the last few hundred million years have also had their highs and lows: the origins of frogs and trees, the end-Permian extinction when 90 percent of all species perished, and the rise and fall of the dinosaurs.

These vignettes are a few of many waypoints in the evolutionary chronicle of life on earth. Evolutionary biologists try to understand this history, explaining how and why life has taken its particular path. But the study of evolution involves more than looking backward to try to understand the past. Evolution is an ongoing process, one possibly operating at a faster rate now than in times past in this human-dominated world. Consequently, evolutionary biology is also forward looking: it includes the study of evolutionary processes in action today—how they operate, what they produce—as well as investigation of how evolution is likely to proceed in the future. Moreover, evolutionary biology is not solely an academic matter; evolution affects humans in many ways, from coping with the emergence of agricultural pests and disease-causing organisms to understanding the workings of our own genome. Indeed, evolutionary science has broad relevance, playing an important role in advances in many areas, from computer programming to medicine to engineering.

1. WHAT IS EVOLUTION?

Look up the word *evolution* in the online version of the *Oxford English Dictionary*, and you will find 11 definitions and numerous subdefinitions, ranging from mathematical ("the successive transformation of a curve by the alteration of the conditions which define it") to chemical ("the emission or release of gas, heat, light, etc.") to military ("a manoeuvre executed by troops or ships to adopt a different tactical formation"). Even with reference to biology, there are several definitions, including "emergence or release from an envelope or enclosing structure; (also) protrusion, evagination," not to mention "rare" and "historical" usage related to the concept of preformation of embryos. Even among evolutionary biologists, evolution is defined in different ways. For example, one widely read textbook refers to evolution as "changes in the properties of groups of organisms over the course of generations" (Futuyma 2005), whereas another defines it as "changes in allele frequencies over time" (Freeman and Herron 2007).

One might think that—as in so many other areas of evolutionary biology—we could look to Darwin for clarity. But in the first edition of *On the Origin of Species*, the term "evolution" never appears (though the last word of the book is "evolved"); not until the sixth edition does Darwin use "evolution." Rather, Darwin's term of choice is "descent with modification," a simple phrase that captures the essence of what evolutionary biology is all about: the study of the transformation of species through time, including both changes that occur within species, as well as the origin of new species.

2. EVOLUTION: PATTERN VERSUS PROCESS

Many people—sometimes even biologists—equate evolution with natural selection, but the two are not the same. Natural selection is one process that can cause evolutionary change, but natural selection can occur without producing evolutionary change. Conversely, processes other than natural selection can lead to evolution.

Natural selection within populations refers to the situation in which individuals with one variant of a trait (say, blue eyes) tend to leave more offspring that are healthy and fertile in the next generation than do individuals with an alternative variant of the trait. Such selection can occur in many ways, for example, if the variant leads to greater longevity, greater attractiveness to members of the other sex, or greater number of offspring per breeding event. The logic behind natural selection is unassailable. If some variant of a trait is causally related to greater reproductive success, then more members of the population will have that variant in the next generation; continued over many generations, such selection can greatly change the constitution of a population.

But there is a catch. Natural selection can occur without leading to evolution if differences among individuals are not genetically based. For natural selection to cause evolutionary change, trait variants must be transmitted from parent to offspring; if that is the case, then offspring will resemble their parents, and the trait variants possessed by the parents that produce the most offspring will increase in frequency in the next generation.

However, offspring do not always resemble their parents. In some cases, individuals vary phenotypically not because they are different genetically, but because they experienced different environments during growth (this is the "nurture" part of the nature versus nurture debate). If, in fact, variation in a population is not genetically based, then selection will have no evolutionary consequence; individuals surviving and producing many offspring will not differ genetically from those that fail to prosper, and as a result, the gene pool of the population will not change. Nonetheless, much of the phenotypic variation within a population is, in fact, genetically based; consequently, natural selection often does lead to evolutionary change.

But that does not mean that the occurrence of evolutionary change necessarily implies the action of natural selection. Other processes—especially mutation, genetic drift, and immigration of individuals with different genetic constitutions—also can cause a change in the genetic makeup of a population from one generation to the next. In other words, natural selection can cause adaptive evolutionary change, but not all evolution is adaptive.

These caveats notwithstanding, 150 years of research have made clear that natural selection is a powerful force responsible for much of the significant evolutionary change that has occurred over the history of life. Natural selection can operate in many ways, and scientists have correspondingly devised many methods to detect it, both through studies of the phenotype and of DNA itself.

3. EVOLUTION: MORE THAN CHANGES IN THE GENE POOL

During the heyday of population genetics in the middle decades of the last century, many biologists equated evolution with changes from one generation to the next in gene frequencies (*gene frequency* refers to the frequencies of different alleles of a gene). The "Modern Synthesis" of the 1930s and 1940s led to several decades in which the field was primarily concerned with the genetics of populations with an emphasis on natural selection. This focus was sharpened by the advent of molecular approaches to studying evolution. Starting in 1960 with the application of enzyme electrophoresis techniques, biologists could, for the first time, directly assess the extent of genetic variation within populations. To everyone's surprise, populations were found to contain much more variation than expected. This finding challenged the view that natural selection was the dominant force guiding evolutionary change, and further directed attention to the genetics of populations. With more advanced molecular techniques available today, the situation has not changed. There is much more variation than we first suspected.

The last 35 years have seen a broadening of evolutionary inquiry as the field has recognized that there is more to understanding evolutionary change than studying what happens to genes within populations— though this area remains a critically important part of evolutionary inquiry. Three aspects of expansion in evolutionary thinking are particularly important.

First, phenotypic evolution results from evolutionary change in the developmental process that transforms a single-celled fertilized egg into an adult organism. Although under genetic control, development is an intricate process that cannot be understood by examination of DNA sequences alone. Rather, understanding how phenotypes evolve, and the extent to which developmental systems constrain and direct evolutionary change, requires detailed molecular and embryological knowledge.

Second, history is integral to understanding evolution. The study of fossils—paleontology— provides the primary, almost exclusive, direct evidence of life in the past. Somewhat moribund in the middle of the last century, paleontology has experienced a resurgence in recent decades owing both to dramatic new discoveries stemming from an upsurge in paleontological exploration and to new ideas about evolution inspired by and primarily testable with fossil data, such as theories concerning punctuated equilibrium and stasis, species selection, and mass extinction. Initially critical in the development and acceptance of evolutionary theory, paleontology has once again become an important and vibrant part of evolutionary biology.

Concurrently, a more fundamental revolution emphasizing the historical perspective has taken place over the last 30 years with the realization that information on phylogenetic relationships—that is, the *tree of life*, the pattern of descent and relationship among species—is critical in interpreting all aspects of evolution above the population level. Beginning with a transformation in the field of systematics concerning how phylogenetic relationships are inferred, this "tree-thinking" approach now guides study not only of all aspects of macroevolution but also of many population-level phenomena.

Finally, life is hierarchically organized. Genes are located within individuals, individuals within populations, populations within species, and species within clades (a clade consists of an ancestral species and all its descendants). Population genetics concerns what happens among individuals within a population, but evolutionary change can occur at all levels. For example, why are there more than 2000 species of rodents but only 5 species of monotremes (the platypus and echidnas), a much older clade of mammals? One cannot look at questions concerning natural selection within a population to answer this question. Rather, one must inquire about properties of entire species. Is there some attribute of rodents that makes them particularly prone to speciate or to avoid extinction? Similarly, why is there so much seemingly useless noncoding DNA in the genomes of many species? One possibility is that some genes are particularly adept at mutating to multiply the number of copies of that gene within a genome; such DNA might increase in frequency in the genome even if such multiplication has no benefit to the individual in whose body the DNA resides. Just as selection among individual organisms on heritable traits can lead to evolutionary change within populations, selection among entities at other levels (species, genes) can also lead to evolutionary change, as long as those entities have traits that are transmitted to their offspring (be they descendant species or genes) and affect the number of descendants they produce. The upshot is that evolution occurs at multiple levels of the hierarchy of life; to understand its rich complexity, we must study evolution at these distinct levels, as well as the interactions among them. What happens, for example, when a trait that benefits an individual within a population (perhaps cannibalism—more food, fewer competitors!) has detrimental effects at the level of species?

Although evolutionary biology has expanded in scope, genetic change is still its fundamental foundation. Nonetheless, in recent years attention has focused on variation that is not genetically based. Phenotypic plasticity—the ability of a single genotype to produce different phenotypes when exposed to different environments—can itself be adaptive. If individuals in a population are likely to experience different conditions as they develop, then the evolution of a genotype that could produce appropriate phenotypes depending on circumstances would be advantageous. Although selection on these different phenotypes would not lead to evolutionary change, the degree of plasticity itself can evolve if differences in extent of plasticity lead to differences in the number of surviving offspring. Indeed, an open question is, why don't populations evolve to become infinitely malleable, capable of producing the appropriate phenotype for any environment? Presumably, plasticity has an associated cost such that adaptation to different environments often occurs by genetic differentiation rather than by the evolution of a single genotype that can produce different phenotypes. Such costs, however, have proven difficult to demonstrate.

Differences observed among populations may also reflect plastic responses to different environmental conditions and thus may not reflect genetic differentiation. However, if consistently transmitted from one generation to the next, such nongenetic differences may lead to divergent selective pressures on traits that are genetically determined, thus promoting evolutionary divergence between the populations. One particular example concerns behavior, which is highly variable in response to the environment—an extreme manifestation of plasticity (see chapter 21). Learned behaviors that are transmitted from one generation to the next—often called *traditions* or *culture*—occur not only in humans but in other animals, not only our near relatives the apes, but also cetaceans, birds, and others. Such behavioral differences among populations would not reflect genetic differentiation, but they might set the stage for genetic divergence in traits relating to the behaviors. One can easily envision, for example, how chimpanzee populations that use different tools—such as delicate twigs to probe termite mounds, or heavy stones to pound nuts—might evolve different morphological features to enhance the effectiveness of these behaviors. A concrete example involves human populations that tend cattle—surely a nongenetically based behavior—and have evolved genetic changes to permit the digestion of milk in adults.

4. IN THE LIGHT OF EVOLUTION

In a 1964 address to the American Society of Zoologists, the distinguished Russian-born biologist Theodosius Dobzhansky proclaimed that "nothing makes sense in biology except in the light of evolution." Ever since, evolutionary biologists have trotted out this phrase (or some permutation of it) to emphasize the centrality of evolution in understanding the biological world. Nonetheless, for much of the twentieth century, the pervasive importance of an evolutionary perspective was not at all obvious to many biologists, some of whom considered Dobzhansky's claim to be self-serving hype. One could argue, for example, that the enormous growth in our understanding of molecular biology from 1950 to 2000 was made with little involvement or insight from evolutionary biology. Indeed, to the practicing molecular biologist in the 1980s and 1990s, evolutionary biology was mostly irrelevant.

Now, nothing could be further from the truth. When results of the human genome sequencing project first appeared in 2000, many initially believed that a thorough understanding of human biology would soon follow, answering questions about the genetic basis of human diseases and phenotypic variation among individuals. These hopes were quickly dashed—the genetic code, after all, is nothing more than a long list of letters (A, C, G, and T—the abbreviations of the four nucleotide building blocks of DNA). Much of the genome of many species seems to have no function and is just, in some sense, functionless filler; as a result, picking out where the genes lie in this 4 billion–long string of alphabet spaghetti, much less figuring out how these genes function, is not easy.

So where did molecular biologists turn? To the field of evolutionary biology! Genomicists soon realized that the best way to understand the human genome was to study it in the context of its evolutionary history, by comparing human sequences with those of other species in a phylogenetic framework. One method for locating genes, for example, is to examine comparable parts of the genome of different species. The underlying rationale is that genes evolve more slowly than other parts of the genome. Specifically, nonfunctioning stretches of DNA tend to evolve differences through time as random mutations become established (the process of genetic drift), but functioning genes tend to diverge less, because natural selection removes deleterious mutations when they arise, keeping the DNA sequence similar among species. As a result, examination of the amount of divergence between two species relative to the amount of time since they shared a common ancestor can pinpoint stretches of DNA where evolution has occurred slowly, thus identifying the position of functional genes. Moreover, how a gene functions can often be deduced by comparing its function with that of homologous genes in other species and by using a phylogeny to reconstruct the gene's evolutionary history.

And thus was born the effort to sequence the genomes of other species. At first, the nascent field of comparative genomics focused on primates and model laboratory species such as mice and fruit flies, the former to permit comparisons of the human genome with that of our close evolutionary relatives, the latter to take advantage of the great understanding of the genomic systems of well-studied species.

More recently, the phylogenetic scope has broadened as it has become evident that useful knowledge can be gained by examining genomes across the tree of life—knowledge of the genetic causes of Parkinson's disease in humans, for example, can be gained from studying the comparable gene in fruit flies, and much of relevance to humans can be learned from understanding the genetic basis of differences among dog breeds.

Dobzhansky would not have been surprised. Evolutionary biology turns out to be integral to understanding the workings of DNA and the genome, just as it is key to understanding so many other aspects of our biological world.

5. CRITIQUES AND THE EVIDENCE FOR EVOLUTION

Unique among the sciences, evolutionary biology's foundation—that species evolve through time—is not accepted by a considerable number of nonscientists, especially in the United States, Turkey, and a few other countries. Public opinion polls repeatedly reveal that most Americans are either unsure about or do not believe in evolution. One yearly poll conducted for more than 30 years, for example, consistently finds that about 40 percent of the US population believes that God created humans in their present form in the recent past.

Yet, the scientific data for evolution is overwhelming. Just like the composition and structure of genomes, many other biological phenomena are explicable only in an evolutionary context. Why, if evolution had not occurred, would seagoing manatees, descended from four-legged land dwellers, have toenails on their flippers? Why would cave fish and crickets have eyes that are missing some parts and could not function even if there were light? Why do human fetuses develop, and then lose, fur and a tail? All these, and many other phenomena, are easily understood as a result of the evolutionary heritage of species, but are inexplicable in the absence of evolution.

The case for evolution is built on two additional pillars. First is the fossil record, which documents both the major and minor transitions in the history of life; each year, exciting new discoveries further narrow the gaps in our understanding of life's chronicle. Second is our understanding of evolutionary processes—in particular, natural selection, the primary driver of evolutionary divergence. Studies in the laboratory and in human-directed selective breeding clearly demonstrate the efficacy of selection in driving substantial genetic and phenotypic divergence; one need look no further than the enormous diversity of dog breeds to appreciate the power of sustained selection. Moreover, scientists are increasingly documenting the occurrence of natural selection in nature and its ability to transform species, sometimes over quite short periods of time.

The public debate is ironic given that manifestation of evolution has so many important societal consequences (see chapter 7). Evolutionary adaptation of disease-causing organisms has rendered many drugs ineffective, leading to a huge public health toll as diseases thought to have been vanquished have reemerged as deadly scourges (see chapter 10). A recent example is the evolution of resistance to antibiotics in the bacterium *Staphylococcus aureus,* which leads to more than 100,000 infections and 19,000 fatalities a year in the United States. A similar story exists about insect pest species that devour our crops and spread diseases. In the United States alone, the evolution of pesticide resistance results in agricultural losses totaling between $3 billion and $8 billion per year. Perhaps most scary is the realization that the human population is an enormous resource to many organisms and that natural selection continually pushes these species to become more adept at making use of this potential bonanza. Ebola, AIDS, influenza—all are diseases caused by viruses that adapt to take advantage of us; a particularly worrisome concern is that some form of avian flu could evolve to become more virulent to or transmissible between humans, with the potential to produce a pandemic that could kill millions (see chapter 9). All these problems are the result of evolutionary phenomena, and all are studied using the tools of evolutionary biology.

6. THE PACE OF EVOLUTION

For more than a century after the publication of *On the Origin of Species,* biologists thought that evolution usually proceeded slowly. To a large extent, this thinking was a result of Darwin's writing—"We see nothing of these slow changes in progress, until the hand of time has marked the long lapse of ages" (Darwin 1859, chap. 4). Darwin was, after all, right about so many things, big and small, from accurately deducing the manner in which coral atolls form to correctly predicting the existence of an unknown moth with a 12-inch proboscis from the morphology of a Malagasy orchid. Hence, biologists have learned that it doesn't generally pay to disagree with what Darwin said.

Nonetheless, Darwin was not right about everything. One major mistake was the mechanism of heredity—not surprising, as Mendel's work was unknown to him, and the discovery that DNA is the genetic material was still a century in the future. A second error concerned the pace at which evolution occurs. Darwin expected that natural selection would be weak and consequently that evolutionary change would happen slowly, taking many thousands or millions of years to cause detectable change. Of course, in his day there were no actual data underlying this conclusion. Rather, this expectation sprang from Darwin's appreciation of the view promulgated by his mentor, the geologist Charles Lyell, that the slow accumulation of changes caused by weak forces would lead in the fullness of geologic time to major changes. This position, of course, was in agreement with the prevailing Victorian wisdom about the slow and gradual manner in which change occurs—or should occur—in both nature and human civilization.

Darwin's view influenced evolutionary biologists for more than a century—well into the 1970s, most thought that evolution usually occurred at a snail's pace. Spurred by the results of long-term field studies of natural selection that began in earnest around that time, we now know that Darwin was far off the mark. Many studies now clearly indicate that selection in nature is often strong and, as a result, evolutionary change often occurs very rapidly.

One important consequence of this realization is that we can observe evolution in real time. Pioneered by the study of Galápagos finches by Peter and Rosemary Grant, who documented rapid evolutionary change in these birds from one generation to the next in response to weather-induced environmental changes, the study of real-time evolutionary change in nature has become a cottage industry, with hundreds, or perhaps now thousands, of well-documented examples. This work not only clearly demonstrates the occurrence of evolution, but also provides great insights into the processes that cause it (usually, but not always, natural selection).

Perhaps most exciting, the rapidity with which evolution can occur has opened the door to experiments in which researchers can alter environmental conditions and test evolutionary hypotheses over a several-year period. Work at the forefront in this area involved studies on the color of guppies in Trinidad. Observing that the fish were generally much more colorful when they occurred in streams without predators, John Endler moved some fish from a stream with predators to a nearby area lacking them; very quickly, the population evolved exuberant coloration, apparently a result of a female preference for brighter males, which, left unchecked by the absence of predators, led to rapid evolution over 14 generations. Subsequent studies have shown that the guppies freed from predation evolve many other differences, such as in growth and reproductive rates. Many similar studies are now ongoing, and it is a safe prediction that field experiments will be an important tool for understanding evolutionary processes in the future.

7. EVOLUTION, HUMANS, AND SOCIETY

Evolution has important implications for humans in a number of ways. Some have already been discussed: humans have used evolutionary principles to alter many species to our own ends (see chapter 12); conversely, wild species are responding to human-caused changes in the environment, adapting to our efforts to control them and responding to new opportunities (see chapter 16). Consequently, it's no surprise that knowledge of evolution is important for efforts to improve artificial selection and combat our evolutionary foes. What is more surprising, perhaps, is the diversity of areas in which an understanding of evolutionary processes is relevant to human society. These include not only medicine (see chapters 7 and 9), biodiversity conservation (see chapter 15), and criminal forensics (see chapter 11), but also important human pursuits such as creating new molecules in the laboratory (see chapter 13) and devising algorithms to solve analytically intractable problems (see chapter 14).

Beyond purely utilitarian functions, an understanding of evolution can tell us much about ourselves: where we came from and where we may be going, perhaps even shedding light on what it means to be human. In recent years, a series of important fossil discoveries have brought into focus many aspects of the human evolutionary story, from our early primate roots to our recent past. Sequencing of the genomes of humans past and present and of our close primate relatives has complemented these findings in important ways and in some cases has led to unexpected discoveries, such as evidence of lineages, like the Denisovans, for which few fossil data exist (see chapter 3).

But what about our evolutionary future? When I was a boy, the public service television station ran short filler promos speculating that in the future, humans would have a bulbous, brain-packed head with tiny eyes and nostrils. Where this idea came from I have no idea, but it probably represented a mixture of orthogenetic thinking—human evolution has been marked by rapid increase in brain size and so must continue in that direction—with a misguided notion that evolution equals progress, and because intelligence is the hallmark of the human species, it would surely continue to evolve into the future. Even then, I could sense that something was not quite right about this prediction, and today, in fact, many believe that human evolution has ended because selection no longer operates on phenotypic traits: not only has medical care ameliorated the negative consequences of many genetic traits, but human cultural practices such as birth control may have severed the positive link between beneficial traits (e.g., physical strength, intelligence) and reproductive output.

Although these points have validity, they are not absolute. In much of the developing world, selective agents such as malaria can still exert strong selective pressure in the absence of adequate medical care. Moreover, new diseases, such as AIDS, for which, at least initially, no treatment exists, continue to emerge and may impose selection on populations in all parts of the world. Even in the developed world, evidence suggests that some genetically based traits are correlated with survival and reproductive success and thus that natural selection is still leading to evolutionary change (see chapters 5 and 23). Finally, natural selection is only one of several evolutionary processes. Surely, the increased mobility of humans is increasing the homogenizing effects of gene flow and diminishing the diversifying effects of genetic drift that acts in small and isolated populations. Human populations never existed as discretely identifiable genetic "races" (see chapter 22), but ongoing genetic exchange is diminishing the geographic variation that was the result of our past evolutionary history (see chapter 23).

Although selection has been important in shaping human evolution, that does not mean natural selection can explain all aspects of the human condition. Many human traits—our large brain, altruistic behavior, keen sense of smell—may have evolved as adaptations, but others may represent phenotypic plasticity or may have evolved for nonadaptive reasons. The field of evolutionary psychology focuses particularly on human behavior and is very controversial; some see in most human behavior evidence for adaptation to conditions past or present, but others are more skeptical (see chapter 6).

Many look to evolution to help address issues about what it means to be human. Those questions are primarily in the realm of philosophy, rather than evolutionary biology, and for the most part do not fall within the purview of this book or this chapter. Nonetheless, I will end with two observations. First, recent advances make clear that plants and animals occupy only a small part of the evolutionary tree of life; a great variety of microbial species constitute most of life's diversity. As a result, the human species is just one of millions of tiny branches on the evolutionary tree, and these microbial species are as well adapted to their ecological niches as we are to ours. It is easy for humans to view life's history anthropocentrically as a great evolutionary progression leading ultimately to us, but microbial species adapted to a great diversity of extreme environments—Yellowstone's hot springs, deep-sea hydrothermal vents—might see things differently. Second, the dinosaurs—members of the class Reptilia—dominated the earth for more than 150 million years. For most of that time, they cohabited with our mammalian ancestors, who were generally small-bodied, minor players in Mesozoic ecosystems. Conventional wisdom has it that our mammal ancestors, thanks to their large brains and warm-blooded physiology, outcompeted dinosaurs, and ultimately would have displaced them. However, evidence for this view is slender; right before the end of their reign, dinosaurs were thriving and showed no evidence of being pushed out by mammals. It is thought provoking to contemplate what the world would be like—where we would be today—had an asteroid not slammed into the earth 65.3 million years ago, wiping out the dinosaurs and clearing the way for the evolutionary diversification of mammals, including our own species.

FURTHER READING

Coyne, J. A. 2009. Why Evolution Is True. New York: Viking. *An excellent review of the evidence that evolution occurs and that natural selection is its primary cause.*

Darwin, C. 1859. On the Origin of Species by Means of Natural Selection, or the Preservation of Favoured Races in the Struggle for Life. London: John Murray.

Dawkins, R. 2009. The Greatest Show on Earth: The Evidence for Evolution. New York: Free Press. *Another excellent review of the evidence for evolution.*

Futuyma, D. J. 2013. Evolution. 3rd ed. Sunderland, MA: Sinauer. *One of the best textbooks on evolutionary biology for majors-level college courses.*

Grant, P. R., and R. Grant. 2008. How and Why Species Multiply: The Radiation of Darwin's Finches. Princeton, NJ: Princeton University Press. *A brilliant exposition of one of the most influential research programs in evolutionary biology, the study of Darwin's finches in the Galápagos Islands.*

Reznick, D. N. 2009. The "Origin" Then and Now: An Interpretive Guide to the "Origin of Species." Princeton, NJ: Princeton University Press. *A detailed explanation of Darwin's magnum opus in the light of twenty-first-century science.*

Zimmer, C. 2009. The Tangled Bank: An Introduction to Evolution. Greenwood Village, CO: Roberts & Company. *An excellent evolution textbook for non-majors biology courses.*

CHAPTER 3

HUMAN EVOLUTION

John Hawks

OUTLINE

1. Origin of the hominins
2. Early *Homo*
3. Neanderthals and the origin of modern humans
4. Recent human evolution

Living humans are the sole living representatives of a lineage, the *hominins*, that diverged from other living apes 6 million to 8 million years ago. Hominins remained limited to Africa for two-thirds of their history. With chimpanzee-sized bodies and brains, early hominins diversified into several lineages with different dietary strategies. One of these found a path toward technology, food sharing, and hunting and gathering, giving rise to our genus, *Homo*, between 3 million and 2 million years ago. As populations of *Homo* spread throughout the world, they gave rise to regional populations with their own anatomical and genetic distinctiveness. Within the last 100,000 years, a massive dispersal of humans from Africa absorbed and replaced these preexisting populations. In the time since this latest emergence from Africa, humans have continued to disperse, interact, and evolve. The rise of agricultural subsistence shifted human ecology, fueling further evolution.

GLOSSARY

Acheulean. A style of stone tool manufacture associated with early humans (*Homo erectus*) during the Lower Stone Age era across Africa and Eurasia. Acheulean technology is derived from the older, Oldowan technology and is a progenitor of the more complex stone tools that characterize the Middle Stone Age.

Australopithecines. Members of the hominin clade with the bipedal gait and dentition of modern humans, but lacking the enlarged brains of the genus *Homo*. Australopithecine species have been assigned to a diversity of genera, but most are now included within *Australopithecus*.

Hominins. Modern humans and extinct species more closely related to humans than to chimpanzees or gorillas.

Oldowan. The earliest stone tool industry, which emerged about 2.6 million years ago and persisted until about 1.6 million years ago, when it was replaced by the more sophisticated Acheulean technology.

Orthograde. An upright posture associated with a bipedal gait, such as occurs in modern humans.

Pronograde. The posture of holding the body parallel to the ground, such as is typical of most quadrupedal vertebrates.

While Darwin avoided discussion of the evolution of humans in *On the Origin of Species*, he soon tackled the issue in *The Descent of Man*, which defined the starting point for modern evolutionary anthropology. In the nineteenth and early twentieth centuries, the main theme of anthropology was a perceived lack of fossil progenitors, prompting a much-hyped search for a "missing link." Gradually, this concern diminished as paleoanthropologists, especially over the last half century, succeeded in uncovering thousands of fossil specimens, representing diverse human ancestors and collateral relatives. While questions still remain, these fossil data provide a rich history of the origin of many of humanity's distinctive physical traits. Furthermore, archaeological finds have provided information on the behavior of hominins during the latter half of human evolution, giving details about diet and social organization. Today, geneticists can add evidence from whole-genome comparisons of living humans, other primates, and some ancient hominins. Through all these lines of evidence a remarkably clear picture of human evolution is now emerging.

We can roughly consider human evolution in three parts. The first, from 7 million up to around 4 million years ago, saw the origination of the hominin lineage and the initial appearance of our bipedal pattern of locomotion. The second, from 4 million up to around 1.8 million years ago, was the age of the australopithecines. This group of species had a stable set of adaptations in body size and locomotion but showed substantial dietary and geographic diversity. Our own genus, *Homo,* arose from australopithecine ancestors during this period and by 1.8 million years ago had begun to spread out of Africa into the rest of the world. The spread of *Homo* throughout the world, along with many later dispersals and population expansions within and outside Africa, laid the foundation for today's human populations.

1. ORIGIN OF THE HOMININS

Chimpanzees and bonobos are our closest relatives among living primates. Whole-genome comparisons suggest that our common ancestors with these apes lived between 5 million and 8 million years ago. Our common ancestors with gorillas lived a bit earlier, within the last 10 million years, and with orangutans even earlier, before 12 million years ago. Hence, it is during the period between 10 million and 4 million years ago that paleontologists look for the immediate precursors of the human lineage. The fossil forms more closely related to living humans than to living bonobos or chimpanzees are placed within the biological tribe Homininae and are called *fossil hominins.*

Although genetic comparisons have produced good evidence of when our common ancestors with chimpanzees and bonobos lived, paleontologists have not yet identified these ancestors from any fossil evidence. A diverse record of fossil apes has been recovered from the Miocene geologic epoch, which lasted from 23 million to 5.2 million years ago. Before 15 million years ago, all known apes lived in Afro-Arabia. Early in the Middle Miocene, some apes dispersed into Asia and Europe, including the Asian ancestors of orangutans. Miocene apes varied extensively in body size and adaptive niche, and evolved a diversity of locomotor strategies. The ancestral apes were *pronograde* quadrupeds, similar to living Old World monkeys such as macaques and baboons, including several forms of early African apes as well as some early relatives of orangutans found in South Asia. The living apes, including gibbons, orangutans, chimpanzees, and gorillas, share adaptations for suspending their body below branches, with long arms relative to their legs, a more mobile shoulder joint with a long clavicle, and a spine and hips that support a broader range of postures and vertical climbing, when compared with Old World monkeys. Although they are all well adapted to below-branch suspension and vertical climbing, chimpanzees, bonobos, and gorillas are also highly terrestrial. Knuckle-walking in these primates is a solution to the trade-off between effective suspensory climbing and quadrupedal gait on the ground. Most specialists today accept that the great apes evolved convergently toward their current body plan, so that our common ancestors with chimpanzees and bonobos may not have had the extremely long arms and reliance on knuckle-walking found in these living apes. Convergent locomotor evolution seems to be a recurrent feature of ape evolution, making it challenging to establish which fossil apes may have been closely related to the living African apes and to humans.

Humans are obligate bipeds, with pelvis, foot, and vertebral adaptations that impede effective quadrupedal gait and climbing. All living apes can move bipedally, and some Miocene apes such as *Oreopithecus* may even have specialized their foot anatomy toward terrestrial bipedality. But clear evidence of our own pattern of obligate bipedality has been found in fossils only from later than 4.2 million years ago, well after the hominin clade diverged from the chimpanzee-bonobo clade.

Several fossils from different parts of Africa are candidates as early hominins, recognized by fossil evidence of more orthograde posture and by dental similarities to later hominin forms. The dental features include small canine teeth, low-crowned molar teeth, and thick molar enamel. *Sahelanthropus tchadensis* from north central Africa is the earliest known, from around 7 million years ago. Represented by a nearly complete skull and jaw, it shows an orthograde placement of the skull atop the spinal column. *Orrorin tugenensis*, from western Kenya dating to 6 million years ago, also has a femur consistent with bipedal weight bearing. *Ardipithecus kadabba*, 5.5 million years old from Ethiopia, combines hominin-like teeth with a toe bone, suggesting that the toe generated force during bipedal walking, as occurs in modern humans. It remains unclear whether these fossil taxa lived before or after the divergence of the human and chimpanzee lineages, and if after, whether they are on the human or chimpanzee side of this evolutionary split.

The best fossil evidence for a prebipedal hominin is *Ardipithecus ramidus*, dating to 4.4 million years ago from Ethiopia, including a large sample and one partial skeleton. From its limb proportions, grasping feet, and apelike hands, *Ardipithecus* was a habitual quadruped that also had good climbing abilities. But several of its features are similar to those of hominins, including a shortened pelvis and aspects of the skull of a more orthograde posture. The teeth and jaws of *A. ramidus*, like those of earlier *A. kadabba*, are among its most hominin-like features. It is often interpreted as the earliest well-documented member of our lineage. However, the data do not rule out the possibility that it may represent another ape lineage that shares some convergent features with later hominins.

Australopithecines

The first fossils to show clear evidence of adaptation to terrestrial bipedal locomotion are assigned to *Australopithecus anamensis*. Between 4.2 million and 3.9 million years ago, this species existed in East Africa. This species is differentiated only by relatively subtle characteristics from the later species, *Australopithecus afarensis*, which is present in more than a dozen fossil-bearing localities representing hundreds of known specimens, all dated to between 3.9 million and 2.9 million years ago. The teeth of these two closely similar species show several temporal trends, toward larger molar and premolar teeth and a loss of the canine-premolar cutting anatomy. Because of these trends, most paleoanthropologists regard *A. anamensis* and *A. afarensis* as successive members of a single evolving lineage.

Other lineages of hominins may have been present at the same time, including *Kenyanthropus platyops* from Kenya, *Australopithecus bahrelghazali* from Chad, and *Australopithecus deyiremeda* from Ethiopia, all dated to the interval between 3.5 million and 3.2 million years ago. These are possibly distinct from *A. afarensis* because of cranial and dental peculiarities, but in each case the fossil evidence is based on fragmentary specimens. Likewise, a partial foot skeleton from Burtele, Ethiopia, may represent yet another lineage with a distinct locomotor strategy, possibly a direct descendant of earlier *Ardipithecus*.

A diversity of contemporaneous forms is much clearer among the hominins near the Plio-Pleistocene boundary 2.5 million years ago. From approximately 2.8 million to 2.3 million years ago, South Africa was the home of *Australopithecus africanus*, also represented by large fossil samples and in most respects similar in cranial anatomy and teeth to *A. afarensis*. Additionally, by 2.5 million years ago, a lineage of large-toothed hominins known as the "robust australopithecines" appeared in East and later in South Africa. *Robust* refers to the chewing mechanics of these hominins, which combined powerful jaw muscles with extraordinarily large molar and premolar teeth. The robust australopithecines, referred to their own genus, *Paranthropus*, had approximately the same body size as other australopithecines but clearly had a different diet, featuring many more leaves and hard seeds. *Paranthropus robustus* was a South African form, possibly descended from *A. africanus*. *Paranthropus boisei* was the apex of this trend toward dietary specialization, found in East Africa between 2.5 million and 1.5 million years ago.

Like humans, all australopithecines were obligate bipeds, meaning their skeletal adaptations to bipedality precluded effective quadrupedal movement. Their feet had a first toe aligned with the other toes, minimal opposability or grasping ability, and arches similar to the feet of living people. Their knees were angled to promote effective weight support in a bipedal stance and did not rotate to facilitate grasping with the feet. In contrast with nonhuman apes, humans and australopithecines have short hip bones that make a broad bowl-shaped structure to support the viscera when upright. In addition, the broader hip and shorter ischium enable effective muscle control of the lower limbs during bipedal walking and running. Our bipedal form of locomotion is not as fast as chimpanzee or gorilla knuckle-walking, but it is highly energetically efficient.

However, despite their clear bipedality, australopithecines had relatively long, heavily muscled arms, curved toes and finger bones, and a long clavicle and apelike shoulder blade, all suggesting that climbing remained important to *A. afarensis* and *A. africanus*, even as these hominins moved into more open grassland settings. Still, with hands and legs ill suited for suspension or above-branch quadrupedal walking, early hominins must have climbed in a manner analogous to that of recent humans. With female masses around 35 kg and males up to 50 kg, they approximated living chimpanzees in body size. Small, presumably female, skeletal individuals, such as the "Lucy" skeleton of *A. afarensis*, had statures of 100 to 140 cm, much shorter than the average of any recent human population. The largest australopithecine skeletons, such as the Kadanuumuu skeleton from Woranso-Mille, Ethiopia, were as tall as 165 cm, within the range of body sizes of small living humans. All australopithecines had small brains, approximately 450 ml on average, which contrasts strongly with the 1350-ml brains of living humans, but were slightly larger than those of living chimpanzees of comparable body size.

2. EARLY *HOMO*

By 1.8 million years ago, a very different kind of hominin had emerged and spread into Eurasia. *Homo erectus* was the size and stature of recent human hunter-gatherer populations, bigger than any known australopithecine. The skulls of *H. erectus* also contained disproportionately larger brains than those of australopithecines, initially between 600 and 900 ml, and relatively small teeth. The earliest clear fossil evidence of *H. erectus* occurs at Dmanisi in the Republic of Georgia and Modjokerto, Java, with additional fossil discoveries in East and South Africa. In each of these areas, remains of *H. erectus* existed along with evidence of stone tool manufacture and transport of stone. The evidence indicates that *H. erectus* relied on a higher-quality diet including meat, which imposed greater demands on technical abilities and social organization but created opportunities for dispersal and range expansion, explaining the species' extra-African distribution.

At present, identifying the population that gave rise to *H. erectus* is one of the most engaging problems in the study of human evolution. *Homo erectus* and later forms of *Homo* relied on stone tools to cut meat from animal bones, to smash bones to retrieve the marrow inside, to process plant foods, and to shape wood. The earliest evidence of stone tool manufacture comes from Lomekwi, Kenya, from approximately 3.3 million years ago. By 2.5 million years ago, many sites preserve evidence of stone tools and cut marks on animal bone, indicating that these tools were often used for butchering animals. But these early tools are not directly associated with any hominin fossils, making it difficult to know how this cultural evidence may have been related to hominin species. Most anthropologists accept that the genus *Homo* arose during this time period, and a handful of fragmentary fossils, including parts of skulls, jaws, and teeth from Ethiopia and Kenya, may represent the earliest members of *Homo*. The oldest of these currently known is a fragment of jaw from Ledi-Geraru, Kenya, which is approximately 2.8 million years old.

Much better known are the later species, *Homo habilis* and *Homo rudolfensis*, both dating to between 2.1 million and 1.6 million years ago. These species lack many of the distinctive features of *H. erectus* and have brain sizes intermediate between those of australopithecines and *H. erectus*, ranging between 500 and 800 ml. Their teeth and jaws are smaller than those of earlier australopithecines. The anatomy of the postcranial skeleton is not known for *H. rudolfensis*, but *H. habilis* post-cranial evidence shows an anatomy very like that of *A. africanus* and other australopithecine species. Later examples of *H. habilis* coexisted with *H. erectus* in Kenya and Tanzania, and scholars disagree about whether *H. habilis* or *H. rudolfensis* gave rise to *H. erectus*, or whether all three lineages may derive from earlier hominin species.

Australopithecus sediba is an exceptionally interesting sample, dating to 2 million years ago from Malapa, South Africa. Two very complete skeletons of this species combine *Homo*-like teeth and hands with the body proportions, brain size, and possible arboreal adaptations of earlier hominins. Whether this species could be ancestral to *H. erectus* or *H. habilis* or both could be influenced by analysis of a handful of fossil fragments from East Africa. In the past, these have been assigned to *Homo*, but until more is known about their anatomy, it will be difficult to test hypotheses about their relationships.

A very large sample of fossil hominins from the Rising Star cave system in South Africa may also inform our understanding of diversity in *Homo*. First described in 2015, these fossils belong to a new species, *Homo naledi*, which shares many anatomical features with early members of the genus, including *H. habilis*, *H. rudolfensis*, and early fossil representatives of *H. erectus*. The large array of postcranial remains documents a mosaic of features including humanlike feet and lower limb anatomy; a more primitive, australopithecine-like pelvis and shoulder;

and a wrist that is more humanlike than that of either *H. habilis* or *A. sediba* but with curved fingers like those of earlier hominins. Despite the overall pattern of similarities in the skull and teeth to those of *H. habilis* and early *H. erectus*, *H. naledi* had brains that were equivalent in size to those of australopithecines, between 450 and 560 ml.

At the time of writing, there is no geologic age estimate for the Rising Star fossils, so it is not yet clear whether *H. naledi* may be an ancestor of other forms or whether it may represent the long survival of a lineage stemming from near the origin of our genus. However, the fossils do show that the anatomical evolution of *Homo* was not a simple progression from australopithecine anatomy to more humanlike anatomical form but instead must have included some adaptive convergence or parallelism in some anatomical regions.

The expansion of brain size from *Australopithecus* to *Homo* is correlated with many aspects of life history and behavior. Neural tissue imposes a high metabolic cost, which humans meet by adopting dietary and behavioral strategies that provide high caloric returns. The first postnatal year of human brain development includes a rapid expansion of brain size and concomitant shape changes, in contrast with the developmental trajectories of other primates. Neural development in humans extends across a long childhood, with late sexual maturation and an adolescent growth spurt. These ontogenetic patterns appeared in concert with increasing brain size in Pleistocene humans. An increased dependence by *Homo* on hunting and meat scavenging compared with other primates yielded a net increase in diet quality but imposed several risks, such as competition with large carnivores, unreliability of game, and long training necessary for skill development. Modern humans mitigate these risks by food sharing, sexual division of labor, and gathering of plant foods and animal resources, including honey.

Hunter-gatherer social groups are relatively egalitarian, with decision making regulated by a coalition of many group members. In this setting, learning of social rules and communication about social norms are fundamental determinants of survival and reproduction. This social environment is thought to be a major selective driver of larger brains, allowing for more sophisticated communication and inferences about the intentions of other social actors. Whereas australopithecines had vocal tracts similar in form to those of chimpanzees and gorillas, early *Homo* had both vocal and auditory traits that could have supported humanlike sound production and reception.

After its origin, *Homo* diversified into regional populations with some morphological differences. In East Asia, *Homo erectus* occupied a range from north China to Java, which was connected to the Asian mainland during periods of low sea level. Across this range, populations developed regional variations in the shape of the brow ridge and forehead, extent of muscle development of the jaw and neck, and shape of the teeth. Some of these people made a deepwater crossing to the island of Flores by 1 million years ago, where later they may have evolved into a late-surviving isolated dwarf population called *Homo floresiensis*.

In Africa, the fossil record is sparser but supports the idea that *Homo* increased in variability in the period after 1.2 million years ago. The West and South Asian archaeological records show that these regions were also occupied by early human populations, but scant fossils remain. Europe was inhabited by 1.2 million years ago, but the skeletal record represents chiefly the last 800,000 years.

Everywhere they lived, humans used stone tools. The basics of production involved the procurement of stone raw material either from rocky outcrops or from rounded cobbles in streambeds. People were selective about material, choosing fine-grained stone with predictable fracture dynamics, which they sought and transported over kilometers. Removing a sharp flake by itself yields a reliable cutting edge; removing several flakes from a rock, or "core," can shape an edge suitable for chopping or piercing bone. This basic technological pattern is called *Oldowan*. After 1.6 million years ago, however, mainly in Africa and later in Europe and West Asia, people shaped core tools into symmetrical tools with long edges, called hand axes. The resulting *Acheulean* stone industry persisted for some 1.3 million years. Archaeologists know that Pleistocene humans also often used fire, wooden spears, and other implements, and sometimes tools made of bone.

By 300,000 years ago, brain size in *Homo* had increased to between 800 and 1300 ml. Most paleoanthropologists refer these later remains to species other than *H. erectus*. In Africa and Europe, they are often called *Homo heidelbergensis*, although many scientists prefer to call them "archaic *Homo sapiens*." Whatever they are called, these people began to experiment with different technical forms, including a process of stone tool manufacture known as a *prepared core* technique. The result was a greater control over the shape of end products, sometimes yielding blades and points that were attached (hafted) onto spears as compound tools. These stone industries are called Middle Stone Age (MSA).

3. NEANDERTHALS AND THE ORIGIN OF MODERN HUMANS

Genetic evidence has greatly clarified our understanding of the human populations of the last 250,000 years. Archaeology and skeletal remains help complete the story, adding perspective on the causes and timing of the key events. This was a time of vast migrations and mixture of distant populations with one another.

By 250,000 years ago, MSA people had developed regional tool industries with little evidence of interregional movement or exchange. A small skeletal sample represents these MSA populations from across Africa. These represent the earliest humans with modern anatomical characteristics, including a high forehead, face tucked beneath the front of the braincase, and a rounded cranial vault. The functional import of these changes is not yet understood, but they seem to reflect a basic shift in developmental patterning.

The later MSA peoples, after 120,000 years ago, became regionally differentiated. In both southern Africa and the Maghreb, people collected shells and marked objects, for example, with natural pigments and ostrich eggshells. In Mozambique, people gathered large stores of wild grains; in Ethiopia, they transported obsidian over hundreds of kilometers.

Some African populations dispersed into West Asia by 105,000 years ago, taking with them a subset of the genetic variation present in Africa. In western Asia and Europe they encountered the Neanderthals, whose remains are dated to between 200,000 and 300,000 years ago. Beginning from a common anatomical background with modern humans, Neanderthals evolved a number of traits that appeared nowhere else: long, barrel-shaped skulls with a rearward projection called an "occipital bun," thick curving long bones with large joints, and at least in the European part of their range, body proportions now associated with inhabitants of very cold environments. Neanderthals were a small population dispersed over a large space, and even more so than their contemporaries in Africa, depended heavily on meat from large prey animals. The Neanderthals were probably a minor component of the overall Pleistocene human population, but their skeletal and archaeological remains are numerous, so we understand their lifeways better than those of other populations. Additionally, it has proved possible to obtain a partial genome sequence from Neanderthals, which has shed great light on the genetic ancestry of modern humans outside of Africa.

One notable triumph of paleogenetics is the Denisova genome, from the Altai Mountains of southern Siberia. This genome represents a population living at the same time but to the east of the Neanderthals, but substantially distinct from the known Neanderthal genetic sample. Living people in Australia and New Guinea derive around 5 percent of their ancestry from a population similar to the Denisova individual. Neanderthals themselves contributed between 1 and 4 percent of the ancestry of present populations throughout the world (including Australasia), except within Africa itself. These genetic results may help explain morphological features that imply some degree of regional continuity of human populations in Europe, East Asia, and Australasia; however, the spread of Africans within the last 100,000 years accounts for more than 90 percent of the ancestry of living people, but a small multiregional component of ancestry has remained in the face of this and subsequent migrations.

4. RECENT HUMAN EVOLUTION

After modern human populations became established throughout the world, evolution continued to shape our biology. Early human populations in Europe and northeast Asia likely found themselves poorly suited to the low temperature and insolation of these regions. The tropical regions of Asia had a physical geography similar to Africa's but very different floral and faunal communities. Watercraft allowed people to colonize Australia, Melanesia, and other island regions, and facilitated the migration of people from the Bering Land Bridge into the southern parts of the Americas before 14,000 years ago. Rapid evolution by natural selection in all these novel environments was inevitable.

As humans dispersed throughout the world they also increased vastly in numbers. At the end of the last glaciation, people expanded their dietary breadth to a greater number of plant and animal species, a process called the Broad Spectrum Revolution. Some experimented with planting and keeping seed crops; others began managing herd animals more intensively. Over many generations, these processes led to domestication of former wild species, settlement of many human groups into villages and cities, and the rise of political and economic elites. Pastoralists sustained large populations on formerly less hospitable plains and steppes, sometimes migrating over long distances. Civilization was one result of this agricultural revolution; warfare and serfdom were others.

Human skeletal traits (and, by inference, genes) have changed during the last 20,000 years at a rate unmatched in earlier periods. Humans became more gracile as cranial muscle attachments and structures such as the brow ridge became lighter. After the introduction of agriculture, smaller teeth and jaws became common, and a higher proportion of individuals failed to develop third molars, or "wisdom teeth," entirely. Along with such evolutionary changes, skeletal samples document the catastrophic health effects resulting from agriculture and village life.

Pathogens have been among the most obvious causes of recent human evolution. For example, more than 20 different alleles that protect to some extent from *Plasmodium falciparum* malaria are known from different human populations, many of which have arisen within the last few thousand years. Diet is another important cause of recent evolutionary changes, as some human groups have specific genetic adaptations to starchy grains and milk consumption. The physical environment has exerted its own selection on populations at high altitude, with selection affecting oxygen transport, and at high latitude, with recent strong selection on genes associated with pigmentation.

Industrial populations of the last 200 years have undergone further radical changes in longevity, residence patterns, and family size. Nevertheless, selection and evolution of modern human populations is ongoing, with documented selection on quantitative traits of medical and biometric interest. The future direction of human evolution cannot be predicted from our past history (see chapter 23), but the pace of recent evolution suggests that our species may have many more changes ahead.

FURTHER READING

Aiello, L. C., and J.C.K. Wells. 2002. Energetics and the evolution of the genus *Homo*. Annual Review of Anthropology 31: 323–338. *A summary of the impacts of a transition to an energy-dense (high-meat) diet and its significance to some of the main differences between* Homo *and its progenitors.*

Antón, S. C., W. R. Leonard, and M. L. Robertson. 2002. An ecomorphological model of the initial hominid dispersal from Africa. Journal of Human Evolution 43: 773–785. *Analysis and discussion of the dispersal dynamics of* Homo.

Boehm, C. 1993. Egalitarian behavior and reverse dominance hierarchy. Current Anthropology 34: 227–254. jstor.org/stable/2743665. *An interesting discussion of the conditions under which egalitarian human societies are expected to arise, with implications for the social structures in archaic humans.*

Brumm, A., G. M. Jensen, G. D. van den Bergh, M. J. Morwood, I. Kurniawan, F. Aziz, and M. Storey. 2010. Hominins on Flores, Indonesia, by one million years ago. Nature 464: 748–752. doi.org/10.1038/nature08844. *Description of a diminutive fossil hominin from the Indonesian island of Flores.*

Dunbar, R.I.M. 2003. The social brain: Mind, language, and society in evolutionary perspective. Annual Review of Anthropology 32: 163–181. *A review of the hypothesis that the unusually large brains of* Homo *evolved in response to the cognitive demands of living in social groups with complex bonds.*

Green, R. E., J. Krause, A. W. Briggs, T. Maricic, U. Stenzel, M. Kircher, N. Patterson, et al. 2010. A draft sequence of the Neandertal genome. Science 328: 710–722. doi. org/10.1126/science.1188021. *Description of the genome sequence of Neanderthals and a comparison with modern human variation.*

J. Hawks, E. T. Wang, G. Cochran, H. C. Harpending, and R. K. Moyzis. 2007. Recent acceleration of human adaptive evolution. Proceedings of the National Academy of Sciences USA 104: 20753–20758. doi.org/10.1073/pnas.0707650104. *An analysis showing that, contrary to common perception, humans have experienced recent, rapid adaptive evolution that is likely ongoing and accelerating.*

Leakey, M. G., F. Spoor, F. H. Brown, P. N. Gathogo, C. Kiarie, L. N. Leakey, and I. McDougall. 2001. New hominin genus from eastern Africa shows diverse middle Pliocene lineages. Nature 410: 433–440. *Description of* Kenyanthropus *and discussion of its relationship to* Australopithecus.

McBrearty, S., and A. S. Brooks. 2000. The revolution that wasn't: A new interpretation of the origin of modern human behavior. Journal of Human Evolution 39: 453– 563. *A discussion of the origins of many modern human traits, such as sophisticated Middle Stone Age tools and other artifacts; argues that these traits emerged gradually within Africa but then rapidly spread to the rest of the world 40,000–50,000 years ago.*

McHenry, H. M., and K. Coffing. 2000. *Australopithecus* to *Homo*: Transformations in body and mind. Annual Review of Anthropology 29: 125–146. *A review providing useful perspectives on the origin of the genus* Homo.

Patterson, N., D. J. Richter, S. Gnerre, E. S. Lander, and D. Reich. 2006. Genetic evidence for complex speciation of humans and chimpanzees. Nature 441: 1103–1108. *A ge-*

nomic analysis that dates the human-chimpanzee divergence and argues for gene flow between the two lineages after their initial split.

Reich, D., R. E. Green, M. Kircher, J. Krause, N. Patterson, E. Y. Durand, B. Viola, et al. 2010. Genetic history of an archaic hominin group from Denisova Cave in Siberia. Nature 468: 1053–1060. doi.org/10.1038/nature09710. *Description of the genome sequencing of archaic human remains that were contemporaneous with Neanderthals but geographically separated.*

White, T. D., B. Asfaw, Y. Beyene, Y. Haile-Selassie, O. C. Lovejoy, G. Suwa, and G. Wolde. 2009. *Ardipithecus ramidus* and the paleobiology of early hominids. Science 326: 75–86. doi.org/10.1126/science. 1175802. *Description of* Ardipithecus ramidus *and discussion of its significance for understanding of human origins.*

Wood, B., and T. Harrison. 2011. The evolutionary context of the first hominins. Nature 470: 347–352. doi.org/10.1038/nature09709. *A review of the challenge of identifying fossil hominins and determining whether they belong to the lineage leading to modern humans.*

CHAPTER 4

HUMAN COOPERATION AND CONFLICT

Joan E. Strassmann and David C. Queller

OUTLINE

Humans live in groups where they cooperate so extensively that it is sometimes easy to forget that cooperative actions evolve only if they benefit the individual or cause the genes underlying the action to proliferate. Conflict, by contrast, is comparatively easy to understand, because different individuals can profit from the same resources. The advantage to group living in humans probably comes from protection against predation and attack, including by other groups of humans. Co-operation within groups can be called fraternal or egalitarian, with the former being between relatives and the latter between unrelated individuals. Both kinds of cooperation evolve only if genes underlying the actions proliferate as a result of the actions they cause. With fraternal cooperation that proliferation can happen through kin selection, with a relative reproducing more, while for egalitarian cooperation the cooperator must benefit directly. Conflict can destroy the advantages of group living and undermine cooperation, so its control within groups is crucial. Conflict can be limited by relatedness, common interest, power, and policing. Cooperation and conflict may seem like strictly behavioral attributes, but they can be found in the physiological struggle for nutrients between mother and fetus on the battleground of the placenta.

Interactions with other organisms—from those with our symbiotic bacteria to those with our domesticated food animals and plants—are also largely mutualistic. The essence of humanness, our intelligence, may have come from the tension between cooperation and competition in mixed groups of relatives and nonrelatives. In this chapter we give an overview of cooperation and conflict in humans from an evolutionary perspective.

GLOSSARY

Altruism. A behavior that is costly to its performer but that benefits the recipient

Cheating. A behavior that benefits oneself at a cost to others.

Cooperation. An interaction that benefits the genes of both the actor and the recipient.

Fitness. In the evolutionary sense, the relative ability to pass on one's genes to the next generation relative to the population.

Hamilton's Rule. A mathematical formula for the inclusive fitness of an action, so that the action should evolve. For an altruistic act benefiting the fitness of a relative, Hamilton's rule would be $rb - c > 0$: the benefit to the relative times relatedness must exceed the cost to the altruist.

Inclusive Fitness. The sum of the effects of an individual's actions on its own and other's reproduction, each devalued by relatedness. Ultimately, the measure of an individual's actions on the representation of its genes in the next generation.

Indirect Reciprocity. Payback for an act benefiting others that does not come from that individual but comes from someone else, or society at large. Considered to be a basis of morality.

Kin Selection. Selection that comes from affecting relatives.

Policing. Related to power, but in this case the actors are acting for the good of the group not just themselves. Not limited to the vernacular meaning of policing.

Power. Ability to determine the outcome of interactions, for example, through force or deception.

Reciprocity. Direct reciprocity is a costly act performed by an individual that benefits another but is directly repaid to that individual by an equivalent act or gift.

Relatedness. A measure of gene sharing relative to the population mean, often calculated through pedigrees.

Social Brain Hypothesis. Theory that humans have evolved large brains to deal with complex social interactions, particularly multiple pair-wise interactions requiring coordination.

1. OVERVIEW OF COOPERATION AND CONFLICT THEORY

Blue jays gang together and mob a raiding hawk. Burying beetles carefully tend a rotting squirrel for their babies. Honey bee workers protect their colony by suicidally repelling a raccoon tens of thousands of times larger. A lowly bacterium explodes itself, releasing chemicals that kill invading clones to benefit its clone. Nature is sometimes said to be red in tooth and claw, and genes are said to be selfish, so we can see Darwin's struggle for existence in most of these examples. And yet each of them also shows that organisms sometimes cooperate, performing actions that benefit other individuals. Cooperation is especially common in humans and includes actions like a mother nursing her baby, a hunter sharing meat, or a village damming a stream for crop irrigation. But conflict is also common and includes stealing food, mating outside a pair bond, and participating in between-group raids.

It is easy to understand how conflict evolves: a hawk's genes get propagated when they cause the hawk to catch blue jays, while a blue jay's genes get propagated when they enable the blue jay to chase hawks away. Cooperative actions are harder to understand, but the underlying principle is the same: cooperation is favored by natural selection when it consistently benefits reproduction of the actor's genes.

There are two main pathways by which this happens, sometimes called *fraternal* and *egalitarian*. Fraternal interactions are those among relatives, whereas egalitarian cooperation occurs among nonrelatives (even of different species). The evolutionary logic of both fraternal and egalitarian interactions is summarized by *Hamilton's rule*, which states that a behavior will be favored when the sum of all its fitness effects on different individuals, including itself, each multiplied by the relatedness of the actor to that individual, exceeds zero. This sum is also known as the *inclusive fitness* effect of the behavior, and it recognizes that the fitness of an action must include effects not only on the actor but also on others who might share the gene. Each effect is multiplied by relatedness, because this measures the degree of genetic similarity at the locus causing the behavior, above random similarity in the population. Such similarity is mainly generated by kinship, reflected in standard population-genetic measures such as ½ for parent-offspring, ½ for full siblings, ¼ for half-siblings, and zero for random population members. Hamilton's rule shows that cooperation has to benefit either one's own fitness or sufficiently benefit the fitness of relatives.

Kin selection explains fraternal cooperation and even the extreme form of cooperation called *altruism*, in which actors sacrifice their own fitness for kin. According to Hamilton's rule, genes for altruism are favored evolutionarily when the fitness costs of altruism to the actor are more than compensated for by the benefit to the recipients multiplied by their relatedness, or $-c + rb > 0$. Kin selection, the rb part of the equation, explains the suicidal stinging of honey bees and the self-detonation of bacteria. The loss of their own life is compensated for by the increased fitness of their relatives.

Because relatedness of an actor to a beneficiary is never greater than one, the circumstances that favor fraternal cooperation generally are those in which there is a synergy inherent in cooperating; that is, the results of two or more acting together are greater than the sum of the actions of the individuals. For example, two can defend a home 100 percent of the time, perhaps taking turns foraging when away, a qualitative difference from what one alone could do. A social wasp might stay and help her mother rear young, because if either the mother or the daughter should fall victim to a predator while foraging, the other can take over brood care—the life insurance advantage to cooperation. Beginning with such simple advantages, social insects have evolved highly elaborate cooperative interactions in colonies, with many workers caring for the progeny of their mother, the queen, to great evolutionary and ecological success. Kin selection has been studied in many organisms and is responsible for many kinds of cooperation within nonhuman species.

Cooperation among nonkin, in contrast, is egalitarian in the sense that it requires direct benefits to both actors. Such benefits sometimes arise simply as a consequence of group living. Schools of fish, flocks of birds, colonies of sea lions, or nesting bank swallows benefit not because of any direct cooperative action but because the simple presence of the others dilutes vulnerability to predators. Eavesdropping on actions intended to benefit relatives, like alarm calls, is another kind of advantage that can occur in groups.

Egalitarian cooperation is not restricted to incidental benefits of being in a group. A baboon may groom a nonrelative and then wait to be groomed itself. Fish may take turns in the risky act of predator inspection. This kind of cooperation requires payback to work, so the problem is how to avoid the sucker's payoff of giving a costly benefit and then not receiving a return. The sucker's payoff is most easily avoided if each individual act is of small cost and immediately followed by the same act by the partner. Cooperative acts that are temporally separated and costly to the actor require strategies like tit-for-tat, in which the partner's cooperative or noncooperative behavior is mimicked in the next interaction with that partner. This strategy can limit the sucker's payoff to one interaction per noncooperator while allowing repeated beneficial interactions with cooperators.

However, interactions that require accounting for and remembering past actions like tit-for-tat need cognitive abilities unlikely to be found outside of humans, so many kinds of reciprocity are rare outside of humans. In humans, egalitarian cooperation can even extend to something called *indirect reciprocity*, in which cooperators benefit from others' witnessing their actions and thus regarding the cooperators as good partners for future cooperation.

Egalitarian cooperation between individuals of different species is usually called *mutualism* or *symbiosis* if one partner lives in or on the other and can operate through similar mechanisms, such as incidental benefits, tit-for-tat, and partner choice. However, with individuals of two different species, it is common that goods or services easily produced by one partner are exchanged for different goods easily produced by the other partner. A milkweed gives a bumble bee nectar in exchange for transport of pollen to another milkweed. A cleaner shrimp eats parasites on a grouper fish, cleaning it in the process. Aphids, cicadas, and termites eat poor but abundant food sources they digest and use with the aid of bacteria. In these kinds of interactions both parties benefit from the different abilities of each. But since the individuals are unrelated, the potential for conflict is ever present.

The more the interests of two parties are aligned, the less common are cheating and conflict. Genetic relatedness within families is a powerful but not complete way to align interests. Among nonrelatives like hosts and their bacteria, interests can be aligned by processes such as common inheritance, as when a host passes symbiotic bacteria on to its progeny. Tightest among these alliances is that of the eukaryote cell, in which nuclear and mitochondrial DNA are coinherited through females. Yet, even in these cases conflict potentially remains. Whether cooperation is based on kin selection among relatives or on mutualistic or symbiotic interactions, cheaters are likely to arise, where cheaters are those that reap the benefits of an interaction without contributing sufficiently. These cheaters can potentially be controlled by the group or by a partner, with control generally falling to the more powerful over the less so.

In sum, cooperation can evolve among relatives or among nonrelatives, even of different species, and cooperative alliances are remarkably successful. The theories for understanding cooperation, cheating, conflict, and control of conflict are based on theories of inclusive fitness, mutualism, and group living. We now consider their application to human behavior. Because modern societies exist in such radically altered environments, researchers often focus on the small groups that may be closer to the environments in which human behavior evolved: hunter-gatherers and traditional small-scale farmers and herders.

2. COOPERATION WITH RELATIVES

A mother nursing her baby, an uncle teaching his nephew to fish, a child carrying her younger sister on her back, and a family pooling resources to farm are all examples of human fraternal cooperation. When choices exist, humans are repeatedly shown to cooperate more with relatives than with nonrelatives, and with closer relatives over more distant ones. This cooperation can be within or between age classes or residential groups. It can involve countless small interactions that occur many times a day, and costly ones that are more rare. Humans rely on relatives for support of many kinds in many ways. For example, hunter-gatherers generally give more food to family members. Default inheritance laws indicate a society's belief that an individual would favor its kin, even when there is no will making this desire explicit. In Missouri, for example, half the estate of a person with a spouse and children who dies intestate will go to the spouse, and the rest will go to the children. Similarly, half the estate of a person without a spouse or children will go to his or her parents, and half to his or her siblings, in manners predictable by relatedness.

One of the most challenging things that people do is reproduce and rear children to independence. In this most important task for fitness, relatives help each other so much that our life histories have deviated from those of our closest relatives. Probably because of the assistance of relatives, humans have shorter interbirth intervals than do chimpanzees.

However, which relatives help and how they do so is complicated in humans and varies with ecological conditions. A review of 16 studies of populations with natural fertility across the world focused on the easily quantifiable variable of child survival. A relative could have had a positive effect on increasing child survival, or their very presence could actually have decreased child survival. In the studies reporting statistically significant results, there were positive (100% of studies) effects of mothers; positive (47%) and negative (7%) effects of fathers; positive (64%) and negative (9%) effects of maternal grandmothers; positive (60%) and negative (13%) effects of paternal grandmothers, and positive (83%) effects of older siblings, to list some of the most important categories they examined. These numbers suggest helping is widespread but are likely to underestimate it, because relatives often also compete for food or other resources, canceling out some of the beneficial effects of helping.

A study of 32 hunter-gatherer societies found that bands average around 28 individuals. For the average individual, about a quarter of bandmates are genetic relatives, so an action by an individual that helps the group overall will provide some genetic return even without specific kin preference. This also means 75 percent of individuals in the group are not genetic relatives, so there is an incentive to recognizing relatives. Humans recognize their relatives through observation of associations with known relatives, like mothers, as when a baby nurses from its own mother. Physical resemblance is generally an unreliable cue, but can be shown to influence sharing. Other relationships are largely known because of information given by older people. Relationships through the maternal line are more easily identified than those through the paternal one, because genetic paternity may not match social paternity.

Fraternal cooperation in humans with relatives is important, frequent, and takes many forms, just as it does in some nonhuman species. Perhaps more surprising, in comparison with other species, is the extent of egalitarian cooperation among humans. (See also chapter 5.)

3. COOPERATION WITH NONRELATIVES AND INDIRECT RECIPROCITY

A hunter teaches boys his techniques. A nighttime patrol guards a village. Men cooperate to raid a distant village, gaining territory and women. These are all examples of human cooperation among unrelated individuals that has to benefit the actor (or relatives) to be evolutionarily stable. Cooperation with nonrelatives is particularly important if human groups typically include unrelated individuals. Since one-quarter of individuals in hunter-gatherer bands are related, most band members are unrelated (though a fraction of those are in-laws). These groups have high levels of sharing not limited to kin, including child care, food, and defense.

Direct reciprocity or bartering, in which an exchange takes place over a short time for roughly equivalent values, is ubiquitous among humans. Monetary transactions make the payback immediate, so these exchanges are relatively easy to understand. But other interactions do not involve immediate payback and instead set up long-term cooperative networks in which individuals give and exchange without immediate benefit. A great deal of effort has gone into explaining such interactions.

Interactions among nonrelatives that lack immediate payback seem largely restricted to humans, most likely because only humans have the cognitive abilities to keep track of past interactions, past cooperative or uncooperative actions, and the values of goods or actions traded. Reciprocity works best when resource possession changes frequently, making sharing cheap for the giver who currently has much, and precious for the receiver who currently has little. Tallies need not be exact when the cost is much less than the gain for such cooperation to benefit both parties. Interactions may be most exaggerated in extreme environments like the Arctic, where the Inuit share extravagantly through potlatch.

But it is also common that cooperative acts are directed toward individuals unlikely ever to repay. What might explain such behaviors? These actions can be explained as *indirect reciprocity*. People constantly observe interactions among others, then choose their own subsequent partners according to perceived trustworthiness and value of others. Indirect reciprocity explains apparently random acts of kindness and forms the foundation of our moral systems. We cooperate because others will view us as cooperative and choose us as partners when it really matters. Indirect reciprocity is difficult in other animals that lack the cognitive abilities of humans for witnessing, calculating, and choosing partners. Of course, in humans these calculations are generally not founded on direct conscious choice.

4. CONFLICT AND CONTROL AMONG RELATIVES

Cain kills his brother, Abel. Oedipus kills his father, Laius, as prophesied. Despite the prevalence of cooperation among humans, we are masters of conflict, such that even murder of kin is enshrined in powerful myths. This is not surprising from an evolutionary point of view. Individuals can often gain fitness using the resources of others, and selection normally favors this behavior. Yet, in some ways the real question is, why is conflict comparatively rare in humans? There are two main ways in which overt conflict is controlled: alignment of interests, so conflict itself is reduced, and control by a party sufficiently powerful to suppress overt conflict.

Alignment of interests can occur, as we have seen, through gene sharing or kinship. Every individual has a stake in the reproduction of its relatives. But this alignment of interests is imperfect. Even the closest relatives, such as parents and offspring or full siblings, share only half their genes. The biological explanation for conflict is based on the other half of the genes, the ones not shared.

For example, though mothers care for their offspring, a mother and offspring can be in conflict when the mother's ability to provide for one offspring comes at a cost to her other existing or future offspring. The mother is equally related to all her offspring and should favor resource transfers from one to another—for example, weaning one to nurse a newborn—if the fitness benefit to one exceeds the cost to the other. Each offspring, however, should value its own fitness twice as much as its siblings' fitness and can therefore gain inclusive fitness by getting more than the mother is selected to give. This difference explains conflict over nursing and weaning. When resources are limited, these conflicts can be severe. Our name for severe cases of child malnutrition, *kwashiorkor*, comes from a word in the Ga language for the nutritional stress of the toddler when a new baby is born.

Mother-offspring conflict is not only behavioral but begins before birth. The mother supplies nutrients to the fetus through uterine arteries. But this is insufficient from the fetus's point of view; fetal cells attack the walls of these arteries, making them unable to constrict and limiting the mother's ability to restrict blood flow. Mothers have evolved ways of reinstating maternal control, including lowering overall blood pressure and making the arteries spiral in form, which increases resistance and decreases blood flow. Other maternal fetal conflicts can be found in the supply of blood sugar to the fetus, evident in the interplay of genes controlling insulin. However, it is important to remember that there is also a large area of overlapping interest in which the mother is selected to behave in ways that enhance the survival of her child.

Other kinds of conflict within families follow the same lines as mother-fetus conflict. When genes are shared, individual interests overlap but are not entirely coincident. Siblings, particularly same-sex siblings, often compete for resources or position. For example, older brothers are often able to marry at younger ages and may deplete family resources, as in the Dogon of Mali.

Power is the other limiter of conflict among relatives. If two parties are unequally matched, it may not pay for the weaker party to persist in conflict. In weaning conflict, the ultimate power seems likely to lie with the mother, who has absolute ability to cut off nursing. Nevertheless, to the extent that she relies on offspring signals of hunger, the offspring might use that power to manipulate the mother into continuing nursing for longer.

The more powerful can both prevail in conflict situations and also act to reduce conflict among others. Older individuals may be particularly effective in controlling conflict among youngsters both because of their power and because their relatedness to the young is more symmetrical than the relatedness among the young themselves. Parents are equally related to their children, grandparents are equally related to all their grandchildren, and great-grandparents are equally related to all their great-grandchildren, so each ascending level of ancestry provides a basis for still more inclusive cooperation. One argument for the origin and adaptiveness of religion begins with the respect given to living ancestors as arbiters of conflict and extends into veneration of more ancient deceased ancestors that bring together even wider circles of kin.

5. CONFLICT AND CONTROL AMONG NONRELATIVES

A child hits a friend. A man steals. A group ambushes and kills members of another village. A war involves millions across the globe. If conflict evolves among kin, it is expected even more so among nonrelatives that do not share genes. We have already noted that humans show uncommon degrees of cooperation among kin. How can conflict be limited? Again, the two paths involve alignment of interests and power.

In unrelated individuals, some alignment of interest can come through joint reproduction. A mother and father share reproductive interests in their offspring and so, to some degree, do the relatives of the mother and father. Thus, natural allies are made through marriage and reproduction. A good share of the three-quarters of hunter-gatherer relationships that are unrelated are actually bound by marriage, either directly or to kin. The joint interest in common offspring explains why marriage often results in broad alliances between lineages. A nonhuman example of joint reproduction is vertical transmission of symbionts through the eggs of the host; because the symbionts can reproduce only through the host's eggs, the symbiont will be selected to do everything it can to aid the egg production.

As with kinship, this alignment of interests is imperfect, even for the closest bond, the mated pair. Humans are not strictly monogamous, so not all offspring are shared. In the case of unshared offspring, each party will favor its own. And, of course, for the many unrelated individuals in groups that are not spouses or in-laws, this kind of alignment of interests does not work.

Power also matters here. A strong individual can sometimes keep a weaker one from contesting disputes. Dominance hierarchies in many organisms, once established, help prevent continual conflict. However, the asymmetries between individuals of similar ages and sexes may often not be strong enough for one party to easily suppress another. Nevertheless, collective power can sometimes reduce conflict. Indirect reciprocity relies on a form of unorganized collective power that depends on behavior being witnessed by others, which affects social standing and the likelihood of being chosen for cooperative interactions. Stronger sanctions include actual punishment, not just by stronger individuals in the group but also by group-sanctioned actions that are typically called *policing*. This use of the term refers to actions taken by multiple individuals, not necessarily the one affected. Policing also occurs among organisms like honey bees, when worker bees remove the eggs laid by other workers, leaving room for the queen's eggs, to which they are collectively more closely related. Perhaps the ultimate consequence for conflict behavior is expulsion from the group, which likely led to death among ancestral humans.

The high cooperation in human societies makes some people argue that societies are virtually organismal, much like honey bees in a superorganismal colony, or cells in a multicellular organism. We think this is unlikely, because of serious levels of conflict in human societies that are still not controlled by alignment of interests or power, which do not occur in organisms.

6. HUMAN SOCIALITY AND THE BRAIN

Kinship is important in human sociality, but humans stand out from other organisms in the amount, diversity, and intensity of cooperation among nonkin. Both reciprocity and indirect reciprocity require extensive mental bookkeeping about individuals, their actions, and their reactions to others, which is presumably facilitated by the large brains of humans relative to body mass. In fact, causality may work both ways; perhaps the most compelling hypothesis for the rapid expansion of our cerebral cortex is the *social brain hypothesis*. According to this hypothesis, the origins of selection for large brains come from the complications of group life: sorting out friend from foe and relative from nonrelative, keeping track of past interactions, detecting deception, building coalitions, and synchronizing tasks. Such cognitive abilities may be particularly favored in large groups of mixtures of different categories of individuals. Selection for large brains is facilitated in ecologically dominant species whose biggest foes are other groups of the same species. Thus, sophisticated within-group cooperation is favored for success in between-group conflict. If either or both hypotheses are correct, it puts cooperation and conflict at the core of what makes us human.

However, any evolutionary hypothesis on human brains and intelligence requires caution, because neither the comparative method nor experimentation works well for hypothesis testing with the evolution of such unique characteristics.

7. ARE HUMAN COOPERATION AND CONFLICT MALADAPTIVE?

The general approach we have taken in this chapter is to explain human cooperation and conflict from an evolutionary perspective, in which we seek reasons that behavioral acts augment the reproductive success or inclusive fitness of the actors. However, the success of this approach can be undermined by several other consequences of the large human brain. Humans are unique in having language, greatly increasing the potential for culture, which is a nongenetic form of inheritance and evolution. Culture in turn has rapidly modified human environments, which can make previously favored actions disfavored. Evolved proximate mechanisms may no longer result in the advantages they once did. We tend to choose fatty and sugary foods, following predilections that might have been great for picking ripe fruits and nutritive cuts of meat in a regime of scarcity. Under current conditions of excess, however, these preferences now threaten our health, leading to diabetes and heart disease.

Even more basic is the uncoupling of sex and resources from reproduction. Historically, sexual pleasure must have led to more births and resource acquisition to successfully rear more children. Yet, in industrialized societies, though people continue to be rewarded by the pleasures of sex and of acquiring and consuming resources, they no longer regularly turn these activities into more babies. Instead wealthier people tend to have fewer children. The evolved proximate mechanisms are still there, but they operate to new ends.

Controlled experiments suggest that humans are more cooperative than expected in laboratory one-shot interactions with nonrelatives, where there are no possible gains from kin selection and reciprocity. This might indicate evolutionary "mistakes" in a novel artificial environment if our behavioral mechanisms evolved purely in human bands that always included relatives and where interactions were never one-shot. Alternatively, some cooperation might have evolved through nongenetic cultural means. One possibility is that there was cultural group selection, in that groups with more cooperative cultures replaced those with less effective ones. Cultural norms can reduce within-group variation and increase between-group variation, making selection between groups stronger, in contrast with other kinds of group selection. Whether or not that is true, it is clear that cooperation can be greatly increased (and sometimes decreased) by cultural institutions such as religion, government, and markets. It is beyond the scope of this chapter to fully evaluate these issues, but it is important to remember that not all actions are genetically adaptive or adaptive in all environments. Nevertheless, biological evolution does provide insights into our social past and into the origins of the social traits that make us human.

FURTHER READING

Alexander, R. D. 1987. The Biology of Moral Systems. Hawthorne, NY: Aldine de Gruyter. *This is a masterful book that investigates how morality evolved. It is also the best explanation of indirect reciprocal altruism.*
Dunbar, R.I.M., and S. Shultz. 2007. Evolution in the social brain. Science 317: 1344–

1347. *This article explains the social brain hypothesis, which posits that keeping track of complex social interactions is what led to our unique brains.*

Haig, D. 1993. Genetic conflicts in human pregnancy. Quarterly Review of Biology 68: 495–532. *This article presents the argument and evidence for genetic conflicts between mothers and their fetuses that can compromise the health of both.*

Hill, K. R., R. S. Walker, M. Božičević, J. Eder, T. Headland, B. Hewlett, A. M. Hurtado, et al. 2011. Co-residence patterns in hunter-gatherer societies show unique human social structure. Science 331: 1286–1289. *This study brings together leaders in the study of hunter-gatherers to show the surprising frequency of alliances among nonrelatives.*

Hrdy, S. B. 2009. Mothers and Others: The Evolutionary Origins of Mutual Understanding: Cambridge, MA: Harvard University Press. *This book makes a strong case for the importance of nonrelatives in child rearing.*

Kurzban, R., M. N. Burton-Chellew, and S. A. West. 2015. The evolution of altruism in humans. Annual Review of Psychology 66: 575–599. *This review makes clear cases for kin selection and reciprocal altruism and clears up a lot of fuzzy thinking from other sources.*

Macfarlan, S. J., R. S. Walker, M. V. Flinn, and N. A. Chagnon. 2014. Lethal coalitionary aggression and long-term alliance formation among Yanomamö men. Proceedings of the National Academy of Sciences USA 111: 16662–16669. *The Yanomamö are perhaps the best-studied group on warfare among bands.*

Sear, R., and R. Mace. 2008. Who keeps children alive? A review of the effects of kin on child survival. Evolution and Human Behavior 29: 1–18. *The information on the impact of relatives on child survival comes from this review.*

Strassmann, B. I. 2011. Cooperation and competition in a cliff-dwelling people. Proceedings of the National Academy of Sciences USA 108: 10894–10901. *This is a vivid description of social cooperation and conflict in the Dogon, a tribe of agriculturalists in Mali, Africa.*

Summers, K., and B. Crespi. 2013. Human Social Evolution: The Foundational Works of Richard D. Alexander. New York: Oxford University Press. *This book reprints the foundational papers arguing for an evolutionary approach to human behavior along with more recent pieces that address the latest advances.*

CHAPTER 5

HUMAN BEHAVIORAL ECOLOGY

Virpi Lummaa

OUTLINE

1. Development of human behavioral ecology
2. Problems and criticism
3. New focus on evolution in modern societies
4. What can human behavioral ecology contribute to the general study of evolution?

Human behavioral ecology applies the general theories and mathematical models developed for understanding variation in traits across species to test similar questions in humans. The focus is on studying the consequences of particular traits or behavioral strategies for an individual's success at passing on its genes to the following generations, given the ecological and social environment of that individual. Humans experience a wide global range of living conditions and lifestyles, from traditional communities to extreme urbanization, and human behavioral ecologists today use a range of study designs and data sources to investigate all these populations from an evolutionary perspective. The type of data available on humans makes it possible to investigate the details of many central questions in evolutionary biology.

GLOSSARY

Cohort Studies. Longitudinal study designs commonly used, for example, in medical and social science research, and increasingly also in human behavioral ecology. Such studies record the life events of a group (cohort) of individuals sharing a common characteristic or experience (e.g., born during the same year or exposed to a famine in utero) and compare these individuals with other cohorts or the general population.

Demographic Transition. The transition from high birth and death rates to low birth and death rates as a country develops from a preindustrial to an industrialized economic system.

(Historical) Population Records. Registers of births, deaths, marriages, and migrations that have been maintained in many countries over long periods of time (e.g., by the church or governmental departments) and that are now a frequent source of data in human behavioral ecology.

Hunter-Gatherer. Ancestral subsistence mode of *Homo* in which most or all food was obtained from wild plants and animals, in contrast with agriculture, which relies on domesticated species. All humans were hunter-gatherers at least until approximately 10,000 years ago.

Intervention Studies. Procedures used to test a cause-and-effect relation in epidemiological studies by modifying the suspected causal factor(s) affecting health outcomes (e.g., by supplementary feeding of a group of subjects or treating them with a given medicine) and recording their future life events in comparison with those of subjects not receiving the treatment.

Microevolution. A change in gene frequency within a population over time.

Optimality Models. Simulations that weigh the costs and benefits of a given trait or behavior compared with another trait or behavior for maximizing fitness.

Pleistocene. A time period 2,588,000 to 12,000 years before the present when key events in human evolution took place.

Twin Registers. A type of data often used in human behavioral genetics recording various traits of up to thousands of twin pairs from a given country or cohort. Such data sets are most commonly used to estimate the relative importance of environmental and genetic influences on particular traits and behaviors in humans by comparing individuals in identical and fraternal twin pairs.

Human behavioral ecology is an evolutionary approach to studying human behavior that applies methods virtually identical with those used by behavioral ecologists studying other species. The focus is on studying the consequences of particular traits or behavioral strategies for an individual's success at passing on its genes to the following generations. The most successful behavior from the viewpoint of evolutionary fitness may vary among individuals depending on attributes such as their wealth, age, living environment, family support available, or set of genes. Empirical studies in human behavioral ecology use data from different human populations to test predictions produced by the general theories and mathematical models developed for understanding variation in traits across species. One of the most widely studied questions is whether variation among individuals in partner choice and reproductive patterns in humans is adaptive: Does mate choice capitalize on reproductive prospects in the future? Does age at first reproduction reflect the "best age" for the given man or woman to start a family to maximize his or her overall number of children reared over a lifetime? Or is there an adaptive explanation for women going through menopause before the end of their life span? For example, it is postulated that women living in an environment with a high mortality hazard benefit from giving birth at a young age to ensure reproducing before dying, despite the risks to both maternal and baby survival associated with early motherhood. In contrast, a woman living in a more stable environment might maximize her overall number of surviving offspring by delaying the onset of motherhood until she has finished growing and maturing.

Application of evolutionary theory to understanding human behavior has grown increasingly popular since the publication of *Sociobiology* by Edward O. Wilson (1975), often considered as "giving birth" to the field. An evolutionary approach to explaining variation among individuals in traits such as mate preferences, marriage patterns, and childbearing—or even differences in hunting patterns, diet, language, diseases, and personality—has gained popularity in disciplines besides biology, such as anthropology, psychology, and more recently, medicine. This approach has also been applied in economics, where—much as in evolutionary thinking—maximization and self-interest are central concepts. In contrast, sociologists, for example, have traditionally been slower at integrating evolutionary theory into their approach to explaining human behavior. Consequently, scientists applying evolutionary theory to understanding human behavior have backgrounds and training in an extraordinary diverse range of disciplines. They often disagree about how evolutionary theory can be applied to understanding human behavior and how such attempts should incorporate any influence of culture, modernity, inheritance of wealth, and other factors often considered particularly relevant in humans as compared with other species.

This chapter focuses on discussing the success of the behavioral ecological approach in explaining variation among humans. The first part introduces the key approaches and assumptions traditionally used in the study of human behavioral ecology and lists the main areas of research and their findings. The second part discusses the difficulties and criticism faced by such studies. The third part highlights the recent developments in the field that arose in response to such criticism and points out the areas in need of further investigation. Finally, although studies on humans suffer from many unavoidable methodological difficulties, the last section highlights the particular benefits that working with humans offers for advancing our understanding of evolutionary processes in general.

1. DEVELOPMENT OF HUMAN BEHAVIORAL ECOLOGY

Human behavioral ecology began by testing predictions formulated largely from optimality theory. *Optimality models* weigh the costs and benefits of alternative traits or behaviors for maximizing fitness and have been successfully used to further our understanding of behavioral variation in other animals. In humans, short birth intervals, for example, could be associated with the benefit of producing many offspring over the limited reproductive life span, but such benefits must be weighed against the costs of short birth intervals to both mother and child in terms of mortality risk. The best (optimal) strategy thus involves a trade-off between such factors to maximize the overall possible number of offspring reared in a lifetime. The approach typically considers human behavior to be highly plastic and likely to produce adaptive outcomes in different environmental settings. Such a black-box approach assumes that there is a link between genes and behavior, but the existence of this linkage was for a long time not studied in detail.

In humans, most quantitative data to test the models have been collected studying contemporary "traditional" societies, such as extant hunter-gatherer, agropastoral, or horticultural groups, for example, in southern Africa (!Kung San), Kenya (Kipsigis), Amazonia (Yanomamö; Tsimane), and Tanzania (Hadza) (see Hawkes et al. 1997). Only the hunter-gatherer lifestyle (e.g., that of the traditional !Kung San) is usually, strictly speaking, expected to be similar to that during Pleistocene, when human evolution is thought to have been rapid; however, because of the current rarity of such groups, research has expanded to other populations little influenced by globalization and with "natural" mortality and fertility rates, with the idea that studying such tribal groups is close to studying our ancestors. Thus, the traits that increase reproductive success among the currently living traditional populations have also done so in the past and can inform us about selection pressures operating in past environments. Because of the desire to correlate given traits or behaviors with measures of reproductive success, such as the number of living children or grandchildren, the data analyzed on these populations have largely been correlational in nature; that is, they have involved collection of anthropometric, behavioral, and demographic data on individuals without the possibility—available for other shorter-lived organisms—to conduct experiments.

One of the first areas of focus was research on foraging behavior to show that, on the whole, human foragers select food sources that maximize nutrient acquisition, as predicted by optimal foraging theory. Further research has applied the optimal theory framework to investigating mating patterns (e.g., to test whether females may gain higher fitness by mating with a male who already has a mate), life history variation (e.g., age at maturation and first reproduction, birth spacing, and senescence), and parental investment according to the prevailing social and environmental conditions. Overall, although these studies cannot necessarily show that the traits in question are the products of past selection, they have proven that applying the same framework as scientists use in working on similar questions in other species can indeed produce convincing support for the tested hypothesis and provide insight into how natural selection maintains variation in the trait.

For example, one of the greatest mysteries in human life history has been the existence of female menopause, a complete and irreversible physiological shutdown of reproductive potential, well before the commonly achieved overall life span in all human populations. This phenomenon is evolutionarily puzzling, because all organisms are predicted to seek to maximize their genes in the following generations, a goal that is normally achieved by breeding throughout life. The problem is that adaptive benefits of menopause are difficult to test empirically, because all women experience it; we will never know whether in our evolutionary past, women experiencing menopause produced significantly more and/or superior offspring than women who continued to reproduce until death. What is better understood, however, is that whatever the cause for menopause itself, the extended life span after menopause gives an evolutionary advantage to women. A woman with genes for living beyond her decline in fertility produces more grandchildren (and hence forwards more genes to the following generation) than a woman who dies at menopause, because postreproductive women can have positive effects on their offspring's reproductive success—they help rear their own grandchildren. Among the Hadza of Tanzania, child weight is positively correlated with grandmother's foraging time (see Hawkes et al. 1998 for details). The presence of a grandmother has also been linked to increases in grandchild survival chances in many contemporary traditional as well as historical populations around the world. Finally, research using data available for farming/fishing communities of eighteenth- and nineteenth-century Finnish and Canadian people has shown that mothers indeed gained extra grandchildren by surviving beyond menopause until their mid-seventies. These data show that life span can be under positive selection at least until this age. This effect arose because offspring in the presence of their living postreproductive mothers bred earlier, more frequently, for longer, and more successfully. Such benefits were not present if the mother was alive but lived farther apart from her adult offspring, which suggests that the findings are not a mere artifact of better overall survival of both grandmothers and grandchildren in some families (see Lahdenperä et al. 2004 for details). An additional discussion of the evolution of menopause in humans can be found in chapter 8.

Another main interest in human behavior ecology has been to investigate the effect of environmental conditions on the fitness benefits of different traits. For example, costs of reproduction to females need to be analyzed in relation to the energy budget of the woman: high costs of reproduction do not have the same effects on women who have good diets and low levels of physical activity compared with women in poor energetic condition. Such physiological consequences of reproduction for women with differing food access are well documented in humans. Further evidence that resource availability may affect selection on life history traits in humans comes from studies showing a negative relationship between number of offspring and postmenopausal life span among poor landless women, whereas for wealthier women, the relationship between fecundity and postmenopausal life span is often positive. A negative relationship between fecundity and longevity may therefore be expected in women who owing to multiple pregnancies and breastfeeding pay high costs of reproduction that cannot easily be compensated for by increases in dietary intake and reduction in physical activity. In contrast, wealthier women can more easily "afford" both large family size and long life span. Comparable differences in the costs of reproduction could also be created, for example, by differing amounts of help available from other individuals with raising the offspring, such as partners, grandparents, or other helpers in the nest, that affect the level of investment made by the mother, but few studies have investigated such effects.

2. PROBLEMS AND CRITICISM

The downside of the original focus on traditional populations with high fertility and mortality rates is that sample sizes tend to be limited; groups are rapidly disappearing or are affected by globalization; collection of multigenerational data often essential for addressing evolutionary questions is time consuming or impossible; and ages are merely estimates. Focusing preferentially on hunter-gatherers also ignores the fact that human evolution has been most rapid, in terms of generation-to-generation changes in gene frequencies, since the invention of agriculture. Investigating modern populations is equally interesting, because differences in reproductive and survival rates among individuals still lead to selection favoring certain heritable traits over others, albeit the alleles being favored might also be influenced by culture (see chapter 21), in particular, modern medical care. Moreover, modern populations lend themselves to current genomic and population genetic analyses.

First, recent analyses of the human genome have revealed that human genetic makeup has responded to the domestication of plants and animals and the spread of agriculture, numerous genes have experienced recent positive selection, and overall considerable selection has occurred in the past 10,000 years (see chapter 23 for more details and examples). These results are at odds with the claims that natural selection affecting humans stopped with the spread of agriculture or at least with recent modernization, and investigating only those humans exhibiting lifestyles comparable to those practiced during the Pleistocene is relevant for understanding human evolution. Clearly, agriculture has been a powerful selection force whose effects should be more rigorously investigated, and the continued evolution of humans should be better documented.

Second, analyses of the human genome have also revealed that significant genetic differences both among and, in particular, within human populations have arisen from recent selection events. Many scientists who apply natural selection to understand human behavior have traditionally been uncomfortable with assigning any role for genes in explaining variation among individuals or populations, perhaps because of social Darwinism and racially discriminatory perspectives on human evolution put forward during the early half of the 1900s (see also chapter 22). In contrast, a modern approach to investigating the role of genes in human behavior should focus on studying the effects of mating and reproductive patterns on genetic variation, and genetic constraints on trait evolvability in different populations, as well as on how the documented selection on traits together with their underlying genetic architecture predict responses to such selection.

Third, early attempts to apply an evolutionary framework to contemporary Western populations sparked criticism on the ground that some aspects of the modern industrialized world are too novel, and humans may be responding nonadaptively to them, making studies on adaptive traits in such populations pointless. This view ignores the fact that in both industrialized human societies with easy access to modern contraception and medical care and traditional societies there is a large variance in the reproductive success of both sexes. In other words, although survival to old age is high among all individuals, not everyone has the same family size, and many individuals even forgo reproduction altogether. Such a variance provides material to natural selection that will capitalize on any heritable trait variation linked with higher reproductive success. Thus, even if many behaviors in novel modern environments turn out to be maladaptive, the large opportunity for selection coupled with heritable traits linked with differences in reproductive output of individuals might lead to rapid changes in the genetic makeup of the population over generations, and selection against any traits genetically linked to maladaptive behavior, because any genetically variable traits associated with the variance in reproductive success will experience selection and evolution regardless of the mechanism by which reproductive variance is affected. Consequently, while social Darwinism should not be tolerated, the reality that humans can continue to evolve should not be negated. Yet, because of the trend in human behavioral ecology to focus on the past, and the previous criticism for using other than hunter-gatherers (or to some extent horticulturalists, agropastoralists, or farmers with high mortality and fertility rates) as model populations, only recently have scientists started investigating the behavior of people living in industrialized societies from an adaptationist viewpoint. Even fewer studies have been undertaken to examine how the modern environment itself continues to fuel evolution by favoring or disfavoring certain alleles of the genes, and how the drastic demographic shifts in many populations to low birth and death rates during the recent centuries has affected the overall opportunity for selection or specific trait selection.

Human behavioral ecologists are also criticized for seeking adaptive explanations for behaviors even when such explanations are unlikely. Such criticism applies to all behavioral ecology, but pointing out flaws and factors not correctly considered in the evolutionary models of behavior is obviously easier when the study subject is our own species. It should, however, be stressed that human behavioral ecologists investigate not only how human behavior "fits" the given environment with adaptive benefits but also how environmental conditions constrain individual success. For example, poor early environmental conditions for developing individuals, such as unfavorable month or season of birth, reduce longevity and reproductive performance, yet women commonly reproduce during such times. Social norms, cultural practices, and traditions often lead to reproductive outcomes that are not necessarily beneficial in terms of evolutionary fitness—the study of cultural evolution represents an entire field of research investigating such topics but is not discussed further in this chapter (see chapter 21 for details). Furthermore, poor dietary intake during gestation that leads to reduced birth weight of babies has been shown to be associated with their subsequent risk of adverse health, age at sexual maturation, ovarian function, and life span, which suggests that poor early-life conditions influence development and produce adverse effects later in life. The implications of such effects for evolutionary processes should be considered in more detail.

3. NEW FOCUS ON EVOLUTION IN MODERN SOCIETIES

Recent methodological improvements in the ability to measure selection, heritability, and response to selection in natural populations of animals have inspired many human behavioral ecologists. The central focus of human behavioral ecology has recently begun to shift from asking how the behavior of modern humans reflects our species' historical response to natural selection, to measuring current selection in contemporary populations as well as investigating how that might (or might not) cause evolution. Calculations that incorporate a measure of selection and heritable variation in traits allow us to predict how traits under selection could change over time. Such evolutionary changes in human populations are likely, because natural selection operates on several morphological, physiological, and life history traits in modern societies through differential reproduction or survival, and variation in many of these traits has a heritable genetic basis. This change of focus has led to several important changes in methods and approaches used in the field.

First, the type of information that can be analyzed has become more diverse, allowing researchers to take full advantage of the exceptional data available only for humans. Historical demographers, population geneticists, and evolutionary biologists are making increasingly better use of (*historical*) *population records* of agricultural or industrialized populations. Such data sets have the benefit of large multigenerational samples, although the type of data available is usually limited to demographic information such as births, marriages, reproductive events, and deaths. Recently, there have been promising attempts to make better use of extremely large and versatile *cohort studies* and *twin registers* collected by epidemiologists and social scientists on representative samples of people living in contemporary Europe, the United States, and Australia. *Medical intervention studies* that have collected long-term data on their subjects (e.g., after supplementing mothers' diet during pregnancy) offer a much-needed experimental framework for human behavioral ecologists. These data sets are only now making their way into evolutionary studies. Many scientists are also beginning to use noninvasive manipulations, especially in questions related to sexual selection and mate choice, but also when studying life history strategies. For example, subjects can be exposed to images ("environment") associated with high versus low mortality risk and then asked questions about reproductive investment intentions and preferences. Primatologists have conducted between-species comparisons across primates to draw conclusions on human patterns, and worldwide ethnographies and encyclopedias provide an opportunity to perform similar tests among the large variety of human societies, too. All in all, humans experience the widest global range of living conditions and lifestyles—from traditional communities to extreme urbanization—and human behavioral ecologists today ought to use a wide selection of study designs and data sources to investigate all these populations from an evolutionary perspective.

Second, the focus on studying *microevolution* in contemporary populations has made it necessary to reexamine the old assumption among behavioral ecologists that the details of trait inheritance do not seriously constrain adaptive responses to ecological variation. Estimating heritability of human traits is often considered problematic: an estimation of heritabilities and genetic correlations requires large multigenerational samples and sample sizes often not available in traditional anthropological studies. Furthermore, effects of a common environment shared by close relatives, and cultural transmission, can inflate estimates of heritability. Nevertheless, a review by Stephen Stearns and colleagues (2010) of studies investigating heritability of life history and health traits in humans suggested that although the heritability levels vary considerably among traits and among study populations, many human traits, such as age at first and last reproduction, cardiovascular function, blood phenotypes, weight, and height have measurable heritability and will respond to selection if they are not constrained by genetic correlations with other traits. Fewer studies have investigated such genetic correlations between traits (caused, for example, by the same gene affecting variation in several traits), but there is some suggestion that such effects can set genetic constraints on trait evolution in humans. For example, a study using the historical pedigree records available on rural Finnish people showed significant negative genetic correlations between reproductive traits and longevity (see Pettay et al. 2005). The existence of this genetic variation and covariation implies that females who reproduced at faster rates also had genes for relatively shorter life span, supporting the hypothesis that rate of reproduction should trade off with longevity. Overall, investigation of genes underlying behavioral differences is only beginning in humans, but studies so far suggest that detailed knowledge of the genetic architecture and its dynamics with environmental conditions can provide helpful information on the current evolutionary processes.

Third, an increasing number of studies show that both the opportunity for selection (variation among individuals in fitness) and selection on particular traits can be strong in contemporary populations (see, e.g., Courtiol et al. 2012). The important question is, do these results predict any phenotypic changes taking place in the mean trait values or the genetic makeup of the population over generations? Understanding such responses to selection reveals how the rapidly changing culture, such as medical care, is changing the biology of humans. A recent study by Sean Byars and his colleagues (2009) measured the strength of selection, estimated genetic variation and covariation, and predicted the response to selection for life history and health traits in the current US population. Natural selection appears to be causing a gradual evolutionary change in many traits: the descendants of the study women were predicted to be on average slightly shorter and stouter, to have lower total cholesterol levels and systolic blood pressure, to have their first child earlier, and to reach menopause later than they would in the absence of evolution. A similar study on a preindustrial French-Canadian population found natural selection to favor an earlier age at first reproduction among women, a trait that was also highly heritable and genetically correlated to fitness in this population. Age at first reproduction declined over a 140-year period and also showed a substantial change in the breeding value (part of the deviation of an individual phenotype from the population mean due to the additive effects of alleles), suggesting that the change occurred largely at the genetic level. These studies demonstrate that microevolution might be detectable over relatively few generations in humans. It must, however, also be borne in mind that phenotypic changes may not always provide robust evidence of evolution, as they may not reflect underlying genetic trends. Many traits such as height, weight, mortality, age at first reproduction, and family size have shown strong secular changes during a *demographic transition* (the change from high birth and death rates to low ones as a country develops from a pre-industrial to an industrialized economic system) that may mostly be associated with rapid changes in diet, medicine, and contraception availability. Further studies focusing on how selection interacts with changing early- and later-life environment of individuals and is associated with changes in specific sections of the genome are thus needed.

4. WHAT CAN HUMAN BEHAVIORAL ECOLOGY CONTRIBUTE TO THE GENERAL STUDY OF EVOLUTION?

Evolutionary studies on humans are said to suffer from many drawbacks compared with investigations on model animals, because the data are "correlational"—given the difficulty in conducting experiments, and the study objects are exceptionally long-lived, which complicates the collection of lifelong data in the field. Nevertheless, humans make it feasible to investigate the details of many central questions in evolutionary biology.

Only in humans is it possible to work on databases that contain the lifetime vital records, medical history, and a range of physical and psychological details for up to millions of recognizable individuals that can in some cases be traced back for several generations. Such data sets allow researchers to investigate selection on and evolutionary change in physiological and health-related traits that could never be feasibly collected for any other animal in natural conditions. Moreover, such data sets also allow studies in selection on personality and cognitive abilities, which have become popular among behavioral ecologists working on animals, but in humans these can be explored in greater detail than in other species and can be linked to lifetime reproductive success. Furthermore, huge investments in documenting the human genome shadow those available for most other species, and genetic data are sometimes available alongside historical pedigree data. In addition, ongoing large research programs to unravel developmental origins of health and disease in humans should offer excellent opportunities to investigate the evolutionary implications of interplays between developmental conditions and genetics in a much longer-lived species than those studied so far.

Data available on humans also allow investigations of fitness in a more reliable way than is often possible in similarly long-lived other species, or even in short-lived species in the wild. Many registers allow accurately determining the numbers of grandchildren for each individual, and these provide a far better measure of fitness than simply the number of offspring born, given the considerable trade-offs detected between offspring quantity and quality in humans (and likely in many other species, too, in which large parental investment improves offspring survival and mating success). Importantly, population-based registers allow inclusion of those individuals who never reproduce into the calculations of variance in fitness, which appears crucial given that in the past as well as present human populations, a large fraction of each birth cohort fail to contribute their genes to the next generation, and selection is often strongest through recruitment differences rather than differences in the family size among those who do reproduce.

Many "natural experiments" also offer opportunities to investigate evolutionary questions. Such events involve well-documented famines such as the Dutch Hunger Winter during the Second World War (see, e.g., Roseboom et al. 2001 for details), sex-ratio biases created by wars, documented long-term year-to-year variation in crop success and local ecology linked with individual fitness data, or large-scale changes in the demographic parameters of the population that have occurred repeatedly across the world but at different periods in different countries.

Given that humans exhibit all mating systems documented in the animal kingdom (monogamy, polygyny, polyandry, and even promiscuous mating), they also offer interesting opportunities for investigating how changes in mating system affect selection. For example, over the reproductive lifetimes of Utahans born between 1830 and 1894, socially induced reductions in the rate and degree of polygamy corresponded to a 58 percent reduction in the strength of sexual selection, illustrating the potency of sexual selection in polygynous human populations and the dramatic influence that short-term societal changes can have on evolutionary processes.

Finally, humans are also exceptional in that it is possible to reliably study individual variation in complex cognitive traits. Researchers have used methodology relying on simple experimental settings to collect quantitative data on traits such as mating preferences, cooperativeness, and personality. Similar studies are virtually impossible to conduct on animals, because the methods involve a certain degree of abstraction. For example, the same individuals can be asked to choose between large numbers of fictive alternatives, such as hypothetical partners. These preferences for mate characteristics can then be further compared with real-life partner characteristics, and the ecological and individual causes and fitness consequences of the degree of mismatch between preferences and actual pairings can be examined.

FURTHER READING

Alvergne, A., and V. Lummaa. 2010. Does the contraceptive pill alter mate choice in humans? Trends in Ecology & Evolution 25: 171–179.

Byars, S. G., D. Ewbank, D. R. Govindaraju, and S. C. Stearns. 2009. Natural selection in a contemporary human population. Proceedings of the National Academy of Sciences USA 107: 1787–1792.

Courtiol, A., J. Pettay, M. Jokela, A. Rotkirch, and V. Lummaa. 2012. Natural and sexual selection in a monogamous historical human population. Proceedings of the National Academy of Sciences USA 109: 8044–8049

Dunbar, R., L. Barrett, and J. Lycett. 2005. Evolutionary Psychology: A Beginner's Guide. Oxford: OneWorld.

Hawkes, K., J. F. O'Connell, N. G. Blurton Jones, H. Alvarez, and E. L. Charnov. 1998. Grandmothering, menopause, and the evolution of human life histories. Proceedings of the National Academy of Sciences USA 95: 1336–1339.

Hawkes, K., J. F. O'Connell, and L. Rogers. 1997. The behavioral ecology of modern hunter-gatherers, and human evolution. Trends in Ecology & Evolution 12: 29–32.

Lahdenperä, M., V. Lummaa, S. Helle, M. Tremblay, and A. F. Russell. 2004. Fitness benefits of prolonged post-reproductive lifespan in women. Nature 428: 178–181.

Laland, K. N., and G. R. Brown. 2011. Sense and Nonsense: Evolutionary Perspectives on Human Behavior. 2nd ed. Oxford: Oxford University Press.

Moorad, J. A., D.E.L. Promislow, K. R. Smith, and M. J. Wade. 2011. Mating system change reduces the strength of sexual selection in an American frontier population of the 19th century. Evolution and Human Behavior 32: 147–155.

Pettay, J. E., L.E.B. Kruuk, J. Jokela, and V. Lummaa. 2005. Heritability and genetic constraints of life-history trait evolution in preindustrial humans. Proceedings of the National Academy of Sciences USA 102: 2838–2843.

Roseboom, T., J. van der Meulen, C. Osmond, D. Barker, A. Ravelli, and O. Bleker. 2001. Adult survival after prenatal exposure to the Dutch famine 1944–45. Paediatric and Perinatal Epidemiology 15: 220–225.

Stearns, S. C., S. G. Byars, D. R. Govindaraju, and D. Ewbank. 2010. Measuring selection in contemporary human populations. Nature Reviews Genetics 11: 611–622.

Wilson, E. O. 1975. Sociobiology: The New Synthesis. Cambridge, MA: Harvard University Press.

CHAPTER 6

EVOLUTIONARY PSYCHOLOGY

Robert C. Richardson

OUTLINE

1. The Darwinian background for evolutionary psychology
2. The modern-day program of evolutionary psychology
3. Psychological evidence
4. The application of evolutionary models in evolutionary psychology
5. Evolutionary alternatives

Evolutionary psychology is an approach to cognitive psychology that aims to inform work in psychology with evolutionary ideas and to re-form cognitive science by placing it in an evolutionary context, that is, by focusing on how psychological traits such as aggression, mate se-lection, and social reasoning were adaptive in ancestral environments. This methodology involves a variety of psychological and behavioral evidence that is relatively independent but may be interpretable in evo-lutionary terms; in other cases, it involves psychological models that depend on evolutionary models. One such example is incest aversion, which can be interpreted in terms of kin selection or inclusive fitness. There are problems in integrating the two domains. More specifically, the evolutionary interpretations often lack empirical evidence. In gen-eral, it seems evolutionary psychology could benefit from a more inclu-sive and contemporary infusion of evolutionary theory.

GLOSSARY

Computational Mechanisms. Algorithms that compute determinate input-output functions, dependent only on the structure of representations involved; sometimes called *Turing computability*.

Ecological Rationality. A hypothesis characteristic of evolutionary psychology that what counts as a rational procedure is relative to the ecological context in which it is applied, and cannot be determined without knowing the context in which it is applied.

Environment of Evolutionary Adaptedness (EEA). The environment characteristic of human evolution, both physical and social; sometimes the EEA is thought of as the environment of Pleistocene ancestors in the African savanna; sometimes it is treated as a statistical composite of ancestral environments.

Incest Taboos. General social prohibitions against sexual relations among more closely related individuals; among humans, the paradigm is the prohibition of sexual activity between individuals closer than second cousins. This is often held to be a human universal.

Modules. Cognitive mechanisms that operate in relative independence from other mechanisms that govern other domains (e.g., face recognition is a capacity in humans that is relatively independent of other capacities).

Social Exchange. Any exchange of value among individuals, with costs and benefits attached; in more interesting cases, these involve iterated exchanges in which reciprocal altruism can be effective.

Evolutionary psychology (EP) is a field of research that seeks to rely on evolutionary biology in the development and elaboration of specifically human psychological hypotheses or psychological mechanisms; more generally, EP looks to the integration of cognitive psychology with evolutionary biology in explaining and interpreting human behavior. EP developed from sociobiology, with the perspective that because humans, like all other organisms, have evolved, and the principles of evolutionary biology are universal, evolutionary theories will help us understand our own origins and features (see chapters 3 and 5). Yet, the application of principles of evolution to humans still engenders debate and skepticism (see also chapter 5), as it did when the idea of evolution was first introduced.

The primary range of the cognitive models in EP includes such issues as attraction to mates, patterns of jealousy, reasoning applied to *social exchange*, probabilistic reasoning, and *incest taboos*; it also includes more controversial topics such as the evolution of rape, differences between male and female aggression, and the patterns of child abuse. Psychological mechanisms are assumed to be subject to shaping by natural and sexual selection, and as a result, current behavioral patterns can be understood in terms of human evolutionary history. EP assumes that current psychological mechanisms are adaptations to ancestral environments and not to contemporary environments. Dietary preferences, for example, that may have been adaptive in ancestral environments, such as a preference for sweet food, are not adaptive in our current environment. The same disparity should apply to other psychological patterns and mechanisms.

Interpreting psychological hypotheses and mechanisms in terms of evolutionary principles is not a simple matter. In one view, evolutionary theory may be used primarily as a heuristic for defining and elaborating psychological hypotheses. Some cases seem to fit in this category, such as gender differences in the sorts of traits that are attractive in potential mates. This use seems relatively unproblematic but makes no substantive use of evolutionary theory.

In contrast, evolutionary models may be integrated into the evidence for psychological hypotheses, supposedly contributing to their evidential credentials. For example, incest taboos are argued to have evolved by natural selection and function to avoid inbreeding. Examples of this sort are the most controversial. Here, the specific evolutionary models are often not supported by adequate evidence. When this is so, the psychological hypotheses are correspondingly uncertain, or at least no more certain than otherwise warranted by the psychological evidence. For example, human language is plausibly an evolutionary adaptation: given the complexity of the underlying structure and function of the mechanisms, the incorporation of recursive grammars, and the complex patterns in the acquisition of children's languages, this is the sort of complex mechanism we should expect to be an adaptation. In this portrayal, human languages are *adaptations* for human communication, which is no doubt true given the importance of language; but we are left in the dark about the specific features of human languages, such as their recursive structure, that likely make them adaptations. Communication, for example, is adaptive, but the connection of a general appeal to communication to recursive structures is not clear. In such cases, there is a disconnection between the supposed adaptation and the adaptive model.

1. THE DARWINIAN BACKGROUND FOR EVOLUTIONARY PSYCHOLOGY

Attempts to unify evolution and psychology date to Darwin, but contrary to the modern pursuit of identifying the mechanisms of evolution such as kin selection or natural selection that shape human psychology, Darwin's focus was more on arguing that human psychology has evolved than on exploring or suggesting how psychology evolves. There are few mentions of human evolution in *On the Origin of Species*, but Charles Darwin did write in the final chapter that the *Origin* would "open fields for far more important researches. Psychology will be based on a new foundation, that of the necessary acquirement of each mental power and capacity by gradation. Light will be thrown on the origin of man and his history" (1859). Out of context, this appears to suggest that interpreting psychology in the light of *natural selection* would put psychology on a "new foundation." However, this passage appears in a section that lists topics that Darwin felt would be transformed by acknowledging evolution. He did not discuss the mechanism of evolution in this section, just its existence. Thus, while he tackled head-on the most contentious area of all—what would seem to differentiate us from all other animals—human psychology, he did so without reference to the mechanism of natural selection.

Darwin developed his ideas on human evolution more fully in *The Descent of Man and Selection in Relation to Sex* (1871). In the opening passage of *The Descent* he writes:

> He who wishes to decide whether man is the modified descendant of some pre-existing form, would probably first enquire *whether man varies*, however slightly, in bodily structure and in mental faculties; and if so, *whether the variations are transmitted* to his offspring in accordance with the laws which prevail with the lower animals; such as that of the transmission of characters to the same age or sex. Again, are the variations the result, as far as our ignorance permits us to judge, of the same general causes, and are they governed by the same general laws, as in the case of other organisms? (Darwin 1871, 9; italics added.)

In both the *Origin* and in this passage from *The Descent*, Darwin is discussing evolution, or common descent, and not specifically natural or sexual selection. There is appeal to variations and to inheritance, and to the "laws" governing each of them, but there is not a hint of competition or the "struggle for existence," much less of natural selection. Here in *The Descent*, he initially recapitulates the argument for common descent from the *Origin*, extending it to what he calls the "mental faculties" of man, saying at the outset that his object "is solely to shew that there is no fundamental difference between man and the higher mammals in their mental faculties" (1871). Darwin is clear that this commitment to evolution is meant to include what he calls the "moral sense." This was crucial for Darwin. It meant, among other things, that our capacities for social interaction and our psychological propensities were meant to be within the purview of his evolutionary theory.

Darwin was neither the first nor the last to bring evolutionary insights to the discussion of our social sentiments and reasoning. In the nineteenth century, Herbert Spencer had placed his discussion of psychology in an explicitly evolutionary setting before the publication of the *Origin*; William James's psychology was inspired by Darwinian insights, as were other important psychologists at the turn of the century. In the twentieth century, there were other ventures into the evolution of human psychology, some of which are in retrospect less well regarded, such as Desmond Morris's *The Naked Ape*. With the elaboration of models designed to capture social behavior in the middle of the last century, *Sociobiology* by E. O. Wilson dealt with the task of capturing animal behavior in evolutionary terms, and almost as an appendix extended that project to the domain of human social behavior. From Wilson's book, the field of sociobiology was born and thrived in the 1980s, with various attempts to extend sociobiology to encompass the human case. This brings us to the most recent approach, evolutionary psychology, which takes up the Darwinian idea that evolution should shed light on human psychology, and which has usurped sociobiology.

2. THE MODERN-DAY PROGRAM OF EVOLUTIONARY PSYCHOLOGY

Contemporary evolutionary psychology is not a homogeneous collection of views, even with respect to its evolutionary commitments, though it is possible to articulate a loose set of claims that are broadly endorsed, and typical of contemporary adherents. In large part, these are commitments consistent with evolutionary theory as it was articulated during the "evolutionary synthesis" years in the first half of the twentieth century, updated by evolutionary models from the 1960s. Not every advocate of EP is committed to precisely the same set of claims, but it is possible to provide a kind of portrait. The following are some characteristic commitments:

- Psychological mechanisms are the result of natural selection and sexual selection. While it is generally acknowledged that evolution has some outcomes that are owing to chance or are by-products of selection for other traits, the focus of EP is on traits presumed to be adaptations and therefore that reflect evolution from selection—traits centered on problems such as finding a mate, cooperative activities like hunting, or the rearing of offspring. The assumption is that natural selection will tend to "solve" problems like this with considerable efficiency. Possible alternatives to selection are rarely considered, nor are alternative selectionist regimes.
- Psychological mechanisms can be thought of as *computational mechanisms*. Among such mechanisms are included cognitive processes (e.g., probabilistic reasoning or problem solving) as well as emotional responses (e.g., jealousy or fear). The idea that psychological mechanisms are computational is a common assumption among a range of cognitive scientists, though its prevalence has faded considerably in the last decade or so. Alternatively, these computational mechanisms can be thought of as exhibited in and causing behavioral *strategies* for responding to environmental challenges, where the strategies are genetically specified.
- Psychological mechanisms evolved in response to relatively stable features of *ancestral* environments, often collectively referred to as the *environment of evolutionary adaptedness* (EEA). EP asserts that because most of human evolution took part during the Pleistocene (roughly 2.6 million to 12,000 years ago), and presumably in the later Pleistocene, what is seen today in terms of psychology evolved to be adaptive in this hypothetical EEA. Often, the EEA is identified with the savanna of the African Pleistocene, with a hunter-gatherer lifestyle. The EEA can also be identified with a kind of statistical aggregate of the total range of ancestral environments.
- Because psychological mechanisms are adaptations to ancestral environments, there is no assumption that they are adaptive in contemporary cir-

cumstances. Social environments are a significant part of the environment and are obviously crucial to human evolution. If we assume with EP that our ancestral social environment consisted of small, nomadic bands of relatives, then the difference between that and our contemporary culture suggests that whatever strategies were adaptive for our ancestors, may not be so for us. Likewise, if we assume that our distant ancestors lived in a sugar-deprived environment, then our fondness for sweets might be "natural" though no longer adaptive. In general, EP assumes that evolutionary responses are too slow to have had any significant effect in the last 12,000 years or so, since the advent of agriculture and sedentary life.

• The human mind is a kind of mosaic of mechanisms, each with some specific adaptive function, rather than merely a general-purpose learning machine. Different adaptive problems will require different solutions and different strategies for dealing with them. So, for example, a mechanism for mate selection is unlikely to be of much use in foraging. At least some of this machinery must be domain specific, specialized for particular tasks, and some of these mechanisms may count intuitively as *instincts*. Several advocates of EP treat these mechanisms as *modules*, though others insist that all that is required is distinct domain-specific mechanisms.

3. PSYCHOLOGICAL EVIDENCE

Evolutionary psychologists make use of an array of techniques to evaluate their psychological models, most of which do not specifically depend on the evolutionary assumptions. These methods include, among others, the use of questionnaires, controlled experiments, observational methods, and brain imaging (functional magnetic resonance imaging [fMRI] and positron emission tomography [PET]). Some also make use of a variety of less standard techniques, including ethnographic records, paleontological information, and life history data. Evolutionary assumptions do come into play in advancing and formulating hypotheses, as suggestive of psychological hypotheses to test. Whether they are more than merely heuristic is sometimes not clear.

Evolutionary psychologists have articulated and tested a wide array of psychological hypotheses inspired by evolutionary thinking. These include human propensities for such matters as cooperation and cheater

detection, differences in spatial memory, and short-term mating preferences. Some simple examples may be sufficient to illustrate the method.
Assume that human memory is sensitive to items that affected fitness
among our ancestors, such as food items, shelter, or possible mates.
Using standard experimental memory probes within psychology that
are concerned with recall and recognition for lists of words, researchers
found that recall for survival-oriented terms was significantly better
than recall for more neutral words. Similarly, theories of parental investment suggest that given monogamous coupling, females will tend to
prefer mates that are more likely to invest in offspring. From an EP perspective, this also suggests that males and females will differ in the patterns of jealousy, with females on average more sensitive to emotional
infidelity (as a risk of abandonment) and males more sensitive to actual
sexual infidelity (as a risk to paternity). These predictions have been
supported by straightforward evaluations of preferences using questionnaires, spontaneous recall, and fMRI.

4. THE APPLICATION OF EVOLUTIONARY MODELS IN EVOLUTIONARY PSYCHOLOGY

Relying on work in paleoanthropology and ethnography relating especially to contemporary hunter-gatherers, evolutionary psychologists
have elaborated a portrait of ancestral social life. While their description
is plausible, it is also controversial among anthropologists. EP assumes
ancestral hominids lived in relatively compact kin-based groups of no
more than 100 members. It is presumed that a sexual division of labor
existed, with males more engaged in hunting and females more engaged
with gathering, and that stable male-female bonds existed, as well as
long periods of biparental care. Within each kin group, there was cooperative foraging. Much more is known about the biotic and abiotic environment that existed. For example, it is known that during this time,
humans were subject to a variety of predators and pathogens and considerable variance in the availability of resources.

Assuming this broad portrait of early human social life and abiotic
influences allows evolutionary psychologists to construct a variety of
evolutionary scenarios. Depending on the case, they use a variety of re-

sources from evolutionary biology, including theoretical models concerning reciprocal altruism (see chapter 4), parental investment, kin selection, and evolutionary game theory. Beginning with the relevant dimensions assumed to be typical in the EEA, evolutionary psychologists construct an account of the adaptive functions that must be satisfied, rather like a design specification. The problem for EP is then to reverse-engineer a solution to the adaptive problem and test it in modern populations.

Reverse engineering is a powerful theoretical tool, but it can lead to difficulties and criticism, especially if the "adaptive problem" is not clear. If it is poorly articulated, then it is not clear whether the evolutionary solution is the right one. To use a nonhuman example, if the ecological "problem" is how insects can walk on water, knowing the adaptive "solution" depends on the surface tension of water, and that in turn depends on knowing the saline content; absent the determinate content of the problem, there is no general solution to crucial issues such as foot structure, though the specific solutions are solved readily. In the human case, language is certainly involved in communication, but this fact offers no explanation for the peculiarities of human language—its recursive structure, for example—which are plausibly adaptations.

EP also raises issues about connecting the psychological hypotheses and the evolutionary interpretations. Consider a Darwinian theory of the evolution of incest avoidance. Incest is a very interesting case of a social prohibition, since psychological studies show that disapproval of it survives even the recognition that it will produce no actual harm. It has been a very significant issue, first for sociobiologists and now for evolutionary psychologists. There is a straightforward case against incest from an evolutionary perspective based on inbreeding. Inbreeding can result in reduced fitness, termed *inbreeding depression*. Where inbreeding depression exists, there should be an evolutionary pressure against inbreeding.

The importance of inbreeding depression leads EP advocates to suggest that there is a natural tendency—sometimes a psychological "module"—for incest avoidance. Debra Lieberman, together with Cosmides and Tooby, has suggested that humans have a specialized kin recognition system (there are such mechanisms in other animals). They observe that incest avoidance could facilitate an avoidance of any deleterious consequences associated with inbreeding depression and suggest that this leads to selection for incest avoidance.

The *Westermark hypothesis* posits a mechanism of the sort Lieberman, Cosmides, and Tooby predicted, suggesting that children raised together develop a sexual disinterest, or even a sexual aversion, to one another. The proposed function is to avoid incest, since those who are raised together are most often closely related. Lieberman assumes, reasonably, that coresidence during periods of high parental investment should be a reliable indicator of kinship or would have been a reliable indicator in the EEA with small kinship bands. Together with Cosmides and Tooby, Lieberman shows considerable support for the conclusion that duration of coresidence is psychologically predictive of sexual aversion.

The evolutionary interpretation that incest avoidance evolved because of selection against inbreeding is plausible on the surface but nonetheless problematic. The association cannot be directly tested in ancestral populations, but it does fit the patterns of some contemporary "hunter-gatherer" populations (which may be considered a proxy for ancestral populations), though not all. It is, of course, true that siblings would typically be associated with one another during childhood, but the proper question is whether in ancestral groups the set of people that an individual may have selected from when choosing a mate included the siblings with which one interacted as a child.

A more straightforward and general problem with EP is assuming a single EEA. We know that our Pleistocene ancestors did not have simply one lifestyle in one region but lived on the African savanna, in deserts, next to rivers, by oceans, in forests, and even in the Arctic, employing very different foraging methods and living off diverse diets, with technologies ranging from the simple chopping tools of *Homo habilis* to the rich and sophisticated stone, bone, and antler toolbox of late Pleistocene *Homo sapiens*. There is little reason to think that there was a single form of social structure associated with the full range of human physical environments, much less that contemporary "hunter-gatherer" populations are typical of ancestral groups. For the hypothesis of incest avoidance as a mechanism to avoid inbreeding depression, it is hard to know whether association will be limited to siblings, or more likely to be with siblings, absent a fairly specific account of social organization, including the relative viscosity of the groups and issues such as group size.

In many animal species, there is a tendency for animals to disperse prior to mating; they move away from their familial unit. This clearly has the effect of reducing inbreeding, though there doesn't seem to be any consensus on whether incest avoidance is particularly significant in supporting dispersal. Among chimpanzees, when males come of reproductive age, they tend to emigrate from the ancestral clan. There is no need for incest aversion, since they move away from their siblings. To know how to apply inbreeding avoidance models to ancestral human groups, it would be necessary to know whether both males and females remained with the ancestral groups or emigrated. There is some evidence that among early *Homo*, the males tended to move out of their ancestral groups once they were reproductive, as with chimpanzees, but it doesn't matter whether this is correct. The important point is that absent such information, the relevance of the evolutionary models of selection to incest aversion is not clear. If reproductives tend to emigrate, then there is little need for incest aversion, and none for incest taboos.

5. EVOLUTIONARY ALTERNATIVES

The preceding general assumptions that form the backdrop for EP are characteristic of only a selected subsample of work in evolutionary biology. More generally, evolutionary biology incorporates a wide variety of disciplinary perspectives that do not feature in the work within EP. Many of the assumptions characteristic of EP may be problematic, because they do not incorporate more recent advances in our understanding of how evolution works. There are many recent developments in genetics, in evolutionary biology, and in developmental biology that might improve EP considerably, bringing it more in line with more recent evolutionary thinking. Several are briefly noted here.

- Natural selection and sexual selection are doubtless potent evolutionary forces. There are alternative evolutionary factors that can, and do, affect evolutionary trajectories. Evolutionary psychologists acknowledge such factors as genetic drift (though it plays no role in their scenarios, and they do not address it as an alternative) but do not incorporate phylogeny or comparative biology. As a result, our primate kin do not typically feature in EP explanations. One salutary change would be to take account systematically of our relatedness to our primate kin and the possibility that features exhibited by humans are inherited from a common ancestor. At the very least, this would provide expectations as to social patterns that feature so prominently in EP, and give testable alternative hypotheses. In particular, it would downplay the commitment to natural selection acting on specifically human social capacities.

- EP typically assumes that the relevant selection forces are relatively ancient and that recent changes are insignificant. From the perspective of EP, modern humans are Pleistocene relics. Yet, we know that there have been substantial changes in the human genome over even the last 10,000 years and that these changes are ongoing (see chapter 23). Many of the evolutionary changes reflect the adoption of agriculture and the domestication of animals, environmental changes that surely impose selection.

- The environment of the Pleistocene is known to have been highly spatio-temporally variable. The environment of the early Pleistocene was very different from, say, the late Pliocene. Moreover, humans came to be widely dispersed, occupying a variety of distinctive environments. Given what we know, it would be reasonable as well to think that social structures would be different in different physical environments—for example, some would be more conducive to sedentary lifestyles, and others to more mobile ones. Though humans are not as genetically diverse as many other animals, there is sufficient genetic variation to support genetic changes in relatively short amounts of time.

- Human behavior is both adaptive and malleable. EP tends to assume, by contrast, that evolved computational programs are species specific and species universal. When there is within-species variation, the assumption is that the strategies are conditional, evoked in different conditions. The validity of this conjecture is questionable. Evolutionary biologists have found that the rate of evolution can be much faster than EP tends to assume. Advances in our understanding of epigenetics and developmental plasticity provide alternatives to conditionality that are typically not incorporated in EP. It is not that EP assumes some form of genetic determinism; rather, the point is that the kind of interplay seen among genetic factors, epigenetic influences, and learning makes universals less likely.

- There are significant alternatives to the typical emphasis of EP on individual- and gene-centered models of evolution. This is an issue beyond the problems of applying EP's preferred modes of analysis. Gene-culture co-evolution may be an important source of evolutionary changes (see chapter 21). This is becoming a well-developed alternative, emphasizing the role of cultural practices in modifying the human brain. In general, gene-culture dynamics can enhance and accelerate rates of evolution, even if we do not yet know how important these are. Multilevel selection models are also being developed. With distinctive, genetically isolated groups that compete as groups, it is possible to develop models for the evolution of social behavior that do not assume the typically individual- and gene-oriented perspective of EP.

There are alternatives that could enrich the work within EP but that typically remain beyond its purview. Darwin was right to think that evolution should reshape our understanding of human psychology. There are many avenues yet to explore in seeing how an evolutionary perspective can contribute to our understanding of human psychology. Most of these avenues are ahead of us.

FURTHER READING

Barkow, J., L. Cosmides, and J. Tooby, eds. 1992. The Adapted Mind: Evolutionary Psychology and the Generation of Culture. New York: Oxford University Press. *A collection that defined the case for evolutionary psychology.*

Bolhuis, J. J., G. R. Brown, R. C. Richardson, and K. N. Laland. 2011. Darwin in mind: New opportunities for evolutionary psychology. PLoS Biology 9: e1001109. doi:10.1371/journal.pbio.1001109.

Buller, D. J. 2005. Adapting Minds: Evolutionary Psychology and the Persistent Quest for Human Nature. Cambridge, MA: MIT Press. *A critique focused on the psychological case for evolutionary psychology.*

Buss, D. M. 1995. Evolutionary psychology: A new paradigm for psychological science. Psychological Inquiry 6: 1–49. *A very important and useful overview from an advocate.*

Buss, D. M., ed. 2005. The Handbook of Evolutionary Psychology. New York: John Wiley & Sons. *A more recent comprehensive collection.*

Laland, K. N., and G. R. Brown. 2011. Sense and Nonsense: Evolutionary Perspectives on Human Behaviour. 2nd ed. Oxford: Oxford University Press. *A criticism of EP from within evolutionary biology and recent psychology.*

Richardson, R. C. 2007. Evolutionary Psychology as Maladapted Psychology. Cambridge, MA: MIT Press. *A philosopher of biology skeptical of the evolutionary credentials of EP.*

PART II

Evolution in Health and Disease

CHAPTER 7

EVOLUTIONARY MEDICINE

Paul E. Turner

OUTLINE

1. Evolution and medicine
2. Pathogens
3. Defense mechanisms
4. Trade-offs in human traits
5. Mismatches to modernity
6. Implications of evolutionary medicine

Evolutionary biology concerns the *ultimate* origins of trait variation within and among populations; human medicine, by contrast, concerns the *proximate* causes of disease, including the effects of individual differences. Of course, knowing *how* disease symptoms arise is essential for practicing medicine, but understanding *why* these symptoms occur is also crucial. The merger of medicine and evolutionary biology represents the new field of *evolutionary medicine*, defined as the use of modern evolutionary approaches to better understand human health and improve disease treatment. A central question in evolutionary medicine is this: Why has natural selection left our bodies vulnerable to disease? This chapter focuses on four possible answers. First, human evolution is too slow to cope with the coevolutionary arms races involving microbial pathogens. Second, our evolved defense systems against these pathogens may, paradoxically, have harmful effects. Third, there are limitations and constraints on what selection can do, and disease often results from constraints including evolutionary trade-offs. Last, human evolution can't keep pace with novel environments, especially those of our own making, leading to diseases that result from a mismatch to modernity.

GLOSSARY

Coevolutionary Arms Race. The sequence of mutual adaptations and counteradaptations of two coevolving species, such as a parasite and its host.

Evolutionary Trade-Off. A balancing between two traits that occurs when an increase in fitness (reproduction and survival) due to a change in one trait is opposed by a decrease in fitness due to a concomitant change in the second trait.

Germ Theory of Disease. The once-controversial idea that certain diseases are caused by invasion of the body by microbes; research by Louis Pasteur, Joseph Lister, and Robert Koch in the late nineteenth century led to widespread acceptance of the theory.

Hygiene Hypothesis. The idea that a lack of early childhood exposure to infectious agents, symbiotic microorganisms (e.g., gut flora), and parasites increases susceptibility to allergic diseases by interfering with the normal development of the immune system.

Mismatch to Modernity. Maladaptation produced by time lags, especially the inability of human adaptation to keep pace with rapid cultural changes (e.g., ready access to high-caloric foods).

1. EVOLUTION AND MEDICINE

In *On the Origin of Species* (1859), Charles Darwin described how natural selection led parasites to exploit their hosts. His examples included wasps whose eggs are laid on or inside the bodies of specific host insects, and whose larvae develop within them; and cuckoos that lay their eggs in the nests of other birds, relying on them to rear their young. Soon after the book's publication, Louis Pasteur's experiments in the 1860s confirmed that some more familiar diseases were caused by parasitic microbes, which motivated improvements in medical hygiene that saved countless human lives. Given these parallel advances in evolution and medicine, it is ironic that the hybrid discipline of evolutionary medicine did not gain traction until the 1990s, when evolutionary biologist George Williams and physician Randolph Nesse argued that the understanding of human illness should be informed by evolutionary thinking.

The historical separation between the two disciplines reflects medicine's proximate focus on restoring the proper functioning of the human body without considering how our evolutionary history has shaped human health and disease. Nonetheless, such evolutionary considerations help explain why humans become ill, and provide knowledge that can be harnessed to suggest new or refined methods of treatment. In particular, evolutionary thinking might generate insights and improve health in myriad ways, including improved practices for the prescription of antibiotics, management of virulent diseases, administration of vaccines, treatment of cancer, advice for couples with difficulties in conceiving and carrying offspring, counters to the recent epidemics of obesity and autoimmune diseases, and insights into how and why aging occurs. The focus of evolutionary medicine is certainly not to provide an alternative to current medical training with an understanding of evolution; rather, it is to demonstrate that evolutionary biology is a useful basic science that poses new medically relevant questions, raises hypotheses and possible answers, and thereby contributes to research while also improving medical practice.

The field of evolutionary medicine is relatively new, but the following sections provide some examples of evolutionary insights that have refined the understanding of medical issues, and even some instances in which medical treatments have been modified owing to evolutionary thinking.

2. PATHOGENS

Parasitism is perhaps the most common lifestyle on the planet. Throughout history humans have suffered widespread morbidity and mortality caused by infectious diseases. Some of these infections are evident even in archaeological remains, such as ancient Egyptian hieroglyphics that depict humans afflicted with limb deformities characteristic of childhood infection by poliovirus (figure 7-1). Other evidence comes from mummified human remains showing, for example, the typical scarring due to skin lesions associated with the variola virus infection that causes

Figure 7-1. Ancient Egyptian carving showing a priest with a shriveled leg typical of a recovered case of paralytic poliomyelitis (polio). Polio is an infectious disease caused by polioviruses that can permanently damage parts of the nervous system. In some cases, as seen here, it can cause paralysis. This bas-relief was carved around 1500 BCE. (Copyright Photo Researchers.)

smallpox. Diseases caused by pathogens constitute a substantial fraction of human mortality, perhaps on the order of 25 percent of all deaths per year. One goal of evolutionary medicine is to promote human health by improving therapies to combat infectious disease agents, based on the knowledge of how pathogens evolve in general, and especially in relation to selection pressures imposed by the human immune system and by current therapies.

The evolution of antibiotic resistance in bacterial pathogens is a popular and important example used to illustrate the process of adaptation by natural selection. The age of antibiotics began with the discovery of penicillin in the late 1920s, followed by small-scale attempts to use it to treat patients in the 1930s, and its mass production and application in the 1940s. Unfortunately, owing to the evolution of resistance, penicillin is no longer effective in combating many pathogens. Moreover, the pharmaceutical industry is lagging in the race to develop new antibiotics that can replace penicillin and other once-effective antibiotics. Widespread therapeutic use of antibiotics (along with the use of antibiotics as prophylactics in agriculture) has led to a very large "experiment": bacteria have been exposed to vast amounts of antibiotics that select for bacterial genotypes with greater resistance to the effects of antibiotics. The large population sizes and short generation times of bacteria virtually assure rapid evolution in response to this strong selection. The unfortunate rise of antibiotic-resistant bacteria reflects not only selection for mutations that confer resistance but also the ability of many bacteria to obtain resistance genes via horizontal gene transfer from other bacteria (including conspecifics and even other species); this has allowed resistance to proliferate even more quickly within and among species of bacteria. The emergence of antibiotic resistance is thus easily understood from an evolutionary perspective; in fact, if any scientific puzzle remains, it is why some bacterial pathogens have not (at least yet) evolved resistance to traditional antibiotics.

The realization that the use, overuse, and misuse of antibiotics have selected for bacterial resistance has begun to influence the practice of dispensing these drugs. Today, most physicians prescribe antibiotics only when necessary, avoiding their use in the case of viral illnesses and for bacterial pathogens against which an antibiotic is known to be ineffective. Recognizing that antibiotic resistance has become a global problem, in 2011 the World Health Organization issued an international call for concerted action to halt the spread of antimicrobial resistance, with specific recommendations on policies in medicine and agriculture to help control the problem. One such recommendation addressed the need for improved measures for preventing the spread of resistant bacteria in hospitals and surrounding communities. For example, highly resistant superbugs, such as the methicillin-resistant *Staphylococcus au-*

reus (MRSA), flourish and spread through hospital-acquired infections. The dangers and costs associated with these superbugs have prompted the development of spatially based evolutionary models that predict the spread of resistant bacteria within hospitals, according to factors such as the number of beds per room and the location of immune-compromised patients. The goal of this evolution-based thinking is to understand and track the spread of infections caused by these superbugs and thereby reduce the risk of acquiring a life-threatening pathogen as an unfortunate consequence of a routine hospital stay.

The rise of antibiotic-resistant superbugs has also necessitated a search for alternatives to traditional antibiotic therapy. To slow or prevent the evolution of antibiotic resistance, three options have been proposed. The first is to continue using the antibiotics in our current drug arsenal but to slow resistance evolution by managing therapy, in particular by prescribing antibiotics in smaller doses for shorter periods. Large doses of antibiotics kill sensitive strains but not the resistant ones, which then are free to multiply without having to compete with the more sensitive strains the antibiotics eliminated. In contrast, lower doses used for shorter periods should allow susceptible strains to survive, thus suppressing the evolution of their resistant competitors. If this therapy can be employed without endangering patient health, it can give the immune system time to clear the infection without promoting the evolution of antibiotic resistance.

A second alternative is phage therapy, in which certain viruses (called *phages*) that attack bacteria are used to target pathogens. In the 1940s, when Western physicians started widely using antibiotics, phage therapy was already common in the former Soviet Union and Eastern bloc nations for some infections (such as wounds). Like their bacterial hosts, phages multiply exponentially—as well as mutate and evolve—and so they are essentially self-amplifying drugs that can persist in the patient's body. Phage coevolution might also counter bacteria that evolve resistance to a given phage. Unlike traditional antibiotics, which typically kill nonpathogenic bacteria (including, for example, those that help digestion), phages can target pathogenic bacteria without harming the beneficial types of bacteria in the microbiome.

The drawbacks of phage therapy include the following: the narrow host range of a typical phage makes them inappropriate for broad-spectrum use on infections caused by uncharacterized bacteria; phages are living drugs that may evolve in unexpected ways; and phages can interact with the immune system, even though healthy people seem not to mount a large immune response. In any case, phage therapy is increasingly seen as a viable option to the growing problem of antibiotic resistance worldwide.

The third alternative strategy proposes to disrupt various signals or substances produced by bacteria that function as *public goods*—that is, products that benefit all members of the infecting bacterial population but impose a metabolic cost on the individual bacteria that produce them. Two examples of these public goods are molecules that signal their local population density, and siderophores produced to scavenge iron. Both of these public goods are relevant as virulence factors because some pathogens attack host tissues only if they detect that their density is sufficient to overcome the immune system, and because bacteria need iron to grow and reproduce. Certain chemicals can be used to quench or otherwise interfere with these extracellular public goods, offering potential advantages over traditional antibiotics. For example, their extracellular action avoids resistance mechanisms that block drug entry. Moreover, fitness benefits that might be generated by mutants producing more of the public goods would be diluted across the pathogens in a given host tissue, thereby limiting a mutant's proliferation. Also, such chemotherapeutic agents are not likely to have been experienced previously by bacteria, making it unlikely that they would harbor preexisting resistance.

Bacteria and fungi evolved antibiotic production and resistance millions of years before humans existed, and so it is not surprising that we are now struggling to circumvent what natural selection discovered long ago. A similar though more recent issue is the challenge of controlling pathogens that evolved elaborate traits to uniquely exploit humans. Some of these pathogens have adaptations to promote their growth within specific target cells inside the body, especially when humans constitute their main host. One example is the protist *Plasmodium falciparum*, which causes malaria. During infection this pathogen enters red blood cells and radically alters them so that the pathogen can reproduce. Owing to the huge burden caused by this disease, people living in sub-Saharan Africa and other regions where malaria is endemic have been strongly selected for resistance, sometimes at an extreme cost in terms of other health consequences. In particular, selection to limit malaria infections has favored alleles that cause abnormal types of hemoglobin. While heterozygous individuals—who carry one copy of the variant allele—are more resistant to malaria, those who are homozygotes—with two copies of the resistance allele—suffer from sickle-cell anemia and other disorders. This severe trade-off suggests that resistance to malaria is difficult to achieve. This challenge, among others, is contributing to the difficulties to date of generating an effective malaria vaccine.

Emerging pathogens are those that have recently entered the human population or are causing disease at increased rates. Such pathogens are a huge concern because high rates of mortality can result when a disease agent jumps into the human population from a nonhuman host, sometimes rapidly evolving to become a human-specific pathogen. For example, phylogenetic analyses clearly show that simian immune deficiency viruses (SIV) entered human populations many times in the last century or so, leading to the emergence and spread of human immune deficiency viruses (HIV) I and II, and high mortality especially in HIV-I infected individuals. Other emerging pathogens may have a long history of infecting humans, but changing circumstances may cause increased mortality, as when human populations previously sheltered from a pathogen—and therefore possessing little immunity or resistance—are exposed to infections. For example, European colonization of the Americas brought along the virus that causes smallpox, which contributed to the devastation of Native American populations. Emerging pathogens can thus shape human demography and influence the course of human history.

One important goal of evolutionary medicine is to predict what types of pathogens might emerge in the future. In addition to possibly forestalling some diseases, such research may help prepare the medical community for future challenges when the pathogens of tomorrow burst on the scene as unexpectedly as did the agents of MRSA, HIV, and SARS (severe acute respiratory syndrome) recently. The ability to predict which pathogens will next emerge in human populations is currently rudimentary at best. But some patterns seem evident, as disease surveillance data and laboratory evolution studies imply that generalist patho-

gens—those that previously evolved to infect multiple host species—
tend to be better poised to spread to yet another host species.

Evolutionary thinking is already changing how we combat influenza
virus A, which causes seasonal epidemics in humans, and perhaps
500,000 deaths in a typical year. Although flu vaccines can be highly
effective in protecting humans against infection, they are strain specific
and take many months to produce. Thus, a flu vaccine that targets a
particular strain must be mass-produced well ahead of the season in
which the virus circulates in humans. In the 1990s, researchers began
to use evolutionary phylogenetics to analyze influenza virus A isolates
and track how successful lineages of the virus undergo *antigenic drift*—
that is, specific changes in their hemagglutinin proteins that promote
escape from human immune surveillance, allowing these strains to
dominate subsequent flu seasons. Refinements of this approach are
now used to predict which circulating strains are most likely to give
rise to the dominant strains in the coming flu season, so that a vaccine
against them can be produced and distributed before they become
widespread. However, the effort to match the vaccine with the most
important flu variants in a given year sometimes fails. These failures are
most problematic when a new strain emerges through *antigenic shift*—
that is, when recombination between virus strains generates a variant
that escapes the immune system. Important goals of evolutionary medicine include more accurate forecasting of seasonal flu variants and
better predictions of the emergence of new flu strains capable of spur-
ring pandemics.

The harm that a given pathogen or parasite causes to its host is often
called *virulence*. An important fact is that virulence often differs markedly, even between closely related pathogens. Thus, another goal of evolutionary medicine is to understand how the virulence of a pathogen is
shaped by natural selection. If a pathogen relies on direct transmission
from an infected host to a susceptible host, then activities of the pathogen that might seem to increase its opportunity for transmission (e.g.,
greater within-host production of infectious particles) may weaken the
current host and thereby actually reduce the probability of successful
transmission. An extreme example would be a pathogen that kills its
current host so quickly that a new host is not likely to be encountered.

Thus, when transmission and virulence are tightly coupled, theory predicts that a pathogen should evolve an intermediate level of virulence, one that balances the cost of harming the current host and impeding its mobility (which limits the pathogen's net transmission) against the benefit of producing more propagules (which allows the pathogen to infect its next host).

However, greater virulence may evolve when pathogens are readily spread by hosts debilitated by infection, as when pathogens are spread by insect vectors, via water, or through needles shared by drug users. These circumstances decouple the success of a pathogen from the mobility of its current host, and selection thus pushes the pathogen population toward more rapidly reproducing variants that leave their hosts more severely debilitated. A similar situation often occurs when hosts are coinfected by multiple pathogen variants, because the advantage goes to the pathogen type that replicates faster, destroying host tissues in the process. However, the evolution of increased virulence may sometimes be "shortsighted" from the pathogen's perspective. That is, within-host selection may favor a more virulent pathogen type, even when this variant is not the most successfully transmitted to new hosts. Yet other scenarios are also possible with coinfections. For example, antagonistic interactions among coinfecting genotypes may reduce the overall load of pathogens, leading to lowered virulence.

Understanding the evolution of pathogen virulence is an active area of research in evolutionary medicine, one in which mathematical models make concrete predictions, based on specific assumptions, that can be tested in the laboratory using appropriate experimental systems. Such efforts will help determine which pathogens might be expected to evolve increased or decreased virulence through time. For instance, selection exerted by the use of new vaccines may inadvertently exacerbate some diseases by favoring more virulent pathogens that can escape the vaccine-induced host response.

3. DEFENSE MECHANISMS

The history of the human species has involved coevolution with various parasites and pathogens that evolve much more rapidly than we can as a result of their short generation times. To keep pace in this *arms race*, we rely on the protection afforded by our innate and adaptive immunities, which in essence provide rapid evolution at the cellular and molecular levels during our individual lifetimes.

However, a drawback to a complex immune system is that it can malfunction in myriad ways. Diseases in which the immune system attacks the body's own cells, mistaking them for pathogens, are termed *autoimmune diseases*. Similarly, allergies and *atopic diseases* (those in which an allergic reaction becomes apparent in a sensitized person only minutes after contact, such as asthma and anaphylaxis) represent situations in which the normal processes of defense are inappropriately or excessively activated.

Genetic differences among individuals affect the potential to develop autoimmune disease. These differences are usually associated with certain alleles of the human leukocyte antigen (HLA) system, which is the major histocompatibility complex in humans. One example is HLA-B27, an allele associated with increased risk of ankylosing spondylitis, a chronic inflammatory arthritis more prevalent in men that can cause fusing of the vertebrae in the spine. However, not all males with HLA-B27 develop the disease, which suggests that some additional trigger, perhaps environmental, is involved. Some evidence hints that the disease may involve an aberrant response to *Klebsiella* bacteria, which are usually benign inhabitants of the human gut. Yet other autoimmune diseases may be related to a lack of exposure to certain microbes (figure 7-2A).

Historically, the human species has evolved in environments in which individuals were frequently exposed to severe, persistent infections; in particular, most people carried parasitic worms most of the time. In developed countries, by contrast, humans live in more hygienic environments, such that few people have worms, and few adults die from infection. However, in these same settings the prevalence of asthma, allergies, and chronic inflammatory disorders such as Crohn's disease has increased dramatically (figure 7-2B). The *hygiene hypothesis* offers a possible explanation: the vertebrate adaptive immune system coevolved with diverse commensal and pathogenic microbes, and exposure to them early in life may be essential for establishing appropriately regulated immunological pathways. In other words, the lack of exposure may cause improper activation of immune responses, which are then manifested as allergic or autoimmune diseases.

Helminth worms are especially implicated as stimulators of proper immune function, but the mechanism by which worms suppress or prevent autoimmunity remains unknown. Worm therapy has been tested repeatedly in mouse models, especially nonobese diabetic (NOD) mice. Most NOD mice reared in sterile environments progress to become diabetic at roughly 20 weeks of age, whereas it takes them 40 weeks to become diabetic when they are reared under conventional breeding conditions. Worms have been implicated as the infectious agents that delay the onset of diabetes in the NOD mice. Other animal studies have also suggested a protective role for worms in preventing other autoimmune diseases, such as inflammatory bowel disease, colitis, and collagen-induced arthritis. As a result, it has been proposed that some helminth worm might be introduced into humans as a therapy to reduce the incidence of autoimmune diseases. A leading candidate is *Trichuris suis*, the pig whipworm, which can establish an infection in humans but is not very pathogenic, causing mild symptoms, if any.

This new understanding of the evolutionary role of helminths in modulating proper immunity has led to changes in medical treatment. Some doctors now attempt to treat autoimmune diseases, such as Crohn's disease, by injecting *T. suis* eggs or proteins derived from these worms, to activate an inhibitory arm of the immune system that is otherwise suppressed in many modern populations. However, the results have been mixed; some clinical trials for Crohn's disease show that worm therapy does not improve patient health relative to a placebo.

Other potentially devastating autoimmune diseases might be treated using worms, notably multiple sclerosis (MS), for which treatment has been difficult. Although natural infection by worms slowed development of brain lesions in MS patients relative to uninfected individuals, there were some adverse effects of long-term worm infection. In some cases, symptoms became so serious that drug therapy to remove worms was required even though it worsened MS symptoms.

Thus, worm therapy shows some promise, but it also involves potentially risky trade-offs. For example, in the tropics, worm infections generally increase the susceptibility of humans to tuberculosis, malaria, HIV, and other infectious diseases. Worm infections can also reduce the

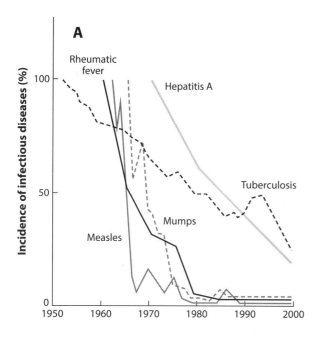

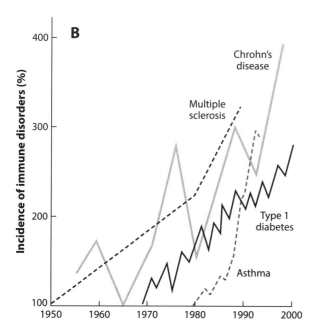

Figure 7-2. As the incidence of prototypical infectious diseases has fallen in high-income countries (A), the incidence of immune-related disorders has increased (B). (From J.-F. Bach. 2002. New England Journal of Medicine 347: 911–920.)

protection offered by vaccines, because the infections interfere with the immune response to vaccination.

Evolutionary medicine studies are also addressing a different hypothesis to account for increasing allergy rates. Modern environments may contain specific risk factors that were absent from previous environments. For example, modern homes are much warmer and drier than the habitations in which our ancestors lived, typically containing fitted carpets, warm bedding, and central heating. This environment is ideal for mites that inhabit house dust and ingest sloughed-off human skin. The feces of these mites are highly allergenic to susceptible people and may contribute to the rise in allergy rates.

4. TRADE-OFFS IN HUMAN TRAITS

One of the most important generalizations that evolution offers medicine is consideration of the body as a collection of trade-offs. No trait is perfect, and when natural selection improves one trait, it might make another worse. If we produced less stomach acid, then we might be less prone to ulcers, but the trade-off would be greater vulnerability to gastrointestinal infections. This line of thinking is particularly important as we gain new ways of altering and engineering our body. Many of us try to get by with less sleep so that we can cram more activities into 24 hours, but natural selection has adjusted the length of our sleep for many thousands of years. Thus, the adverse health consequences of sleep deprivation may exceed any benefits.

As another example, contraceptive use by many women in postindustrial countries causes them to experience an average of 400 menses per lifetime. In contrast, women in cultures that experience historically typical birth intervals of two and a half years, and breastfeed during that time, have about 100 menses per lifetime. This perennial cycling by many postindustrial women causes prolonged elevated levels of hormones, especially estrogen, which can promote the growth of tumorigenic cells often responsible for breast and ovarian cancers. This mechanism may place women at increased risk for certain cancers and might explain why breast cancer rates are often higher for women in postindustrial societies. Contraceptives need not induce a monthly period, and so perhaps a new solution can be found that allows women to experience a level of estrogen sufficient to maintain bone strength and avoid osteoporosis while avoiding increased risk of cancer.

Trade-offs also arise in the context of treating cancers. Chemotherapy strongly selects for drug-resistant cancer cells in the human body, just as antibiotic therapy rapidly selects for drug resistance in bacterial populations. Alternative strategies are being explored that might either slow or prevent the evolution of chemotherapy resistance. One alternative is targeted immunotherapy, which exploits the existing capacity of the immune system to locate and destroy cancer cells. In particular, this approach attempts to stimulate cytotoxic T-cells that specifically react to the cancer while curbing the action of the immunosuppressive cells that tend to increase in cancer patients. Although there have been some dramatic successes with this approach, disappointing failures have also occurred, particularly with highly virulent cancers such as metastatic melanomas and ovarian cancers.

Another alternative in treating cancer based on evolutionary thinking is the use of drug cocktails, as has been successful in treating HIV. The idea is to use a combination of drugs that work through different mechanisms, each targeting a different cancer gene product or pathway. This strategy should greatly reduce the probability that a mutant cancer cell will become resistant to all the drugs in the cocktail. The trade-off is that the mixture of multiple drugs exposes the patient to the combined side effects of all the drugs, each of which is usually highly toxic.

In contrast, so-called adaptive therapy has been proposed as a more promising alternative to traditional chemotherapy. In adaptive therapy lower doses are used to control the cancer to avoid eliminating the less aggressive cancer cells that compete for resources with the more aggressive ones. Results of mouse experiments support the potential value of this approach; tumors in mice receiving the standard high dose of anticancer drugs rapidly evolved resistance, whereas tumors in mice undergoing adaptive therapy did not increase in size.

Another promising alternative targets the public goods of cancer cells to disrupt the cooperation that occurs among cells within tumors. (This strategy is similar to one being explored to control the rise of antibiotic-resistant bacteria.) Cancer cells make growth factors that suppress the immune system, and they stimulate angiogenesis (development of new blood vessels) to feed the tumor. The production of these growth factors is costly to an individual cancer cell, making it vulnerable to other cancer cells that benefit from the growth factors without having to produce them. In the proposed therapeutic approach cells would be removed from a tumor and modified by knocking out those genes that code for growth factors. The modified cells would then be reintroduced into the cancer patient. These modified cells would have a within-tumor advantage by virtue of not paying the cost of producing the growth factors while still benefiting from their production by the unmodified cancer cells. The intended consequence would be that the modified cells would increase within the tumor and eventually cause the tumor to cease growing or even collapse owing to lack of essential growth factors. Preliminary experiments in cell culture suggest that this approach has some potential, but convincing data have yet to be obtained in animal models. In any case, because current cancer therapies are expensive, and resistance to chemotherapy evolves quickly, the search for affordable, evolution-proof alternatives remains a key goal in anticancer therapy.

Trade-offs arise in other contexts as well. Indeed, any trait can be analyzed in terms of its associated costs and benefits. Trade-offs limit the extent to which fitness can be improved when an improvement in one trait compromises some other trait. These compromises can emerge as unpleasant surprises when interventions ignore or overlook potential trade-offs. From that perspective, consider the trade-off between deploying physiological resources toward peak reproductive performance and investing in reparative functions that might sustain health in the postreproductive years. Life span in most human societies has markedly improved owing to large reductions in extrinsic mortality following improvements in public health and medical care. But one consequence of this longer postreproductive life is an increase in chronic diseases of middle to older age that may reflect insufficient physiological investment in self-repair processes including, for example, repair of cellular damage from oxidative stress. The outcomes may include atherosclerosis, arthritis, osteoporosis, cognitive decline, neurodegeneration, and increased susceptibility to infection. An evolutionary perspective might help in development of tests to identify those individuals most at risk of suffering the adverse effects of such trade-offs. Novel therapies might then be based on a better understanding of this variability.

5. MISMATCHES TO MODERNITY

Evolutionary change is a slow process, especially for organisms with generations as long as those of our own species. By contrast, our social and physical environments have changed very rapidly owing to our cultural evolution and the impact of technologies on the environments in which we live. Thus, our ancestors largely evolved under circumstances very different from the ones in which we now live. Many of the health problems we face, including some discussed earlier, likely reflect the consequences of those differences. Moreover, when the differences in the environment exceed an individual's capacity to adapt physiologically there is a disconnect, which has been called a *mismatch to modernity*.

It is ironic that mismatches to modernity result in diseases that stem from humans having access to historically critical, but now often excess, resources. The modern lifestyle allows many people to have an excess of energy intake relative to energy expenditure (figure 7-3), especially given technologies that also limit the need for physical labor. The body's inability to cope with this metabolic mismatch may well explain the increasing global incidence of obesity, type 2 diabetes, stroke, and cardiovascular disease. Lifestyle changes can combat this problem, but they are difficult to achieve because the pleasures of the present often outweigh considerations of the distant future; after all, few of our ancestors lived such long lives that this was a concern. Also, our evolutionary history of living with limited resources may have selected for a physiology that benefited from taking on extra nutrients when they were available. In particular, individuals of lower birth weight are at higher risk of becoming obese and developing metabolic disorders, including hypertension and diabetes. Early nutritional stress, it seems, is a signal whose evolved response sets the individual on a special developmental course with a physiology that is effective for conserving energy but ill prepared for abundant food. This relationship suggests that in utero cues about nutrition may affect the development of metabolic priorities later in life. In particular, the mismatch between early- and late-life nutritional status may contribute to increasing obesity rates, rendering those who are born in poverty and grow into plenty most vulnerable. A better understanding of the global epidemic of metabolic diseases will require

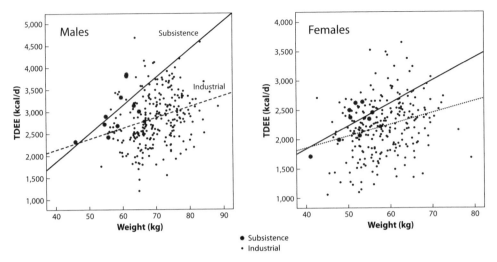

Figure 7-3. Total daily energy expenditure (TDEE; kcal/d) versus body weight (kg) for adult men and women of industrial and subsistence-level populations. Individuals in subsistence-level groups have systematically higher levels of energy expenditure at a given body weight. (Reprinted with permission from S. C. Stearns and J. C. Koella, eds. Evolution in Health and Disease, 2nd ed., Oxford University Press.)

consideration of these early cues and their effects on development, in addition to the interaction between our changing lifestyle and the physiological predilections that we inherited from ancestors who lived under very different conditions.

Diseases caused by the mismatch between our bodies and environments can arise from deficiencies as well as excesses. For example, iodine was often routinely consumed by humans who ate seafood and plants grown in iodine-rich soils; however, endemic goiter and cretinism (a specific syndrome of mental retardation) occur when humans live in noncoastal environments deficient in iodine, such as certain mountainous regions. In developed countries, salt is routinely iodized to reduce the risk of both goiter and cretinism. Similarly, the annals of sea exploration contain many stories of scurvy outbreaks caused by a lack of dietary vitamin C, for which fruits and some vegetables are the only natural sources. Humans, unlike most other species, cannot synthesize vitamin C, and so we are entirely dependent on our food (or, today, supplements). Our ancestors evolved in environments where they had constant access to fruit, and the mutation that caused our inability to synthesize vitamin C was apparently neutral in that environment. But when a person is exposed to a novel environment—say, a long sea voyage—where a dietary source of vitamin C is lacking, scurvy is the result.

6. IMPLICATIONS OF EVOLUTIONARY MEDICINE

Evolutionary medicine should be of interest and use to practicing physicians as well as to biomedical researchers for many reasons. Humans continue to be locked into a *coevolutionary arms race* with pathogens. These professions can therefore benefit from examining infection from the pathogen's perspective and from anticipating how pathogens will evolve in response to treatments such as antibiotics and vaccines. Evolutionary thinking should also help those clinicians and researchers who deal with cancer, reproductive medicine, metabolic disorders, and autoimmune diseases understand how bodies are affected by their mismatch with modernity. More generally, medical researchers gain from evolutionary thinking because it brings new perspectives, by both posing new questions and addressing old questions in new ways, including tough biomedical problems for which new insights might improve health and save lives.

The long-term promise of evolutionary medicine is that this cross-disciplinary science will yield new or improved methods of treatment. However, the field is still young, and it takes a long time for basic-research findings to be translated to changes in medical practice. Even so, some advances in medical treatment have already resulted from evolutionary ways of thinking. Examples include the greater attention to prescription and use of antibiotics, the simultaneous use of multiple antiviral drugs to treat HIV in order to limit resistance, and the use of products derived from helminths to treat certain autoimmune diseases. Also, evidence bearing on the link between HLA genotypes and pathogen resistance suggests that an HLA mismatch between parents may increase the risk of miscarriage, because the embryo is attacked by the mother's immune system. This information has informed the practice of sperm donation to reduce the risk of spontaneous abortion. Another example is the increasing recognition that the genetic variations in human populations can affect the response of individuals and groups to drug treatments as well as the likelihood of developing disorders such as alcoholism. This understanding informs the new field of evolutionary pharmacogenomics and may eventually lead to personalized medicine, in which a patient's preventive and therapeutic care are optimized for his or her unique genotype.

An important educational goal of evolutionary medicine is to incorporate instruction in evolutionary biology into premedical and medical education. Currently, few medical schools have evolutionary biologists on their faculties, and none teach evolutionary biology as a basic medical science. Physicians and medical researchers may learn something about evolution before medical school, but few have the level of knowledge demanded for other basic sciences. An evolutionary view would correct mistaken notions of the body as a designed machine and would provide physicians with a better sense for the organism and what constitutes disease. A recent change in the Medical College Admissions Test (MCAT) requires competency in evolutionary biology, which should improve understanding of evolutionary issues among clinicians more than any other potential measure. In addition to changes in the MCAT, undergraduate institutions should offer courses in evolutionary medicine as part of their premedical curricula. Specific revamps of the medical curriculum could also infuse more evolutionary thinking into medicine. Ideally, professional societies associated with medicine would develop policies to help educate physicians by allowing them to make use of evolution as a basic science for medicine. As future discoveries demonstrate how evolution-based thinking can improve the understanding and treatment of medical disorders, it should become increasingly clear that fundamental knowledge of evolution belongs in the medical toolkit.

See also chapters 9, 10, 21, and 22.

FURTHER READING

Diamond, J. 1997. Guns, Germs, and Steel: The Fates of Human Societies. New York: W. W. Norton. *A Pulitzer Prize–winning book about Eurasian colonization of other societies, sometimes involving infectious diseases that decimated native populations.*
Ewald, P. W. 1980. Evolutionary biology and the treatment of signs and symptoms of

infectious disease. Journal of Theoretical Biology 86: 169–176. *A perspective on the need to understand whether disease symptoms, such as fever and diarrhea, are beneficial for the host or the pathogen.*

Gluckman, P., A. Beedle, and M. Hanson. 2009. Principles of Evolutionary Medicine. New York: Oxford University Press. *A thorough account of evolutionary medicine geared to a medical audience that includes fundamentals of evolutionary biology.*

Merlo, L.M.F., J. W. Pepper, B. J. Reid, and C. C. Maley. 2006. Cancer as an evolutionary and ecological process. Nature Reviews Cancer 924–935. *A review that uses concepts from ecology and evolution to explain why cancer evolves and why it is often difficult to treat.*

Stearns, S. C., and R. Medzhitov. 2016. Evolutionary Medicine. Sunderland, MA: Sinauer Associates. *An introduction to evolutionary medicine, highlighting recent advances and applications of evolution-based thinking in treating human diseases.*

Trevathan, W. R., E. O. Smith, and J. McKenna. 2007. Evolutionary Medicine and Health: New Perspectives. New York: Oxford University Press. *Collected essays on new treatments of mostly noninfectious diseases that have been informed by the field of evolutionary medicine.*

Turner, P. E., N. M. Morales, B. W. Alto, and S. K. Remold. 2010. Role of evolved host breadth in the initial emergence of an RNA virus. Evolution 64: 3273–3286. *A laboratory test of the general hypothesis that pathogens that evolved on multiple hosts are more likely to infect a novel host.*

Williams, G. C. 1957. Pleiotropy, natural selection, and the evolution of senescence. Evolution 11: 398–411. *A classic paper examining the evolution of aging from an evolutionary perspective.*

Williams, G. C., and R. M. Nesse. 1991. The dawn of Darwinian medicine. Quarterly Review of Biology 66: 1–22. *A review of the power of evolutionary medicine to improve understanding of a wide variety of medical disorders.*

CHAPTER 8

AGING AND MENOPAUSE

Jacob A. Moorad and Daniel E. L. Promislow

OUTLINE

1. A natural history of aging
2. Theories for the evolution of aging
3. Menopause
4. Pressing questions on the evolution of aging

Given enough time, organisms lose vigor as they age. Traits that may have once seemed optimized for survival and reproduction degrade, increasing the risk of death and reducing fertility. On the surface, it seems paradoxical that natural selection, which always favors increasing fitness, should permit aging to be nearly ubiquitous. However, evolutionary theory provides us with simple but powerful hypotheses to explain why we senesce and die. At its heart, this theory shows us that fitness depends more on what happens early in life than what happens at old age. In other words, there is more natural selection for early-life function. A basic tenet of aging theory tells us that if the survival or fertility effects of early-acting and late-acting genes are independent but equally distributed, natural selection will favor the evolution of aging in very predictable ways. But do genes really act this way? Humans age, of course, but we are unusual in that our middle-aged females undergo *menopause*. What is so special about our species, and given that men die sooner than women, why is reproductive cessation in males neither as abrupt nor as complete as in women?

GLOSSARY

Aging. *See* senescence.

Antagonistic Pleiotropy. A proposed mechanism for the evolution of senescence. Under this model, selection favors alleles with early-acting beneficial effects that have pleiotropic but deleterious effects at late age.

Disposable Soma Theory. The theory that senescence evolves owing to trade-offs between investment in reproduction and investment in somatic (bodily) maintenance and repair. The optimal strategy is one that favors limited investment in maintenance and repair, such that senescence is inevitable.

Gene Regulatory Network. The complex web of interacting genes, some of which regulate themselves and/or downstream target genes, some of which are regulated by upstream regulatory genes.

Genetic Correlation. The statistical dependence between two traits caused by genes that determine the values of both traits.

Genetic Variance. A measure of phenotypic differences among individuals that are caused by genetic differences.

Inbreeding Depression. The loss of fitness that is associated with the mating of relatives.

Iteroparous. Capable of reproducing multiple times throughout life.

Menopause. The late-onset, irreversible cessation of reproductive capability experienced by women, usually at around 50 years of age.

Mutation Accumulation Theory. Theory based on the notion that the strength of selection declines with age; as a result, late-acting germ-line deleterious mutations accumulate over evolutionary time, leading to age-related declines in fitness.

Programmed Death. The idea, generally rejected by most evolutionary biologists as a principal cause of aging, that natural selection favors senescence, such that genes that actually cause death can spread through, causing catastrophic mortality.

Semelparous. Reproducing just once, and then dying; examples of semelparous organisms include spawning salmon and some species of bamboo.

Senescence. An age-related decline in fitness components, including vital rates (age-specific survival or fertility), behavior, physiology, and morphological traits; used interchangeably with "aging."

1. A NATURAL HISTORY OF AGING

Natural selection is a powerful force. It can shape elaborate developmental pathways that give rise to exquisite morphological characters, it can shape complex behaviors that allow organisms to survive in what to us seem the most inhospitable of environments, and it has even endowed organisms with the ability to heal wounds and to repair themselves. But these characteristics leave us with a puzzle: if selection endows organisms with the ability to repair both genetic and structural damage, why can it not prevent organisms from eventually falling apart as they age?

What Is Aging?

In all human populations, the probability of dying varies across ages, following a bathtub-shaped trajectory (figure 8-1A). Mortality rates are relatively high in utero and in the months immediately after birth. They then decline, reaching a minimum in late adolescence. From this age onward, the probability of dying increases. In humans, the risk of dying doubles every eight years. In our twenties, fertility starts to decline, as another manifestation of the aging process (figure 8-1B).

We measure aging (or *senescence*; here we use the terms interchangeably) as the age-related rate of decline in fitness traits. It is manifested at the level of the population and of the individual. *Demographic senescence* refers to the age-related decrease in the frequency of survival and mean reproductive output of groups of same-age individuals. *Physiological senescence* refers to late-age-related changes in individual phenotypes. Life span per se is not a measure of aging but an outcome of cumulative mortality risks, some of which are constant throughout the life span and some of which vary with age.

Aging in Model Systems

Most of what we know about the biology of aging comes from studies of lab-adapted organisms—yeast, nematode worms, fruit flies, mice, and rats. The general patterns of aging in both survival and reproduction that we see in humans are much the same in these laboratory populations.

From these lab-adapted species, we have learned much about the way that both environmental and genetic factors can shape longevity. Re-

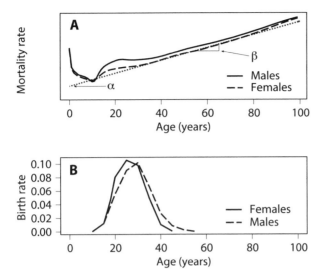

Figure 8-1. (A) Age-specific mortality in the United States, males (solid line) and females (dashed line), 2013 (see cdc.gov/nchs/products/life_tables.htm). The dotted line represents a Gompertz curve fitted to adult mortality, $\mu_x = \alpha e^{\beta x}$, where μ_x is instantaneous mortality at age x, α is the intercept of the line, and β, the slope of the line, represents the rate of aging. (B) Age-specific birth rate in the United States, males and females, 2013 (see http://www.census.gov/nchs/births.htm).

searchers have been able to enhance age-specific survival and late-age physiological function in many species simply by restricting nutrient intake. In most cases, this effect appears correlated with the cessation of reproductive output. The fact that diet restriction can increase life span in species separated by a billion years of evolution has led some to argue that the ability to survive nutrient stress is an ancient adaptation. However, others have suggested that the response may be a side effect of a very modern adaptation to the lab environment. The response to nutrient restriction is much less clear in wild-caught animals.

In the early 1980s, Michael Rose and his colleagues found that artificial selection in fruit flies could dramatically extend mean life span, demonstrating that there was a large amount of genetic variation for longevity. In subsequent work, researchers have shown that changes in the structure or expression level of single genes in many pathways—most notably, genes associated with insulin/insulin-like growth factor signaling—can greatly enhance life expectancy. These results raise an evolutionary question: if altering a gene can increase life span in a worm or a fly, why has nature not already done the experiment? The answer likely follows from the observation that these life-extending mutations generally *reduce* fitness, such as by decreasing early-age fertility.

What can these lab-based studies tell us about patterns of aging in the wild? Until the 1990s, it was commonly assumed that aging did not, in fact, occur in natural populations. Biologists argued that wild animals would not live long enough to manifest signs of aging. It may be true that we rarely see wild animals showing *obvious* signs of infirmity, as they are likely to have been killed by predators. But closer demographic analysis—in particular, measures of age-specific mortality and fertility—have shown clear signs of age-specific declines in fitness components in birds, mammals, and even wild insects.

While almost all species show signs of aging, some live notably longer than others. These differences have led to the recognition that there are certain ecological factors associated with long life span. For example, species that fly (bats and gliding mammals, as well as birds) tend to be long-lived relative to terrestrial species of similar mass. Similarly, species that are well armed against predators (like porcupines) and species that live underground (naked mole rates, queen bees, and ants) have a longer life span.

Some species of animals, such as hydra, appear to avoid aging altogether, with a constant risk of mortality throughout life. Other animals simply live what to humans seems like an extraordinarily long time, including Galápagos tortoises (almost 200 years), some species of rockfish (over 200 years), and even some small clams (over 400 years). Whether these long-lived species show signs of senescence is unknown. However, it is clear that aging can vary and is subject to evolution. This observation is borne out by explicit phylogenetic studies as well.

Sex Differences and Senescence

In the temperate rain forest of eastern Australia, males in the marsupial species *Antechinus stuartii* become sexually active during a brief 1-week period in the middle of winter. By the end of this brief mating season, all the females in the population are pregnant, and half will survive to breed next year. All the males are dead. Such a dramatic difference between *semelparous* males and *iteroparous* females is unusual, but in almost all mammal species, males die sooner than females. In the final section, we consider possible explanations for this widespread pattern.

Aging in Humans

The patterns of aging in humans (and in nonhuman primates) mirror the general pattern seen in model systems, with age-related declines in survival and fertility and higher rates of mortality in males than in females. Human females differ in one important respect in that they show a prolonged period of postreproductive survival (menopause). We discuss the evolutionary explanations for this later (see also chapter 6). Perhaps the most striking pattern in human mortality is that which has taken place over the past 250 years: a 90 percent decline in childhood mortality and a 50 percent increase in life expectancy at birth. The percentage of people living past the ages of 60 or 70 is higher now than at any other time in human history. But to fully understand the evolutionary forces that have shaped human aging, we need to know what human demography looked like prior to modern medicine and sanitation.

In nonindustrial indigenous societies, life expectancy at birth can be as low as 35 or 40 years of age. But for those individuals who make it through the riskiest period of infancy, there is a high probability of surviving to age 60 or 70 or beyond. While the immediate risk of dying is higher for adults in these populations than for those living in developed countries, the general pattern of age-specific mortality and fecundity is very similar. Baseline mortality rates have changed dramatically over time, but rates of aging appear to be quite constant among different human populations.

2. THEORIES FOR THE EVOLUTION OF AGING

The natural world offers up limitless examples of the power of natural selection. Aging differs among populations and organisms, and these differences appear to have been shaped by natural selection. But this observation then leaves us with a puzzle: why does natural selection fail to evolve organisms that can function indefinitely, repairing any damage

(A) Adaptive

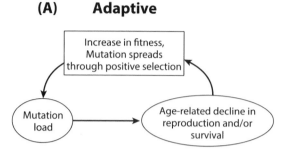

(B) Maladaptive

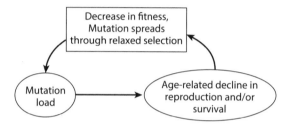

(C) Constrained

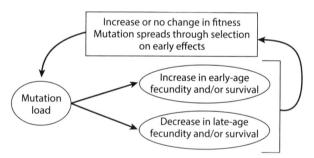

Figure 8-2. Conceptual models of three evolutionary theories of aging. (A) Under an adaptive model, mutations that increase rates of aging are favored by selection because they increase fitness. (B) Under a maladaptive model (Medawar's *mutation accumulation theory*), mutations that increase aging have spread under mutation-selection balance, where selection is relaxed because late-acting mutations have little or no effect on fitness. (C) The constrained model (Williams's antagonistic pleiotropy) refers to genes that decrease or have no fitness effects late in life but spread owing to their beneficial effects early in life. (Reprinted with permission from D. Arbuthnott, P. Promislow, and J. Moorad. Evolutionary Theory and Aging. Handbook of Theories of Aging, Springer, 2016.)

that arises along the way? In fact, this seems to be the case for hydra. So why do we not see this as the dominant pattern? Over the past century, three themes have emerged in attempts to understand aging from an evolutionary perspective—aging as adaptation, aging as maladaptation, and aging as constraint (figure 8-2).

Aging as Adaptation

The German biologist August Weismann is best known for his *germ plasm* theory of inheritance in animals, which recognizes that the germ line and somatic tissue represent two distinct lineages, with inheritance occurring only through the germ line. But Weismann is also credited with formulating the first evolutionary theory of aging in 1881 when he proposed that death in late age is adaptive and that natural selection actively favors the evolution of a death mechanism. This idea has become known as the *programmed death hypothesis*. Weismann emphasized that natural selection acts for the good of the population or species (not the individual) by removing useless individuals from the population and making room for the young.

Researchers have raised many objections to this theory. First, this mechanism requires the preexistence of senescence to explain selection for senescence, because it is predicated on the notion that the old are less fit than the young. Thus, this mechanism certainly cannot explain the *origin* of senescence. Second, the model gives group selection a central role in the evolution of aging but neglects individual-level selection, which should favor longer life span, all else being equal. While recent evolutionary theory considers a role for group selection in the evolution of many kinds of traits (social behaviors, for example), the conditions that allow group-level selection to overwhelm individual-level selection are very restrictive. The evolution of programmed death would require intense group benefits to overwhelm the great loss in individual fitness associated with early death. Third, if aging is genetically programmed, we should be able to identify genes that don't simply reduce age-specific mortality but actually disrupt the programmed mechanism, eliminating senescence. We have not found such mutations.

Aging as Maladaptation

The first modern explanation for the evolution of aging came from biologist Peter Medawar's work in the 1940s and 1950s. Medawar was inspired by evolutionary biologist J.B.S. Haldane's 1941 study of Huntington's disease (HD). Haldane was struck by the fact that HD is a lethal disease caused by a dominant mutation at a single gene. Surely, natural selection should eliminate such a mutation from the population, and yet it struck 1 in 18,000 people in England. Haldane explained its prevalence in terms of two forces, mutation and selection. While selection works to purge the genome of alleles that increase mortality, germ-line mutations that lower survival are constantly being generated. Under mutation-selection balance, natural selection and mutation work in opposition, and an evolutionary equilibrium is met when the two forces are equal in magnitude. Thus, at such equilibrium, lethal alleles can persist in a population. However, the frequency of these alleles should be very low unless selection is very weak.

Consider the case of HD. Affected individuals typically begin to show symptoms around 40 years of age, by which time carriers could have already passed the lethal allele on to their offspring. For most individuals, the lethal consequences of the gene will be seen only after they have finished reproducing. As a consequence, there will be very little natural selection against the allele.

Whereas Haldane saw one particular lethal allele spreading because of its delayed effects, Medawar saw that the same was true for *all* alleles with late-acting deleterious effects. More generally, he recognized that *the later the age at which the effects of a deleterious allele occur, the weaker is the ability of natural selection to eliminate that allele from the population.* This phenomenon creates a *selection shadow* on late-age fitness traits (figure 8-3).

If mutations can have age-specific effects on mortality or fertility (their effect is not manifested at all ages), then at mutation-selection equilibrium, the frequencies of age-specific mutations that decrease fitness will be smallest at early age and greatest at late age. This evolutionary mechanism, usually referred to as the *mutation accumulation hy-*

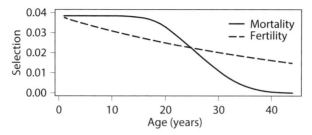

Figure 8-3. Age-related decline in the force of selection against age-specific mortality and for age-specific fertility. Selection is calculated using Hamilton's sensitivities and by using data from figure 8-1. Age-specific fertility values for missing ages are extrapolated.

pothesis, or simply *MA*, was first suggested by Medawar in 1946. This model has been extremely influential because it is both a very general and very simple evolutionary model that assumes only (1) that selection decreases with age (as William Hamilton, also the father of kin selection, was to show with his mathematical models 20 years later; see figure 8-3) and (2) that at least some mutations have effects that are confined to specific ages. We can consider both of these the necessary and sufficient conditions for the evolution of aging.

Aging as Constraint

In 1957, George Williams built on Medawar's insight in one simple but critical respect. Consider a mutation that affects fitness at two different ages, first quite early in life, and then at some much later age. If the direction of the effects at both ages is the same (both beneficial, or both deleterious), then natural selection will lead to the mutant allele's fixation, or loss, respectively. Neither case is very interesting from the perspective of aging, but suppose that the effects of the mutation are in opposite direction at the two ages. Consider a mutation that decreases survival or fertility early in life but increases it later. Selection might favor its spread and contribute to extended longevity of old individuals, but because selection cares more about changes early in life, the early costs of the allele would tend to outweigh its later benefits. In contrast, mutations with benefits early in life and costs late in life are more likely to be favored by natural selection and to spread through a population and lead to senescence. Importantly, this *antagonistic pleiotropy* (AP) model does not view aging as an adaptation, since it is not the aging per se that increases fitness, but as a constraint associated with other adaptations that evolve to maximize early life survival and/ or fertility.

Twenty years after Williams published his AP model, Tom Kirkwood (1977) suggested a plausible model of AP based on how selection is expected to optimize the allocation of limited resources across ages. Kirkwood's *disposable soma* model argued that there is substantial energy demand from both reproductive functions and from functions relating to the maintenance and repair of the individual (the soma). Given that energy is a limited resource, individuals cannot have it all. Selection favors the functions that lead more directly to increased fitness, which, in this case, is the production of offspring at the cost of less energy invested in maintaining and repairing old soma. Thus, the equilibrium level of investment in the soma is one that fails to ward off the effects of senescence.

Genetic Variation and Aging Theory

Starting in the mid-1990s and motivated by quantitative genetic models that seemingly provided diagnostic tests of MA versus AP gene action, evolutionary biologists invested considerable effort in trying to describe the standing genetic variation for aging. These studies compared components of genetic variation and *inbreeding depression* at early and late life with the expectation that they should change if MA causes aging. Work was carried out primarily in fruit flies, but it also included studies of soil nematodes, seed beetles, and even hermaphroditic snails. Most of these studies found putative support for MA. However, Moorad and colleagues have argued recently that quantitative genetic tests of genetic variation and inbreeding depression are not truly diagnostic. From their perspective, the genetic results support the contention that senescence evolved, but they do not favor any one particular model.

Turning to AP, we expect negative *genetic correlations* between early-age and late-age fitness traits. Research in this area began with Michael

Rose's landmark experiments with fruit flies in the early 1980s. When Rose selectively bred from progressively older individuals, he observed not only a dramatic increase in life span but also a *reduction* in fecundity during early life, a result since replicated in other labs. While these results were consistent with predictions from AP, we now recognize that these results can also be interpreted as evidence for MA, because selection against early-acting fecundity mutations in these experiments is relaxed independently of late-life survival. Researchers have also found evidence for negative genetic correlations between traits at different ages in natural populations of swans and red deer, among other species. However, negative genetic correlations cannot distinguish between early-acting advantageous AP mutations that have caused aging to evolve and new, disadvantageous AP mutations that *reduce* aging. The latter kind of gene may have a negative, but transient, effect on life span before its removal from the population by natural selection.

The most compelling evidence for the AP mechanism comes from single genes that are known to extend life span and occur with high frequency (or are fixed in a population). The best example of this sort of gene is the polymorphic *TP53* gene in humans, which appears to create a genetic trade-off between risk of cancer and risk of aging. One variant of the gene (*R72*) reduces the risk of cancer but decreases longevity. The alternative *P72* allele has the opposite effect.

All these studies addressed patterns of standing genetic variation. However, existing genetic variation depends not only on how selection acts but also on the nature of the mutations that enter the genome. By examining the distributions of new mutations, it was hoped that a picture of the raw material for evolutionary change could be resolved without the confounding influences of natural selection. These studies revealed four interesting characteristics of new mutations unanticipated by evolutionary theory:

1. New mutations increase mortality more at early age than at late age.
2. *Genetic variance* caused by new mutation is highest at early ages.
3. New mutations increase mortality at multiple ages (i.e., they generate positive genetic correlations across ages).
4. Effects of new mutations become more age independent (i.e., pattern 3 becomes stronger) as more and more mutations accumulate.

These findings do not support AP (genetic correlations arising from new mutations are positive, not negative). They may also explain two of the aforementioned patterns that are not anticipated by the basic evolutionary models: the occasional reduction in genetic variance (2) with age and mortality deceleration (1).

There is no consensus regarding the primacy of one evolutionary mechanism of aging over the other. Results from quantitative genetic analyses, artificial selection experiments, and mutation accumulation studies are equivocal, which may reflect the fact that MA and AP models are highly idealized and that real aging genes have characteristics of both.

3. MENOPAUSE

Menopause is an unavoidable physiological transition that defines the end of an individual's reproductive capacity. In human females, it lasts between one and three years, occurring at 50 years of age, on average, and is presaged by about 20 years of declining fertility (reproductive senescence). Menopause is marked by a loss of ovarian function (including reduced endocrine production), leading to sterility and a suite of symptoms, including hot flashes, insomnia, mood swings, and increased risk of osteoporosis and coronary heart disease. Human males also lose reproductive function over time, but men lack a similarly well-defined period of fertility loss. Accordingly, there is no upper limit to the age at which men can reproduce (apart from death); the record for extreme male reproduction appears to be 96 years.

While the existence of female menopause is firmly established in humans, little is known about how widespread menopause is in other animals. Some captive animals, such as rats and rhesus monkeys, appear to exhibit female menopause but do so at advanced ages that are believed to be largely unattainable in the wild. Natural populations of cetacean species, such as short-finned pilot whales and orcas, are observed to have large fractions of females that live beyond the age of reproductive cessation. As one can imagine, there are substantial challenges to collecting data to determine how fertility changes with age in natural populations. Nevertheless, this information is critical to comparative efforts trying to understand the forces of selection that cause menopause to evolve.

Why menopause? is one of the more fascinating (and open) questions of evolutionary demography. If menopause is an adaptation, then we are confronted with the challenge of explaining how natural selection can favor the evolution of a trait that ends one's ability to reproduce and transmit genes to the next generation. One obvious explanation is that menopause is simply a manifestation of reproductive senescence brought on by the age-related decline in the strength of selection. But there is a flaw in this reasoning. This argument assumes that historical human populations were characterized by such high rates of adult mortality that adult women rarely lived into their 50s, such that selection could not act with sufficient strength to avoid the accumulation of late-acting sterility mutations. We see menopause in our postdemographic transition world, this thinking goes, because life spans of modern humans are unnaturally long.

As noted earlier, indigenous human populations have very low life expectancy at birth, but for those who make it through those difficult early years, the probability of surviving well past the age of 50 is quite high. In this light, researchers think that menopause predates modern human societies.

Perhaps the greatest problem with the "menopause as modern artifact" argument is that human males tend to die before females and yet they do not undergo the abrupt reproductive cessation observed in females. As mentioned earlier, this pattern of higher male mortality is widespread among mammals. Moreover, evidence from natural populations of baboons and red deer (two species with pronounced elevated male mortality) indicates that the reproductive life spans of males are even more abbreviated than those of females owing to intense male competition for mates (old males do not compete well against their younger counterparts). If menopause were simply reproductive senescence, then we would expect reproductive cessation to be widespread in males and to occur earlier in life than female menopause. We do not observe this pattern.

A more tenable hypothesis for female menopause holds that it is an adaptation. However, the fitness benefits of menopause may be realized by the descendants instead of being conferred to the female in menopause (the benefit to a descendant is indirect, while the cost to the female is direct).

There are two common adaptive hypotheses for explaining menopause—the *mother hypothesis* and the *grandmother hypothesis*. In these models, genes that promote menopause are associated with higher fitness in the children of mothers (or grandmothers) in menopause because these children live longer and/or reproduce more than children descended from females that do not undergo menopause. At their essence, these models imagine that there is conflict between the fitness interest of the maternal ancestor and her descendants (in this sense the evolution of menopause resembles the evolution of altruism). Menopause is expected to evolve at the age at which the benefits of menopause help the children twice as much as menopause hurts the mother (or helps grandchildren four times as much as it hurts the grandmother). Tests of these hypotheses are currently fertile ground in aging research. The first requirement, that the timing of menopause has a genetic basis, has been met. We see in humans that the age of menopause in the mother predicts to some degree the age of onset in daughters, suggesting a heritable basis to the age of onset of menopause. The second requirement, that menopause increases descendent fitness, is discussed later. Note that the validity of both the mother hypothesis and the grandmother hypothesis is subject to lively debate, with supporters and detractors on both sides.

The Mother Hypothesis

The mother hypothesis focuses on the observation that the risk of mortality from complications in pregnancy or delivery increases dramatically as women age. For example, rates of gestational diabetes increase fourfold in pregnant women over 40 compared with those under 30. Problems associated with hypertension may be as much as five times more common among older pregnant women. Complications from both of these, as well as an age-related increase in breech births, increase the frequency of operative births in modern societies. Obviously, this was not an option over human evolutionary timescales, and we can safely assume that many of these age-related problems would have resulted in the death of the mother. In the past, these deaths would have denied existing descendants the care or resource provisioning that a mother would otherwise have provided. In this light, menopause might have mitigated a mortality risk in older women. However, recent analyses suggest the increased risk of mortality for existing offspring owing to losing a mother, other than for those who have not yet been weaned, is actually minimal. In traditional societies, allocare by a mother's relatives may have greatly reduced the cost of her death.

The Grandmother Hypothesis

The grandmother hypothesis argues that at some age the benefits to women of helping their daughters care for offspring outweigh the benefits of continuing to produce more sons and daughters. At this age, selection will favor a shift to care directed from grandmothers to grandchildren. Studies from preindustrial Western societies as well as from indigenous societies show that in at least some populations, older daughters have more children if their mothers survive and that the grandchildren of living women are larger and have lower mortality than those of dead women. See chapter 5 for further details.

Reproductive Competition

A third adaptive argument for the evolution of menopause, recently suggested by Michael Cant and Rufus Johnstone, notes that humans evolved in the context of small groups and that reproduction within these groups came at some expense to the social partners of the mother. Because dispersal among groups seems to have been dominated by young females, females tended to become more closely related to their group as they aged (older females were more likely than young females to have sons in the population). As a result, the social cost of reproduction by older females was borne by her relatives, but the relatives of young females were not affected by her reproduction. Cant and Johnstone reason that the relationship between degree of relatedness and female age caused kin- or group-level selection to favor the cessation of fertility *most* in the older females.

4. PRESSING QUESTIONS FOR THE EVOLUTIONARY THEORY OF AGING

Why Don't All Species Hit a "Wall of Death"?

In 1966, William Hamilton explained how the force of selection would change with age, given explicit values of age-specific fecundity and survival (figure 8-3). Hamilton's model made clear qualitative predictions about the increase of mortality with age. Specifically, there should be three distinct phases. First, age-specific mortality is constant and low from birth until the first age in the population at which reproduction occurs. From then, mortality rates increase with age until the last age of reproduction. Beyond this point, selection can no longer act, and postreproductive populations should hit a sudden "wall of death."

Aside from its occurrence in semelparous species like Pacific salmon and *Antechinus*, this wall of death is generally not seen. How can we explain the fact that populations persist beyond the end of reproduction and that at least in iteroparous plant and animal species we fail to see a wall of death? One reason might be that the effects of individual mutations are less ephemeral than imagined by Medawar or Hamilton, and there are some genes that affect survival at both reproductive and postreproductive ages. As a result, genes that increase postreproductive mortality are not entirely hidden from the purifying effects of natural selection. This sort of gene action may have other, more subtle effects on human mortality, which we discuss in the next section. Another possibility is that even after the age at last reproduction, there will still be selection to survive and help one's offspring reproduce, as discussed earlier.

In contrast with a sudden, climactic increase in mortality, in many populations mortality rates actually level off and can even decline at very late ages. However, scientists have yet to determine definitively whether this pattern of late-age mortality deceleration is a statistical artifact (due to variation in mortality rates among subcohorts within a population) or an evolutionary consequence of selection pressures (or lack thereof) at late ages.

Is There a Limit to Human Life Span?

In a now-famous wager, biologist Steven Austad and demographer S. Jay Olshansky placed a $500 million bet as to whether there will be at least one human who has lived to at least the age of 150, and with mind intact, by the year 2150. Austad and Olshansky don't expect to be around to collect on their wager, so they have each invested $150 and, with careful investment, anticipate that the funds will provide the heirs of the winner half a billion dollars. Austad argues that medical improvements and technological discoveries will lead to 150-year-olds by 2150. Conversely, based on his analysis of existing demographic data, Olshansky argues that we are already close to the limit of human life span.

Is there an evolutionary response to this question? Consider Rose's experiments with fruit flies, in which selection for late age at reproduction doubled the life span of flies in just a few years. We can pursue this reasoning with the following thought experiment (albeit an extreme one). Imagine a human population in which only men and women over the age of 40 reproduced. All else being equal (that is, considering only the effects on life span and assuming everything else stayed the same), would this give rise to, let's say, in 100 generations, or about 4000 years, a significantly longer-lived population? In the case of Rose's fruit flies, we think that selection favored longer life span because only those individuals with genes that promote long life span had a chance to reproduce. In a modern human population, the vast majority of individuals survive to age 40, so there is likely to be little selection on genes affecting survival rates prior to age 40. Rather, we would expect to see a response in an increase in late-age fertility.

Why Do Males Die Earlier Than Females?

There is still no clear answer to why males and females should age differently. However, it is likely that this difference is associated with those traits that define "male" and "female." After all, the fundamental difference that defines two sexes is the size of the gamete that is produced, with female being the sex that produces larger (and typically fewer) gametes. Do sex roles explain why males and females age differently?

Starting with Charles Darwin and Alfred Russell Wallace, evolutionary biologists have been fascinated by the elaborate displays that animals use to attract the opposite sex. Typically, it is the males that bear these traits and/or compete with one another for access to females. We now understand a great deal about the selective forces that have led to the evolution of secondary sexual traits and mating behavior. Less understood are the immediate (proximate) and long-term (evolutionary) costs of these traits.

In thinking about this problem, researchers have created a rich body of literature that falls at the intersection of two conceptually rich fields of study, uniting theories for the evolution of aging with theories of sexual selection and sexual conflict. A 2008 review by Russell Bonduriansky and his colleagues summarized several predictions that have emerged in the literature. These include suggestions that (1) both within and among species, males that invest more heavily in secondary sexual traits should age more quickly; (2) conflicts of interest between the sexes should lead to higher rates of aging, and if one sex has an advantage over the other in this conflict, it should age more slowly than the other; and (3) the influence of sexual selection on sex differences in aging should be mitigated by the degree of genetic correlation between the sexes.

Can We Understand Aging One Gene at a Time?

The evolution of any trait depends on the way in which that trait affects fitness and on the genetic architecture of the trait. The latter component includes the extent to which the focal trait is genetically correlated with other traits that affect fitness, the number of genes that influence the trait, and potential interactions among these genes. With the introduction of high-throughput genomic approaches to the study of aging, researchers are beginning to develop a better understanding of the genetic architecture of aging. At the simplest level, we know that there are a very large number of genes with the potential to influence life span. We are only just beginning to uncover the much deeper structure that relates the genotype to this particular phenotype. For example, consider the *gene regulatory network*, which illustrates how genes are connected to one another if they share a common regulatory mechanism. Research on mice has shown that the number of connections among these genes—the overall complexity of the gene regulatory network—declines with age. We are still at a relatively early stage in the study of gene networks and aging. Just how this complex architecture might have influenced, and possibly constrained, the evolution of aging is an important but unanswered question.

FURTHER READING

Bondurianskly, R., A. Maklakov, F. Zajitschek, and R. Brooks. 2008. Sexual selection, sexual conflict, and the evolution of ageing and life span. Functional Ecology 22: 443–453.

Cant, M., and R. Johnstone. 2008. Reproductive conflict and the separation of reproductive generations in humans. Proceedings of the National Academy of Sciences USA 105: 5332–5336.

Gurven, M., and H. Kaplan. 2007. Longevity among hunter-gatherers: A cross-cultural examination. Population and Development Review 33: 321–365.

Hamilton, W. D. 1966. The moulding of senescence by natural selection. Journal of Theoretical Biology 12: 12–45. *This paper developed the key mathematical framework for modeling evolution in age-structured populations.*

Hawkes, K., J. F. O'Connell, N.G.B. Jones, H. Alvarez, and E. L. Charnov. 1998. Grandmothering, menopause, and the evolution of human life histories. Proceedings of the National Academy of Sciences USA 95: 1336–1339.

Kenyon, C. J. 2010. The genetics of ageing. Nature 464: 504–512.

Masoro, E. J. 2005. Overview of caloric restriction and ageing. Mechanisms of Ageing and Development 126: 913–922.

Medawar, P. B. 1946. Old age and natural death. Modern Quarterly 2: 30–49. *This paper developed the fundamental evolutionary argument that senescence arises owing to an age-related decline in the strength of selection.*

Moorad, J. A., and D.E.L. Promislow. 2009. What can genetic variation tell us about the evolution of senescence? Proceedings of the Royal Society B 276: 2271–2278.

Promislow, D.E.L. 1991. Senescence in natural populations of mammals: A comparative study. Evolution 45: 1869–1887. *This paper established that senescence, measured as age-related increases in mortality rate, is common in natural populations of mammals.*

Rose, M. 1984. Laboratory evolution of postponed senescence in Drosophila melanogaster. Evolution 38: 1004–1010. *The first artificial selection experiment leading to increased longevity and tests of evolutionary theories of aging.*

Sear, R., and R. Mace. 2008. Who keeps children alive? A review of the effects of kin on child survival. Evolution and Human Behavior 29: 1–18.

Williams, G. C. 1957. Pleiotropy, natural selection, and the evolution of senescence. Evolution 11: 398–411. *This paper established the central role of genetic trade-offs in the evolution of senescence. This is one of the few papers in the field whose ideas have been adopted by molecular biologists.*

CHAPTER 9

EVOLUTION OF PARASITE VIRULENCE

Dieter Ebert

OUTLINE

1. Defining virulence
2. The phase model of virulence
3. The trade-off model
4. Vertically transmitted parasites
5. How well do optimality models predict virulence?

Diseases caused by parasite or pathogen infections impair normal functioning in organisms. These impairments can include very diverse symptoms leading to the organism's morbidity and mortality. Studies of the evolution of the virulence of infectious diseases strive to understand the expression of these symptoms as the result of the evolutionary process. This approach is primarily focused on the evolution of parasites (here used to include pathogens), but it may also consider the coevolution of hosts and parasites.

Until about 35 years ago, it was widely accepted that the harmful symptoms of infectious diseases were the side effects of poorly adapted parasites, and that over time virulence would therefore generally evolve toward avirulence. The underlying logic was that a well-adapted parasite should not harm its host, as doing so would deplete the parasite's own resources. This view was challenged in the 1980s by Roy Anderson, Robert May, and Paul Ewald, who argued that virulence is often a necessary consequence when parasites exploit their hosts, and depending on the specific conditions, the optimal level of virulence (i.e., the level that maximizes parasite fitness) may range from low to high virulence. Virulence is now understood as a trait whose evolution can be analyzed within the general framework of evolutionary biology, thus considering the roles of history, chance, and natural selection.

The evolution of virulence is not only of academic interest; its conceptual framework also has implications in various applied fields, such as human and veterinary medicine and agriculture. In particular, public health workers can benefit from considering population-biological aspects of the evolution of virulence.

GLOSSARY

Basic Reproductive Number, R_0. The average number of secondary infections resulting from a primary infection in a population of susceptible hosts.

Horizontal Transmission. The passing of a parasite between two hosts that are not related in direct line, for example, vector-borne transmission and sexual transmission.

Kin Selection. Selection on organisms that share genes by common descent. Fitness is said to be inclusive because it considers the fitness of related individuals.

Myxoma virus. A DNA virus from the family Poxviridae. This virus causes myxomatosis and has been used to control rabbit pests in Australia and Europe. Its virulence evolved rapidly after its release.

Parasite. An infective agent transmitted among hosts and growing or replicating within hosts. Pathogens are included in this definition.

Reproductive Manipulator. Parasites with maternal (vertical) transmission that manipulate host reproduction to increase their representation in the offspring of the next host generation, for example, by killing or feminizing male offspring.

Vertical Transmission. The passing of a parasite from a parent (usually the mother) to an offspring.

Virulence. Parasite-induced morbidity and mortality of a host. More precise definitions, such as parasite-induced host death rate or host fecundity reduction, are used for specific situations.

1. DEFINING VIRULENCE

Virulence is simply defined here as the parasite-induced morbidity and mortality of the host. This definition includes any fitness effect the parasite has on the host, whether that effect is an incidental by-product of the infection or an adaptive trait for the parasite. This definition does not, however, explain how virulence evolves, because it does not specify the link between parasite fitness and virulence. Most attempts to understand the evolution of virulence are based on models of parasite evolution and therefore consider only aspects of virulence that are important for parasite fitness. For an exclusively horizontally transmitted parasite, host mortality is important, as the parasite might die with the host, whereas reduced host fertility or sexual attractiveness—which are important for host fitness—may be of little concern for the parasite. For example, congenital rubella syndrome is a serious illness of babies born to mothers who became infected with the rubella virus during the first trimester of pregnancy. It is a big concern for the human host but is unlikely to have an impact on the evolution of the rubella virus. This picture changes, however, for parasites that rely on *vertical transmission*. For them, host fecundity becomes an important part of the virulence definition. Thus, understanding the evolution of virulence requires detailed knowledge about the host-parasite system in question. In the application of models of optimal virulence to actual diseases, the key factors to consider are the mode of parasite transmission and the trade-offs among parasite fitness components.

Virulence may be further categorized by distinguishing between effects directly beneficial for the parasite (e.g., when host death is required for parasite transmission, as in many parasites of invertebrates) and those effects that are costly for both the host and the parasite (e.g., host death when infections are transmitted among living hosts). Most models of the evolution of virulence consider the latter scenario. Examples of effects with a direct benefit for the parasite include parasitic castration (which liberates resources for parasite reproduction), impaired host mobility (which may increase access to the host by vectors that transmit the parasite further), and parasite-induced changes in host behavior (which may increase chances of transmission).

2. THE PHASE MODEL OF VIRULENCE

The once-dominant view that only novel diseases are highly virulent and that well-adapted parasites are less virulent is based on two related ideas: first, that virulence is a result of a new interaction between a host and a parasite, and second, that virulence changes as the parasite adapts to the new host. A refined version of this two-stage scenario tries to combine the different aspects of virulence evolution and expression into a unified framework (Ebert and Bull 2008). This model distinguishes three successive phases of disease evolution. Phase 1 is the first contact of a parasite with a host that it usually does not infect, often called *accidental infection*. In phase 2 the parasite has only recently established itself in a new host species, at which point the parasite's virulence is not yet the result of adaptive evolution. In phase 3 the parasite has evolved for some time in a particular host and has adapted to the specific conditions of this host population.

The phase model emphasizes that not all aspects of parasite virulence can be understood as a result of adaptive evolution. Consider, for example, the following diseases of humans. West Nile virus does not circulate long enough in humans to evolve an optimal level of virulence. Transmission chains are short, so it remains in phase 1. The human immune deficiency virus (HIV) entered the human population several decades ago. It is clearly able to persist in humans, but it may not have had time to reach an optimal level of virulence. Thus, it can be considered to be in phase 2. Much older human diseases, such as tuberculosis and leprosy, are likely to be in phase 3.

Phase 1: Accidental Infections

Many terrifying human diseases are caused by accidental infections including, for example, bird flu, SARS, anthrax, Lyme disease, Legionnaires' disease, West Nile virus, and echinococcosis. For some of these diseases, untreated infections can approach 100 percent mortality rates. Transmission chains are short, and epidemics do not persist, so the parasites have little opportunity to adapt to their new hosts. At first glance, this case might seem to support the view that novel diseases are highly virulent. However, the most virulent accidental infections are most likely to be recognized, whereas avirulent accidental infections will often go unnoticed, thus producing a strong sampling bias. In fact, avirulent accidental infections may outnumber virulent infections, as experimental transspecies infection trials suggest. Although a few novel infections are highly virulent, most are avirulent. Thus, highly virulent novel diseases are the exception, not the rule, although they may have profound impact on humans and natural populations.

Accidental infections have played and continue to play an important role in applied fields such as medicine and agriculture. In medicine, vaccine development has taken advantage of transspecies host shifts, both by using parasites of closely related species (Jenner's pox vaccine derived from accidental infections with cowpox) and by evolving attenuated parasite lines as vaccines. In agriculture, highly virulent novel infections have been used in pest control, as, for example, to control the very dense populations of European rabbits in Australia and the United Kingdom with the *Myxoma virus*, which is derived from a virus of a related host species.

Phase 2: Evolution of Virulence Following Successful Invasion

After a parasite infects a novel host species, its transmission success will determine whether it will persist and spread in this species. Initial spread is typically epidemic, and the level of virulence of the nonadapted parasite is unlikely to be close to optimum. The recent Ebola virus epidemic in West Africa is an example of this. Selection typically shapes a parasite's life history and virulence during the epidemic and especially over the following period, as it becomes endemic. Every endemic infectious disease has made the successful transition from phase 1 to 2 at some stage in its evolutionary history, but the number of observed cases is very low, despite many more examples of accidental infections (more than 1000 human diseases, so-called zoonoses). Phase 2 can also help us understand the emergence and spread of novel variants (mutants) of a parasite in an established host-parasite association. These mutants may have extreme effects and may spread rapidly, causing an epidemic and being unaffected by existing host defenses. If such epidemics occur frequently, a parasite population may never reach an optimum virulence and thus will remain in phase 2 (Bull and Ebert 2008). In this case, the fitness consequences of having suboptimal virulence are likely to be minor relative to the fitness gains the mutants achieve by evading host immunity and other defenses.

Studying parasites in phase 2 illuminates how the tempo and mode of virulence evolution proceeds in real time. The best example is the *Myx-*

oma virus. A highly virulent strain of this virus was introduced to Australia to control European rabbits, an invasive species that was causing extensive damage. Within a few years the average virulence of the parasite had changed drastically, attaining a new level far below the virulence of the original strain (Fenner and Kerr 1994). Apparently, the original virulence had been far above the presumed optimum, although not so high that it had prevented the spread of the virus. The *Myxoma* example leaves many questions open, as multiple factors changed simultaneously (e.g., host density, host genetic composition). But it does demonstrate that virulence can evolve rapidly, that virulence evolution does not necessarily lead to complete avirulence, and that virulence coevolves with the host (Fenner and Kerr 1994).

Laboratory experiments that follow parasite evolution after a change of environmental conditions or a host shift are in effect creating phase 2 situations. This situation is particularly true for serial-passage experiments, which played an important role in vaccine development, and also for the understanding of virulence evolution. In these experiments parasites are passaged in novel hosts with transmission controlled by the researcher (in some cases, the parasite might go extinct without this intervention). The evolution of the parasite is then monitored over many generations. Evolution typically results in a strong increase in virulence in the host in which the passages take place. At the same time, the parasites display reduced virulence in their former hosts. This attenuation of virulence makes these parasites good candidates for vaccines, as the immune response of the former host still recognizes the parasite, but its low virulence prevents disease or limits its severity. The increased virulence in the novel host is linked to an increase in the parasite's within-host multiplication rate. This increase is likely driven by within-host competition among parasite variants, with the most prolific variants having the highest chance to be transmitted during the next passage.

Phase 3: The Evolution of Optimal Virulence

Parasites that persist for some time in a host are expected to evolve an optimal level of virulence, that is, the level of virulence at which parasite fitness is maximal. It is widely thought that this optimum is characterized by trade-offs among different parasite fitness components. Thus, a key difference between phase 2 and phase 3 is that in phase 2, the parasite is not yet subject to the constraint imposed by the trade-offs. The *Myxoma virus* example was the first to show the existence of a trade-off, in this case between the rate at which rabbits clear the infection (host-induced parasite death) and the rate at which the parasite kills the host (and itself). Highly virulent *Myxoma* strains kill the host too quickly, while strains with low virulence are quickly cleared by the host's immune response. The optimal balance between these two parasite fitness components was shown by Anderson and May (1982) to maximize the parasite's spread in the host population. A mathematical model, later called the *trade-off model*, and the observed data agreed well.

3. THE TRADE-OFF MODEL

When considering models of optimal virulence, it is essential to define virulence precisely. In the mathematical model first applied to analyze the *Myxoma* data, virulence is defined as the parasite-induced host death rate. Other detrimental effects the parasite may have on the host, such as reduced mating success and fecundity, are not considered because they are not fitness components of the parasite. This simplification is acceptable under the assumption that parasite-induced host death rate is positively correlated with the various expressions of morbidity. These positive correlations also justify the use of surrogate measures of virulence in empirical studies, such as host fatigue, sensitivity to stress, fever, or other physiological parameters. However, it is important to keep in mind that correlations between parasite-induced host death rate and other disease-related traits may be weak or may have a negative sign. For example, many parasites of invertebrates castrate their hosts, which eliminates host fecundity but allows the parasite to keep its host alive for much longer than a noncastrator would be able to. In the following discussion of optimal virulence, parasite-induced host mortality is used as a definition for virulence.

The first and still most used model for the evolution of virulence is the trade-off model (Anderson and May 1982). This model is a powerful starting point for analyzing the evolution of virulence, although its simplicity implies certain assumptions and makes it vulnerable to various criticisms. The model stresses the importance of trade-offs between parasite-induced host death and other parasite fitness components. In its simplest form, parasite fitness is estimated as the number of secondary infections that result from a primary infection, R_0:

$$R_0 = \frac{\beta N}{\alpha + \gamma + \mu},$$

where β is the transmission rate in a susceptible host population of density N, α is virulence (parasite-induced host death rate), γ is the rate at which hosts clear infections, and μ is the host background (parasite-independent) mortality rate. Thus, α, γ, and μ are all components of the parasite's overall death rate, while βN indicates the production of new infections. Without trade-offs, parasite evolution would maximize β and minimize the total parasite death rate ($\alpha + \gamma + \mu$), thus driving virulence to zero. A positive correlation between β and α, or a negative one between α and γ (as in the case of the *Myxoma* example), constrains the evolution of virulence. If the increase in β is leveling off relative to the increase in α (i.e., the relationship is asymptotic), intermediate levels of virulence are predicted. This simple trade-off model relies entirely on the assumption that between-host transmission is the quantity maximized by parasite evolution. Although 35 years old, the hypothesis that between-host transmission is crucial for the evolution of virulence is still not strongly supported by empirical evidence. A critical test will be to show that transmission success has the assumed hump-shaped relationship with virulence. Empirical studies that tested this prediction were conducted with unicellular parasites that infect mice, butterflies, and water fleas (*Daphnia*).

Some of the major limitations of the simple trade-off model are that it ignores the role of multiple infections and within-host evolution of the parasites, kin structure of parasites, and host genetic variation. The model also ignores changes over time in the density of susceptible hosts, including those changes that may occur as the parasite itself evolves.

Multiple Infections, Inclusive Fitness, and Virulence

In the 1990s the trade-off model was extended to include within-host competition. Within-host competition describes scenarios in which different variants of parasites compete within hosts, which strongly influences both their likelihood of transmission to the next host and the level of virulence expressed in the multiply infected hosts. Parasite variants that replicate more quickly are assumed to be superior in within-host competition, even if this reduces host survival and thus shortens the period for transmission to take place. Because higher replication rate is associated with higher virulence, average virulence is expected to be higher in populations with frequent multiple infections than in those with single infections, all else being equal. As a consequence, more virulent parasite variants may dominate although they do not maximize the R_0 as predicted by the simple trade-off model. However, more complex models incorporating additional factors that might occur with multiple infections have been developed that predict the opposite result under specific circumstances. In particular, cooperation among parasites to exploit the host, or parasite strategies to exploit one another, may lead to the evolution of lower virulence.

Empirical tests on the role of multiple infections have generally supported the key assumption and prediction of the basic multiple infection model. Thus, higher within-host multiplication rates have been found to be associated with both superior competitive ability and higher virulence. Interestingly, it has also been found that some parasites increase their virulence facultatively when the host has been multiply infected, which implies that the parasites are able to sense, either directly or indirectly, the presence of competing parasite variants.

Multiple infection virulence models make a number of other predictions as well. Ecological and demographic factors such as higher host density, longer parasite survival outside the host, longer host life span (e.g., low, parasite-independent mortality), and less spatial structure of the host population (more mixing) are predicted to lead to increased virulence, because these factors should increase the incidence of infections by multiple parasite variants. These predictions have been explained using inclusive fitness theory, which considers the kin structure of the parasite population. Inclusive fitness is a crucial factor in the evolution of virulence, because within-host competition cannot be considered independently from the genetic relatedness of the competing parasites (Frank 1996). More closely related parasites have more common reproductive interests and thus gain less from competition. Using inclusive fitness theory, one can generalize findings that link ecological features with virulence and place the role of transmission mode into a unified context. For example, in well-mixed host populations, multiple infections result mostly from unrelated parasites, which maximizes competition and thus also virulence. In contrast, in viscose host populations, multiple infections more often arise through infections from the same parasite lineage. In these latter cases, lower levels of virulence are expected to evolve because of *kin selection* in the parasite population. As will be discussed further, kin selection also plays an important role in the evolution of virulence of vertically transmitted parasites.

Host Genetic Variation

Empirical studies have shown that virulence is not only the product of parasite evolution but also the result of the coevolutionary interaction between hosts and the parasites. Some models of virulence evolution have incorporated certain aspects of host genetic variation, but the complexity of the interactions that influence the expression of virulence makes it unlikely that general predictions can be made. Nevertheless, models and empirical data agree that parasite virulence should be lower in genetically diverse host populations. This effect is based on the observation that greater host diversity slows the spread of parasites (parasites spread faster in host monocultures), and this diversity thus reduces multiple infections. Trade-offs in the performance of parasite genotypes across different host genotypes or host species suggest that parasite fitness is compromised (and virulence reduced) in diverse host populations. Finally, host evolution that counters the effects of infections may lead to a reduction in parasite virulence. Incorporating the various effects of host genetic variation into models of disease virulence remains one of the big future challenges for understanding the evolution of virulence.

4. VERTICALLY TRANSMITTED PARASITES

The discussion thus far has focused on those parasites that engage exclusively in *horizontal transmission*, that is, transmission among hosts unrelated in direct line. Many parasites, however, are entirely or partially *vertically transmitted*, usually from mothers to offspring. In this case, the parasite's fitness depends on host reproductive success, and this dependency must be included in models that seek to explain the evolution of virulence. It is best to begin this discussion by focusing on those parasites that are transmitted exclusively from mothers to offspring. Such parasites must either evolve to manipulate host reproduction to their own benefit or evolve to complete avirulence. This necessity can readily be understood when one realizes that a parasite that is transmitted only vertically, and that harms its host's reproductive success without promoting its own transmission, will go extinct as parasite-free hosts outcompete those that are infected.

Some vertically transmitted parasites, such as the bacterium *Wolbachia* and some microsporidians, are *reproductive manipulators*; that is, they have evolved mechanisms to manipulate host reproduction in ways that increase their presence in future generations, despite being virulent. These mechanisms include the killing of host sons, feminizing males to become functional females, and inducing cytoplasmic incompatibility. In some of these cases, only kin selection can explain the observed virulence, because the individual parasites that produce the virulent effects (e.g., killing sons, inducing cytoplasmic incompatibility), in essence, commit suicide because they preclude their own propagation. Other individual parasites are genetically identical or nearly so, and they benefit from these behaviors. Parasites that manipulate host reproduction are very rarely transmitted horizontally.

Parasites with exclusive maternal transmission that do not manipulate their hosts must evolve avirulence, because the parasite's reproductive success is perfectly linked to the reproductive success of its host. In these cases, there is no conflict of interest between host and parasite. Some of the best empirical support for virulence theory has come from the experimental evolution of parasites in diverse systems being propagated under contrasting conditions of vertical and horizontal transmission, with the result that the parasites evolve toward avirulence when they are restricted to exclusive vertical transmission. However, complete avirulence no longer fulfills the definition of a parasite, so that the resulting entities might better be described as symbionts.

Many parasites are transmitted both vertically and horizontally (Ebert 2014). In some cases, the population structure imposes a strict trade-off between the two modes of transmission, such that more frequent horizontal transmission leads to less vertical transmission. Observations and experiments under such conditions have shown that the more horizontal transmission takes place, the more virulent the parasite will be. However, this prediction does not generally hold when the host population is well mixed. In such cases, vertical and horizontal transmission may be positively correlated, making general predictions difficult. Thus, without detailed knowledge about transmission trade-offs and host population structure, the finding that a particular parasite is partially vertically transmitted cannot be used as a predictor of low virulence.

5. HOW WELL DO OPTIMALITY MODELS PREDICT VIRULENCE?

Models of the evolution of virulence are deceptive in their simplicity and power to make testable predictions. Unfortunately, the evidence in support of these models is still rather thin, although many of the key assumptions have been supported experimentally in several systems (e.g., trade-offs have been observed, and multiply infected hosts typically suffer from higher virulence). For example, the field lacks compelling examples of evolutionary changes in virulence associated with trade-offs. Even the frequently cited *Myxoma*–rabbit case is open to some alternative explanations. Those experimental studies that produced the clearest outcomes also had to use rather extreme conditions (e.g., 100 percent vertical versus 100 percent horizontal transmission) that may limit the ability to generalize their findings. Comparative studies employing data from many different host-parasite systems do not explain much of the variation in virulence, suggesting that other effects may overrule general patterns. For example, the type of host tissue affected by the parasite seems to explain more of the variation in virulence than do either trade-offs or transmission dynamics. Next-generation models that take into account host diversity and multidimensional trade-offs might be able to make more accurate predictions, although they are likely also to suffer in terms of generality.

Despite these limitations, models of optimal virulence are important because they provide a starting point for formulating testable predictions. In those cases in which a single environmental factor changes, it may indeed be possible to predict the associated change in virulence. However, changes in one factor often go hand in hand with changes in other factors, which may exert opposing selection on virulence. For example, the trade-off model predicts that an increase in host life span favors low virulence, but this same change may increase the frequency of multiple infections, which favors higher virulence. Therefore, careful evaluations of epidemiological circumstances and host demographic conditions are likely to be necessary before predictions can be made with confidence. It is currently not possible to make simple and robust recommendations for pest management that will favor the evolution of less virulent parasites. Proposals to manage virulence by changing environmental conditions must therefore be evaluated with appropriate care before they are put into practice.

FURTHER READING

Alizon, S., A. Hurford, N. Mideo, and M. Van Baalen. 2009. Virulence evolution and the trade-off hypothesis: History, current state of affairs and the future. Journal of Evolutionary Biology 22: 245–259. *A conceptual review of the trade-off model, mainly from a theoretical perspective.*

Anderson, R. M., and R. M. May. 1982. Coevolution of hosts and parasites. Parasitology 85 (pt. 2): 411–426. *The classic reference on this topic and still readable.*

Bull, J. J., and D. Ebert. 2008. Invasion thresholds and the evolution of nonequilibrium virulence. Evolutionary Applications 1: 172–182.

Ebert, D. 2014. The epidemiology and evolution of symbionts with vertical and mixed-mode transmission. Annual Review of Ecology, Evolution, and Systematics 44: 623–643.

Ebert, D., and J. J. Bull. 2008. The evolution of virulence. In S. C. Stearns and J. K. Koella, eds., Evolution in Health and Disease. 2nd ed. Oxford: Oxford University Press. *The phase model for virulence is here worked out in more detail.*

Fenner, F., and P. J. Kerr. 1994. Evolution of poxviruses, including the coevolution of virus and host in myxomatosis. In S. S. Morse, ed., The Evolutionary Biology of Viruses. New York: Raven. *Much about the biological background for the rabbit–myxoma case.*

Frank, S. A. 1996. Models of parasite virulence. Quarterly Review of Biology 71: 37–78.

CHAPTER 10

EVOLUTION OF ANTIBIOTIC RESISTANCE

Dan I. Andersson

OUTLINE

1. A medical miracle—and how to ruin it
2. Origins of antibiotics and antibiotic-resistance mechanisms
3. Transmission of resistant bacteria
4. Persistence and reversibility of resistance
5. Can resistance evolution be slowed or even stopped?
6. Will antibiotics become a footnote to medical history?

Antibiotics have revolutionized human and veterinary medicine, and over the last 70 years they have made it possible to efficiently treat most types of bacterial infections. Unfortunately, the extensive use—and frequent misuse—of antibiotics has resulted in the rapid evolution and spread of bacteria that are resistant to antibiotics. Arguably, the global use of antibiotics is one of the largest evolution experiments performed by humans, and the frightening consequence is that we are now at the brink of a postantibiotic era in which antibiotics have lost their miraculous power. This problem originates from the strong selection imposed by the extensive use of antibiotics and the resulting enrichment of resistance mutations and horizontally acquired resistance genes. Together these factors have generated high-level antibiotic resistance in the majority of significant human and veterinary pathogens. Several forces act to stabilize resistance in a population once it becomes established, and resistant bacteria may thus persist for a long time even after use of an antibiotic has been reduced. The development of new classes of antibiotics, coupled with more prudent use of antibiotics, will be required to maintain antibiotics as efficient agents for treating bacterial infections.

GLOSSARY

Antibiotic. An antibacterial compound that may have a natural or synthetic origin.

Biological Fitness Cost. The effect that a resistance mechanism has on bacterial fitness (including growth, persistence, and survival within and outside hosts) in the absence of antibiotic.

Compensatory Evolution. Reduction or elimination of the fitness cost associated with a mutation that has a deleterious side effect (e.g., a resistance mutation) by additional genetic changes (compensatory mutations).

Conjugation. Transfer of genetic material between bacterial cells mediated by direct contact between two cells.

Coselection. Process whereby a nonselected gene indirectly increases in frequency by virtue of its genetic linkage (within a genetic element or a bacterial clone) with a directly selected gene; sometimes also called *genetic hitchhiking*.

Horizontal Gene Transfer. A process in which a recipient organism receives and incorporates genetic material from a donor organism without being the offspring of the donor; sometimes also called *lateral gene transfer*.

Minimum Inhibitory Concentration (MIC). The lowest concentration of an antimicrobial drug that inhibits the growth of a bacterial population.

Nosocomial Infection. Infection contracted during treatment in a hospital or other healthcare facility.

Plasmid. A DNA molecule that is separate from and can replicate independently of the chromosomes in bacteria.

Resistome. A neologism that refers to the set of resistance genes and precursors to resistance genes that are present in all pathogenic and nonpathogenic bacterial species combined.

Transduction. Injection of foreign DNA into bacterial cells by a bacteriophage (i.e., a virus that infects bacteria).

Transformation. Uptake of exogenous DNA into a cell through the cell envelope.

1. A MEDICAL MIRACLE — AND HOW TO RUIN IT

Antibiotics represent one of the most important medical advances in modern times, and since their introduction over 70 years ago they have saved countless lives. Today we often take antibiotics for granted in the developed parts of the world, but we have to go back only to our grandparents to find a generation for which common infections such as pneumonia, meningitis, blood poisoning, and intestinal infections were potentially deadly. Charles Fletcher, a young physician who was involved in early clinical trials of penicillin in the 1940s, describes vividly how the introduction of antibiotics changed modern medicine:

> It is difficult to convey the excitement of actually witnessing the amazing power of penicillin over infections for which there had previously been no effective treatment. . . . I did glimpse the disappearance of the chambers of horrors which seems to be the best way to describe those old septic wards . . . and could see that we should never again have to fear the streptococcus or the more deadly staphylococcus.

In addition to being widely used for the treatment of many common community and *nosocomial* (hospital-acquired) *infections*, antibiotics are also an essential component in the treatment and prevention of infections associated with advanced medical practices including chemotherapy of cancers, organ transplantation, implantation and replacement of medical devices and prostheses, neonatal care, and invasive surgery.

Unfortunately, the utility of antibiotics is deteriorating at an alarming rate, and the reason for this change is easily understood in the context of Darwinian adaptive evolution. To put it simply, bacteria adapt genetically to the presence of antibiotics by acquiring various types of resistance mechanisms that prevent antibiotics from performing their inhibitory function. These resistance mechanisms allow the bacteria to grow and reproduce in the presence of antibiotics, and evolution thereby nullifies their efficacy in treating infections. The widespread use—and often the overuse—of antibiotics on a global scale (estimated currently to be at least 100,000 tons per year) for human medicine, veterinary medicine, and agriculture is the main reason for the selection and spread of resistance among both human and animal bacterial pathogens. So, although the introduction of antibiotics is often viewed as one of humankind's greatest achievements, we are now at risk of destroying that achievement. At the very least, we are paying a high price for the increased resistance. The overuse of antibiotics reflects several factors, including poor knowledge among prescribers and patients, profits for physicians and pharmacists from the prescription and sale of antibiotics, aggressive marketing from pharmaceutical companies, and the lack of regulations and guidelines for when and how antibiotics should properly be used. Studies have shown correlations between the amount of antibiotics used and the prevalence of resistance at several levels (e.g., country, hospital), as would be expected if antibiotics select for increased bacterial resistance.

As a society, we are paying a high price for the increased levels of bacterial resistance to antibiotics: resistant bacteria limit our ability to efficiently treat bacterial infections, and they also increase the risk of complications and even death. In addition, antibiotic resistance imposes a large economic burden on the healthcare system owing to increased treatment costs as well as the costs of identifying and developing new, alternative compounds. Worldwide there are areas where bacterial infections have become untreatable as the result of antibiotic resistance, and with the recent spread of gram-negative bacteria that produce multidrug-resistant extended-spectrum β-lactamase (ESBL), the problem is becoming even more acute. This trend toward increasing resistance, combined with diminished research and development of new antibiotics, has led to a dismal situation in which we may face a postantibiotic era.

2. ORIGINS OF ANTIBIOTICS AND ANTIBIOTIC-RESISTANCE MECHANISMS

Antibiotics are compounds that inhibit (bacteriostatic drugs) or kill (bactericidal drugs) bacteria by a specific interaction with some target in the bacterial cell. Some purists limit the definition of antibiotics to only those substances produced by a microorganism, but today all natural, semisynthetic (i.e., a combination of natural and synthetic precursors), and synthetic compounds with antibacterial activity are generally classified as antibiotics. The target for an antibiotic can be an essential enzyme or cellular process such as protein synthesis, cell-wall biosynthesis, transcription, or DNA replication. Most medically relevant and industrially produced antibiotics originate in nature and are synthesized by a variety of species, mainly soil-dwelling bacteria (in particular the genus *Streptomyces* in the phylum Actinobacteria) and fungi. The benefits of antibiotics for microbial producers is a matter of debate; antibiotics might be used as ecological weapons to inhibit competitors, but they might have a more benevolent function as signals for cell-to-cell communication in microbial communities. In any case, the synthesis and release of antibiotic compounds in nature means that many bacteria (both producers and bystanders) have long histories of exposure to antibiotics, and as a consequence, many have evolved various resistance mechanisms. These mechanisms likely evolved to protect against self-destruction (in antibiotic producers), to defend against antibiotics produced by other species, to modulate intermicrobe communication, or to perform metabolic functions unrelated to antibiotics. This vast pool of resistance genes, known as the *resistome*, has the potential to be transferred within and between species, and to confer resistance to any antibiotic that might be used against human and animal pathogens. In fact, many of the resistance problems generated by the use of antibiotics since the 1940s are a consequence of the acquisition of preexisting resistance determinants by pathogens via *horizontal gene transfer* (HGT). Transfer mechanisms include conjugative transfer of *plasmids* (DNA molecules that are separate from and can replicate independently of the chromosomes in bacteria) and conjugative transposons, *transduction* via bacteriophages (viruses that infect bacteria), and *transformation* of naked DNA taken up from the environment by some species. Of these mechanisms, conjugative transfers are the most common mode of acquiring resistance, whereas bacteriophage transfers appear to be rare. Apart from HGT, resistance can also arise by mutations (including point mutations as well as rearrangements and gene amplification) in native resident genes.

For resistance to become a problem, the acquired or mutated resistance genes must be phenotypically expressed in clinically relevant human and animal pathogens. The evolutionary pathways leading to these outcomes are often complex and often not well understood, especially in the case of resistance acquired by HGT. Even when a potential donor of a resistance gene has been identified by genome sequence data (e.g., the CTXM type of ESBL resistance was likely acquired from *Kluyvera* strains in the environment), several conditions must be fulfilled for resistance to emerge in the case of HGT: (1) a resistance mechanism must be present in a donor bacterium; (2) there must be a genetic mechanism for HGT; (3) there must be an ecological opportunity for transfer between the donor and recipient cells (e.g., in the case of *conjugation* direct contact is needed); (4) the transferred gene must be stably inherited, adequately expressed, and confer a resistant phenotype in the recipient; and (5) there must be strong enough selection—typically, a sufficient level of antibiotic—to favor the resistant recipient organisms, even though resistance may impose a fitness cost (as discussed in section 4). In the case of resistance that occurs by mutation in resident genes, the process is simpler and requires only a suitable resistance mutation and sufficient selection to favor the resistant mutants. Despite the relative ease by which bacteria can become resistant by mutations, HGT is the predominant route for generation of antibiotic resistance in most human and animal pathogens. The likely explanation is that pathogenic bacteria can acquire high-level resistance to a given antibiotic—and indeed, simultaneous resistance to several antibiotics—by means of a single transfer event from the relatively accessible pool of resistance genes in the microbial community's resistome. Of special relevance here are genetic elements called *integrons* that can capture and express arrays of resistance genes; when integrons are transferred on a plasmid, they can convert the recipient strain from being antibiotic susceptible to multidrug resistant. In contrast, mutation-based resistance often produces lower-level resistance and may require several mutational steps to produce high-level resistance, thus requiring a longer evolutionary path to achieve a clinically resistant phenotype. A notable exception, however, is *Mycobacterium tuberculosis*, in which all known resistance mechanisms are the result of mutation rather than HGT, and single mutations sometimes produce high-level resistance (e.g., resistance to aminoglycosides and rifampicin arises from mutations in ribosomal protein S12/rRNA and RNA polymerase subunit β, respectively). Mycobacteria have conjugative plasmids and transducing bacteriophages, and it is unclear why HGT is not associated with resistance evolution in this bacterium.

Horizontally acquired genes and mutations in native genes confer resistance to bacteria by a variety of different mechanisms that either protect the normal cellular target from exposure to the antibiotic or alter the target's structure to prevent the drug from binding. (1) The antibiotic may be enzymatically inactivated by hydrolysis (e.g., resistance to β-lactam antibiotics conferred by β-lactamases) or modification (e.g., acetylation, phosphorylation, or adenylylation of aminoglycosides). (2) Uptake of the antibiotic may be reduced by changes in the cell wall (e.g., mutations that confer low-level β-lactam resistance by altering channels called *porins*). (3) Bacteria may express efflux systems that actively pump the antibiotic out of the cell (e.g., efflux pumps that confer β-lactam or aminoglycoside resistance). (4) The target molecule may be modified such that antibiotic is prevented from binding (e.g., mutations in ribosomal proteins or rRNA that inhibit binding of aminoglycosides). (5) Resistance may result from a bypass mechanism whereby the need for the inhibited target is relieved by provision of an alternative target or pathway (e.g., resistance to peptide deformylase inhibitors by inactivation of formyl transferase). Mechanisms 1, 4, and 5 often provide high-level resistance, whereas mechanisms 2 and 3 are typically associated with lower-level resistance.

3. TRANSMISSION OF RESISTANT BACTERIA

Once resistance has evolved in a bacterial pathogen, the extent to which it becomes a medical problem depends on how rapidly and extensively the resistant type is transmitted from its place of origin into the human or animal population and the rate at which it is disseminated among the hosts. Transmission rates of resistant bacteria depend on many factors including host density, patterns of host travel and migration, various hygienic factors (e.g., in hospitals and during food preparation), host immunity (e.g., vaccination), and the intrinsic transmissibility of the resistant pathogens. In principle, humans can influence all these except the last factor.

4. PERSISTENCE AND REVERSIBILITY OF RESISTANCE

Whether antibiotic resistance will persist in a bacterial population after it emerges depends in general on the relative strength of several selective forces. The most obvious of these is the direct advantage to resistant bacteria caused by exposure to concentrations of antibiotics that are lethal or inhibitory to sensitive strains. An opposing force, however, is the fitness cost of resistance, that is, any effect of the resistance mechanism that reduces the ability of the pathogen to grow, persist, or spread in the host population. Such costs will impede the rise of resistant bacteria, and these costs will also affect the likelihood that resistance can be reversed or otherwise eliminated. While these fitness costs offer hope that resistance can be controlled, other forces discussed later can stabilize resistance in a bacterial population, even when the antibiotic is absent or at a low concentration.

Sub-MIC Selection

Selection clearly favors resistant strains when antibiotic concentrations are above the *minimum inhibitory concentration* (MIC) of the susceptible bacteria, but it remains unclear whether levels far below the MIC can also select for resistance. Direct measurements of antibiotic levels in organs and tissues of treated patients and in various natural environments indicate that bacteria are frequently exposed to sub-MIC levels of drugs. In theory, such low antibiotic levels may select for resistance if susceptible bacteria grow even slightly more slowly than resistant strains. Antibiotics can be introduced into the environment in the urine from treated humans and animals, as well as when antibiotics are used in agriculture (for example, on fruit trees). On average, roughly half of all antibiotics (the proportion varies with antibiotic class) consumed by humans and animals enter the sewage system or other environments via urine, and the amount of antibiotics released into the environment is presently on the order of at least 100,000 tons per year. Recent results have shown that the resulting environmental antibiotic concentrations may be important for both the emergence of new resistant strains and the enrichment of existing resistant strains. In competition experiments between susceptible and resistant strains, selection for resistant bacteria can occur at antibiotic concentrations even less than 1 percent of the MIC of the susceptible bacteria; similar antibiotic concentrations can be found in many natural environments.

These findings are important from a public-policy perspective because they suggest that antibiotic releases into the environment through human, veterinary, and agricultural applications contribute significantly to the emergence and persistence of antibiotic resistance. In particular, they indicate the potential benefits of reducing anthropogenically generated antibiotic pollution and avoiding treatment regimens that involve prolonged periods with low levels of antibiotics.

Coselection of Resistance Genes

A resistance gene located on a transmissible element or in a bacterial clone can increase in frequency in a population as a consequence of its genetic linkage to another resistance gene that is under selection. Such linkage and the resulting *coselection* is a common feature of resistance determinants that have been acquired by HGT, including plasmids, transposons, and integrons, and it can occur more generally in any multidrug-resistant clones. As a consequence, the frequency of resistance to a particular antibiotic can remain stable or even increase in environments where the antibiotic is not currently being used. The linked gene that sustains the unselected resistance gene can be any gene that increases the fitness of the bacterial strain, including another antibiotic-resistance gene, a gene that encodes resistance to some heavy metal or disinfectant, or a gene that encodes some virulence-associated function.

Coselection is an important contributor to the long-term maintenance of resistance in bacterial populations, and it may explain why a reduction in the use of a particular antibiotic often has little or no effect on the frequency of resistant bacteria. For example, a recent study reported that an 85 percent reduction in the use of trimethoprim over a two-year period had only a very small effect on trimethoprim resistance in *Escherichia coli*. Similarly minor effects on resistance were recorded in other studies following reductions in use of sulfonamides, macrolides, and penicillin.

Cost-Free Resistance

The *biological fitness cost* of any particular resistance gene can vary depending on environmental conditions and the genetic background in which it occurs. For example, some resistance mutations impose no cost under standard laboratory conditions but have large costs in laboratory animals, and vice versa. Also, the cost of a resistance function often depends on the particular bacterial strain in which it occurs as a consequence of epistatic interactions between the resistance gene and other genes. Interestingly, some resistance genes do not appear to have any measurable fitness cost, at least in the environments and strains in which they have been tested. Of course, there may be other conditions under which these resistance genes do impose some costs, and measuring fitness costs under natural conditions is very difficult and rarely done; even in the laboratory, where genetically marked strains can be directly competed, it is difficult to measure fitness differences below about 0.3 percent per generation. It is also difficult to know what costs are relevant with respect to the persistence of an antibiotic resistance gene in a bacterial population. In principle, a fitness cost as small as 0.001 percent per generation would mean that a resistance gene would eventually be purged from the population by natural selection if the use of an antibiotic was stopped, although it might require many decades or even centuries, given such small fitness costs.

Fitness-Enhancing Resistance

Although antibiotic resistance often has a fitness cost, in some cases it can actually be advantageous, even in drug-free environments. Interesting examples of such fitness-increasing effects of resistance functions have recently been demonstrated in several bacterial species for the fluoroquinolone class of antibiotics. In *E. coli*, fluoroquinolone resistance commonly evolves by a multistep process involving mutations that alter efflux mechanisms and the proteins targeted by the drug. Each resistance mutation alone provides only a small increase in the MIC, so that clinically relevant levels of resistance require the accumulation of several mutations. In laboratory selection experiments, the accumulation of several resistance mutations typically led to reduced fitness in the absence of the antibiotic, but in a few cases an increase in resistance produced higher fitness. *Campylobacter jejuni* provides an interesting example of the background dependence of fitness effects associated with fluoroquinolone resistance. A single mutation in the gene encoding DNA gyrase enhanced the fitness of the resistant strain in a chicken-infection model, but when that same mutation was transferred into a different strain of *C. jejuni*, it imposed a fitness cost. Similarly, recent results show that combinations of certain types of resistance mutations might also increase bacterial fitness above that of the susceptible bacterium. The disturbing implication of these findings is that selection for improved growth may sometimes favor increased resistance even in the absence of drug selection.

Compensatory Evolution That Reduces Fitness Costs

Resistance to an antibiotic may impose a fitness cost because it disrupts the balanced growth of a bacterial cell that has been finely tuned to express genes and functions at levels that maximize fitness. A common process that stabilizes resistance is thus *compensatory evolution*, in which selection favors mutations that restore the cell's balance and thereby reduce or eliminate the cost of resistance, often without any significant loss of resistance. Indeed, several laboratory and animal and human studies have demonstrated the evolution of mutations that restore fitness and, as a consequence, stabilize resistant populations. Whether adaptation in the absence of antibiotic occurs by compensatory mutations or by reversion (loss of resistance) will depend on several factors particular to any given case, including the mutation rates and fitness effects for compensatory and reversion mutations, as well as population size. The genetic mechanisms of compensation vary depending on the particular drug and microbe involved. These mechanisms may include mutations in the resistance gene itself, as well as mutations that alter the expression of the resistance gene or other genes in ways that restore the appropriately balanced gene expression.

Plasmid Persistence

Plasmids typically carry genes that are nonessential and beneficial only under specific environmental conditions. Hence, they are often expendable, and their persistence requires either ongoing selection (e.g., for resistance genes) or other mechanisms that assure their continued carriage. The various selective processes discussed earlier can promote the maintenance of both chromosomal and plasmid-encoded resistance functions; there are also several mechanisms that can promote plasmid persistence even without selection for antibiotic resistance. For example, some plasmids enhance bacterial growth even in the absence of antibiotic. Many plasmids encode resolution and partitioning systems that prevent spontaneous plasmid loss during cell division, and some plasmids even have toxin-antitoxin systems that kill cells that lose the plasmid. Also, plasmids can be maintained in bacterial populations by their conjugation-mediated horizontal transfer between cells even if they impose a fitness cost.

5. CAN RESISTANCE EVOLUTION BE SLOWED OR EVEN STOPPED?

A pressing question is whether society can reduce the rate at which antibiotic resistance emerges and spreads. Various approaches have been suggested in the literature, but only a few are known to work. One approach—perhaps the most obvious but still difficult to implement—is to reduce the use of antibiotics, thereby reducing the strength of selection that favors both the emergence and spread of resistant bacteria. The efficacy of this approach follows from basic evolutionary principles and is also supported by numerous studies showing that the frequency of resistance is correlated with the volume of antibiotics used at various levels, including individual hospitals, communities, and countries. Global restraint in antibiotic use can be achieved only by concerted action and will require the implementation of several strategies, including (1) avoidance of antibiotic use when none is needed (e.g., when the infection is caused by a virus); (2) discontinuance of the use of antibiotics as growth promoters in animal husbandry; (3) discontinuance of the use of antibiotics in the production of crops and in aquaculture; (4) avoidance of economic situations in which the prescription of antibiotics is profitable for the prescriber; (5) appropriate control and regulation of antibiotic marketing by the pharmaceutical industry (in which prescribers, pharmacists, and consumers are targeted); and (6) prohibition of the sale of antibiotics to the public via the Internet or from pharmacies or other outlets without the need for a prescription.

Also, by increasing use of various hygienic and infection control measures, society can reduce the transmission of pathogenic bacteria and thereby reduce the use of antibiotics. The extent to which these measures will work depends on the pathogen and its mode of transmission among hosts. Pathogens for which hygienic measures have been shown to be particularly successful include various food-borne pathogens (e.g., *Salmonella*) and nosocomial infections such as methicillin-resistant *Staphylococcus aureus* (MRSA). For MRSA infections, screening strategies to track and isolate affected patients, coupled with improved hospital hygiene, have been successful in reducing the transmission of these dangerous bacteria.

Other approaches that have been proposed to reduce the rate at which resistance evolves include changes in dosing regimens and use of antibiotic combinations that reduce selection for resistant mutants without affecting treatment efficacy or safety. The use of drug combinations has been shown to be effective in treating many HIV (the virus that causes AIDS) infections because a mutant that becomes resistant to one drug is nonetheless susceptible to others that are provided at the same time. In addition, drugs and drug targets might be chosen during research and development such that the risk of resistance is minimized. For example, new antibiotics might be developed such that (1) resistance is difficult to acquire by mutation or HGT; (2) the resistance mechanism confers a high fitness cost; and (3) the opportunities for compensatory adaptation are limited. It is interesting to note that no clinical cases of resistance have been reported for certain combinations of drugs and bacteria even after decades of use. For example, penicillin has been used successfully to treat *Streptococcus pyogenes* infections for 60 years. An understanding of the reasons for the lack of resistance evolution might allow more rational choice and design of drugs and drug targets.

In addition to limiting the rates at which resistance emerges and spreads, it might even be possible to reverse the existing problem of resistance by reducing the use of antibiotics. Whether this strategy will be successful depends on the strength of the forces driving reversibility. At the levels of the individual and community, the fitness cost of resistance in the absence of antibiotic is probably the main force pushing toward increased sensitivity, whereas in hospitals the main driving force is probably the continuous influx of patients with susceptible bacteria. In hospitals, mathematical modeling and correlative studies suggest that changes in antibiotic use can cause rapid changes in the frequency of resistance. However, when the fitness cost of resistance drives reversibility, the rate of change is expected to be much slower. The main reasons for this are that in addition to the factors described earlier that can stabilize resistance in bacterial populations, the intrinsic dynamics of reversal are expected to be slow because the strength of selection for sensitivity in the absence of antibiotic is generally much weaker than selection for resistance when antibiotics are used. This inference is supported by clinical intervention studies, performed at both the individual and community levels, in which it has been observed that resistant clones are remarkably stable and persistent even when antibiotic use is reduced.

6. WILL ANTIBIOTICS BECOME A FOOTNOTE TO MEDICAL HISTORY?

How will future generations view our ongoing experiment with antibiotics? Will antibiotics retain their therapeutic value for generations to come? Or will antibiotics be viewed as a failed experiment, one that becomes a mere footnote in the history of medicine? The answers to these questions will depend on many factors, of which two challenges are of particular importance. The first is whether society—including medical practitioners, patients, and the pharmaceutical industry—will have the resolve to use antibiotics in a more restrictive and medically responsible way that will slow the emergence and spread of resistance. Success will require global implementation of changes in healthcare systems and practices that are specifically aimed at reducing the overall use of antibiotics that selectively favors resistant bacteria. Many international resolutions to this effect have been put forward, but so far little has been done to implement any global strategies. What is needed now is leadership and coordination that will allow these recommendations to be put into action. If we fail to implement these recommendations, it is certain that resistance will continue evolving to existing antibiotics, as well as to any new ones that are discovered.

The second major challenge is that the pharmaceutical industry has largely abandoned the development of new antibiotics, mainly for economic reasons; as a consequence, few new classes of antibiotics have been introduced for clinical use in recent decades. It is essential that this industry be recommitted to antibiotic discovery and the development of novel drugs. Potential ways forward might include new business models for collaboration between industry and public sectors, including new regulatory rules and funding schemes. Of course, there are real scientific challenges in finding new drugs, including antimicrobials. However, increased knowledge of structural biology, bacterial physiology and metabolism, medicinal chemistry, genomics, and systems biology provides new opportunities for the discovery of novel antibiotics, including ones that might inhibit new targets such that the evolution of resistance is impeded.

FURTHER READING

Andersson, D. I., and D. Hughes. 2010. Antibiotic resistance and its costs: Is it possible to reverse resistance? Nature Reviews Microbiology 8: 260–271. *A comprehensive review on the subject of fitness cost of antibiotic resistance and its influence on the emergence, spread, and persistence of resistant bacteria.*

Davies, J., and D. Davies. 2010. Origins and evolution of antibiotic resistance. Microbiology Molecular Biology Reviews 74: 417–433. *From leaders in the field, this insightful review discusses the environmental origin of antibiotics and resistance genes.*

Freire-Moran, L., B. Aronsson, C. Manz, I. C. Gyssens, A. D. So, D. L. Monnet, O. Cars, and the ECDC-EMA Working Group. 2011. Critical shortage of new antibiotics in development against multidrug-resistant bacteria: Time to react is now. Drug Resistance Updates 14: 118–124. *An important paper that demonstrates the serious shortage of new antibiotics in clinical development against multidrug-resistant bacteria and points to the need for the involvement of the public sector into research and development of new antimicrobial drugs.*

Hughes, D., and D. I. Andersson. 2015. Evolutionary consequences of drug resistance: Shared principles across diverse targets and organisms. Nature Reviews in Genetics 16: 459–471. *A comparison of drug resistance in diverse biological systems (including bacteria, viruses, and protozoa) describing commonalities and differences that could be useful for drug development and treatment.*

Lenski, R. E. 1997. The cost of resistance—from the perspective of a bacterium. In D. J. Chadwick and J. Goode, eds., Antibiotic Resistance: Origins, Evolution, Selection and Spread, 169. Chichester, UK: John Wiley. *Uses mathematical models and experiment findings to discuss how the growth, dissemination, and persistence of antibiotic-resistant bacteria might be controlled.*

Martinez, J. L. 2008. Antibiotics and antibiotic resistance genes in natural environments. Science 321: 365–367. *Discusses the potential biological roles antibiotics and resistance genes might have in natural environments.*

Morar, M., and G. D. Wright. 2010. The genomic enzymology of antibiotic resistance. Annual Reviews Genetics 44: 25–51.

White, D. G., M. N. Alekshun, and P. F. McDermott, eds. 2005. Frontiers in Antimicrobial Resistance: A Tribute to Stuart B. Levy. Washington, DC: ASM Press. *A tribute to one of the leaders in the field of antimicrobial resistance that covers many relevant areas, including mechanisms and epidemiology of resistance as well as public policy and public education programs to use antibiotics appropriately.*

zur Wiesch, P. A., R. Kouyos, J. Engelstädter, R. R. Regoes, and S. Bonhoeffer. 2011. Population biological principles of drug-resistance evolution in infectious diseases. Lancet Infectious Diseases 11: 236–247.

CHAPTER 11

EVOLUTION AND MICROBIAL FORENSICS

Paul Keim and Talima Pearson

OUTLINE

1. Evolutionary thinking, molecular epidemiology, and microbial forensics
2. The uses of DNA in human and microbial forensics
3. Genetic technology and the significance of a "match"
4. The Kameido Aum Shinrikyo anthrax release
5. The Ames strain and the 2001 anthrax letters
6. From molecular epidemiology to microbial forensics and back

The tools of molecular biology coupled with the evolutionary methods of phylogenetics have found powerful applications in tracking the origins and spread of infectious diseases. Microbial forensics is a new discipline focused on identifying the source of the infective material involved in a biological crime and it, too, increasingly depends on evolutionary analysis and molecular genetic tools.

GLOSSARY

Clonal Populations. Populations in which members, called *clones*, have diverged without exchanging any genetic material across lineages. Members of such populations (e.g., many recently emerged pathogens) are genetically identical with the exception of variation generated by subsequent mutations.

Homoplasy. A shared genetic (or phenotypic) characteristic produced by convergent evolution or horizontal genetic exchange between lineages, rather than by descent from a common ancestor that shared the same characteristic.

Match. An identical genotypic profile (often called a *DNA fingerprint*) based on a particular technology.

Membership. A phylogenetic concept more useful than a "match" for describing relationships among bacterial isolates. Two isolates can be members of the same phylogenetic group without being absolutely identical in their genome sequences.

Multiple Loci VNTR Analysis (MLVA). A DNA fingerprinting method widely used to differentiate bacterial types. Here, VNTR stands for *variable number of tandem repeats*.

Single Nucleotide Polymorphism (SNP). A single base-pair difference between the DNA sequences of two individuals including, for example, two closely related bacterial strains.

The investigation of infectious disease outbreaks has a long history and even predates our understanding of the germ theory of disease, which was formulated by Louis Pasteur in the early 1860s. The classic example, a seminal event in epidemiology, occurred in 1854 when John Snow implicated London's Broad Street water pump as the focus of a cholera outbreak. The correlative association of disease occurrence, potential causative infectious agents, and their sources has grown increasingly sophisticated over the years. Today it is common to examine the genomes of bacteria and viruses to precisely define the pathogen subtype, with the aim of identifying specific case clusters that can reasonably be presumed to be a part of the same outbreak. This approach strengthens any correlative study that aims to identify the disease source by eliminating similar disease cases that did not emanate from the same focus.

These same genomic methods became important after the bioterrorism events of October 2001, when letters laden with *Bacillus anthracis* spores were sent through the US Postal Service, and the investigation that followed sought to identify the source of the letters. Evolutionary theory concerning bacterial populations, mutational processes, and phylogenetic reconstruction were essential for this science-based forensic investigation. The development of the field of microbial forensics was greatly accelerated by the anthrax-letter investigation, and it now provides a paradigm for both forensic cases and other public health investigations that involve infectious agents.

1. EVOLUTIONARY THINKING, MOLECULAR EPIDEMIOLOGY, AND MICROBIAL FORENSICS

The fields of molecular epidemiology and microbial forensics are populated by well-educated individuals. Nonetheless, the failure of these fields to employ evolutionary thinking sometimes limits the quality of the evidence and resulting inferences. For example, public health investigations of bacterial diseases have, in recent years, become highly dependent on one particular DNA-based technology called *pulsed-field gel electrophoresis* (PFGE). PFGE has the advantage that it can be applied to any bacteria, but its drawback is that the resulting data preclude more thorough evolutionary analyses. In particular, PFGE generates restriction fragment patterns—often called *DNA fingerprints*—that are analyzed using simple matching algorithms that produce yes/no outcomes, without allowing more sophisticated evolutionary analyses to identify the similarities and differences among the samples of interest. A *match* between fragment patterns is inferred by analysts based on their experience and the rarity of a particular pattern in large databases. Unfortunately, little effort has been made to understand the evolutionary paths that may connect and explain the varying degrees of similarity among these patterns, and probabilistic models to place confidence estimates on relationships (e.g., a match) are rarely used. Most PFGE practitioners appreciate the validity of evolution, but their use of rigorous evolutionary analysis has been stymied by the difficulty in applying theory to such data and by resistance to making the changes necessary to improve on a widely used method.

In contrast with DNA studies of bacterial diseases, no established uniform technology exists in public health investigations of viral diseases; instead, each pathogen is typically analyzed by sequencing a particular, unique target gene. These sequence data are almost always analyzed using phylogenetic methods, and the analyses frequently include probabilistic models to test alternative hypotheses about the sources of the viruses. These DNA sequence data are in a universal digital format, and evolutionary models of sequence evolution are well developed, allowing for the rapid adoption and application of methodologies from other fields. By contrast, DNA fingerprints are poor substitutes for phylogenetic analyses, and the blind application of phylogenetic algorithms is inappropriate without a better understanding of underlying character state changes. The PFGE-based fragment patterns that constitute the DNA fingerprint can be thought of as complex phenotypes determined by the genotype—but following ill-defined rules—which illustrates the weakness of this approach. However, the lack of evolution-driven approaches in bacterial molecular epidemiology is starting to be overcome as sequence-based methods begin to dominate this discipline, and the costs of sequencing genomes keep dropping. The golden age for the molecular epidemiology of bacterial infectious diseases is arriving with the widespread adoption of whole-genome analysis.

2. THE USES OF DNA IN HUMAN AND MICROBIAL FORENSICS

The utility of DNA fingerprinting for human identification in forensic analysis has had a major impact on society and the legal system: it has led to the exoneration of falsely accused individuals and to the conviction of guilty criminals. The primary methodology is similar in some regards to the PFGE method for bacteria described in the previous section. However, in the case of human forensics, after several years of scientific discussion and debate, the statistical methods used to evaluate matches are firmly grounded on population-genetic models and the scientific understanding of human biology, inheritance, and population subdivisions.

But these same statistical models have little utility in microbial forensics owing to the profound differences between bacteria and humans in terms of reproductive biology and modes of genetic inheritance. DNA is the genetic material of both bacteria and humans, of course, but that fact does not mitigate these differences. While DNA analysis in humans and in bacteria may be similar in terms of the molecular methods used, the inferences that can be drawn must reflect their different modes of inheritance, and population structures.

It is equally important to realize that in addition to these differences between bacteria and humans, bacterial species—and even populations of the same nominal species—also differ from one another in ways that can influence the interpretation of genetic relationships. One important variable is the relative extent of vertical and horizontal modes of inheritance. Bacteria reproduce asexually, so their inheritance is primarily vertical (mother cell to daughter cell). However, horizontal gene transfer (HGT) between bacterial cells also sometimes occurs, and when it does so, it can move genes not only within but also between different species. HGT can have important consequences, such as the movement of antibiotic-resistance genes between species, and can leave conspicuous genetic evidence when it occurs between distantly related species. However, at a finer scale, many bacterial populations, including many recently emerged pathogens, show little or no detectable HGT. Thus, in many epidemiological and forensic situations, the relevant models and hypotheses are for lineages that are strictly clonal (asexual) in their derivation. In these cases, evolutionary analyses are focused on phylogenetic relationships and mutation rates.

3. GENETIC TECHNOLOGY AND THE SIGNIFICANCE OF A "MATCH"

The idea of a genetic match between two DNA fingerprints is jargon that has entered the scientific lexicon via the fields of human identification and forensics. Because individual humans are almost always the unique product of two unique gametes (identical twins being the exceptions), almost every person can be uniquely identified based on his or her alleles at a relatively few hyperdiverse regions of the human genome characterized by short tandemly repeated sequences. An exact allelic match between DNA samples from two individuals is so unlikely that a "match" has been used as the only physical evidence needed to link an individual to the scene of some crime. Likewise, a "nonmatch" can be used to exonerate a suspect. The idea of unambiguous matches and nonmatches has thus proven to be very powerful in the justice system. Unfortunately, this same terminology is often applied to scenarios in microbial epidemiology and forensics; however, the interpretations may be very different as a consequence of biological differences between humans and microbes.

With microbes, the technological context is also critical to understanding the significance of a "match." A perfect genetic "match" can be lost using methods with greater resolution and discriminatory power. Low-resolution methods, including PFGE and multiple-locus sequence typing, would show that many bacterial isolates have identical alleles, but these methods see only a small portion of the genome. Greater discrimination can be achieved using *multiple-locus VNTR analysis* (MLVA), a technique that involves screening multiple loci with *variable numbers of tandem repeats* (VNTR), or by sequencing the entire genome of a bacterial isolate. Such whole-genome sequences may seem to be the ultimate standard, but bacterial geneticists have long realized that mutations will generate variation even within a colony of cells separated by only a few generations. Whole-genome sequencing does not detect these mutations, because most applications generate a consensus DNA sequence that ignores rare variants; in fact, the accuracy of current technologies is such that rare sequencing errors obscure such rare mutations. In the future, however, new sequencing methods might detect rare variants directly from their individual DNA molecules. Therefore, a seemingly perfect match between two samples can be broken either by increasing the extent of genomic sampling or by searching more thoroughly for variants within the population. When it becomes possible to discriminate even between two colonies derived from the same progenitor strain, the ideal of seeking a perfect genetic match becomes more problematic than useful.

Rather than a match, a microbe's *membership* in a phylogenetic group or clade is a more meaningful concept for epidemiological and forensics work. In a clonal lineage with little or no horizontal gene transfer, one can define phylogenetic relationships based on informative characters with membership in a particular clade based on shared derived states. *Single-nucleotide polymorphisms*, or SNPs, are now commonly employed in this way because they are produced by rare mutation events and thus are usually stable over appropriately long periods. With sufficient data, such as obtained by sequencing whole genomes, this stability can easily be tested by discriminating between convergent (e.g., homoplastic) and vertically inherited matches at the level of each SNP. Moreover, additional SNPs elsewhere in the genome are not problematic for inferring membership, because diversity is hierarchically nested within clades.

Thus, additional SNPs produce novel genotypes that are still members of the clade. Even a reversion—a mutation to a prior state—does not change clade membership per se, although it can complicate inferences about membership.

In most cases, multiple point mutations will have occurred along most or all evolutionary branches. However, a single canonical SNP can be used to represent each branch, which can simplify phylogenetic analyses. This paring down of the number of characters is not essential, and it may result in less phylogenetic precision if a sample belongs to some subclade that has not been extensively characterized. In such cases, the failure to include all SNPs along a particular branch may lead to the assignment of that sample to the wrong subclade. This mistake may be caught, of course, by including more SNPs. Thus, the hierarchical redundancy of phylogenetically ordered SNPs creates a safeguard against incorrect assignments of samples to subclades.

4. THE KAMEIDO AUM SHINRIKYO ANTHRAX RELEASE

In the summer of 1993, an attack using a biological weapon was carried out in Kameido, a highly populated suburb of Tokyo, by the Aum Shinrikyo (Supreme Truth) doomsday cult (subsequently reorganized into a group called Aleph and the splinter group Hikari no Wa). Although the cult was large and well financed, and had well-educated scientists involved in the planning, the anthrax attack failed to kill or even sicken the targeted population. In fact, it was many years later before scientists realized there had been a failed attack.

In late June 1993, public health officials were notified by Kameido residents of a highly unusual and odoriferous mist emanating from the roof of the Aum's facility. Unsure of what was occurring, government health officials collected samples of the spray and submitted them for chemical analysis. The analyses evidently provided no evidence of toxic chemicals, and the cult discontinued the spraying, so no further actions were taken. Two years later, however, the cult carried out a chemical weapons attack by releasing sarin gas in the Tokyo subways. Ten people were killed and hundreds seriously injured. It was only after the arrest of cult members and during their subsequent questioning that the Kameido anthrax attack was discovered for what it was. The mist coming off their building was, they stated, from a culture of *Bacillus anthracis*— the causative agent of anthrax.

Hiroshi Takahashi was the investigating epidemiologist, and in 1997, he discovered a small tube of liquid that had been collected from the Kameido building at the time of the 1993 attack. He transferred this material to the United States, where *B. anthracis* cells were cultured and then genetically analyzed using MLVA, which was the best available technology at that time. Eight variable loci were analyzed, including six on the chromosome and one on each of two extrachromosomal plasmids that carry virulence factors. Seven of the loci matched a well-known strain of *B. anthracis*, called Sterne, that is used in the production of a vaccine against anthrax. The assay for the eighth locus failed, a result that was also consistent with the Sterne strain because it is missing the pXO2 plasmid that carries this locus. Indeed, the absence of that plasmid is the reason that the Sterne strain is not virulent. Thus, the anthrax attack had failed to kill anyone because the Aum Shinrikyo cult had used a harmless strain of bacteria. The evidence of a vaccine strain raised the question, why had the cult used a harmless strain? Was it a mistake on the part of the cult? Was it a practice run for a possible later attack? This question remains unanswered today.

B. anthracis is a pathogen with very low genetic diversity, reflecting its recent origin. In the pregenomics era, MLVA was one of the only available methods for distinguishing one *B. anthracis* strain from another. The database at that time contained only 89 distinct genotypes, or fingerprints. Even so, the results of the assays supported several important conclusions: (1) the cult had indeed used *B. anthracis*; (2) several commonly studied and virulent strains (e.g., Ames, Vollum) were excluded as the attack material; (3) the failure of one assay was consistent with the strain's lack of one of two virulence plasmids; and (4) that failure, as well as results from the other seven loci, matched the fingerprint of the widely available vaccine strain Sterne. The first two conclusions were robust. The match to Sterne, however, was less so because other strains share the same seven-locus genotype; the null allele for the plasmid-encoded locus produced additional ambiguity. For the reasons discussed earlier, DNA fingerprinting methods such as MLVA are not well suited for evolutionary inferences. Nonetheless, an important forensic principle is evident—one we will revisit in the next section—in terms of the strength of exclusionary versus inclusionary findings.

5. THE AMES STRAIN AND THE 2001 ANTHRAX LETTERS

Only a few weeks after the September 11, 2001, terrorist attacks had killed thousands of people, the United States faced another shocking incident in October, one that employed a deadly biological weapon. The attacker(s) used the US Postal Service to send at least seven letters containing *B. anthracis* spores. These letters were sent to specific targets, but their routing through the postal system resulted in widespread contamination by spores, which disrupted several mail centers and other government facilities including congressional buildings. Molecular genetics and evolutionary approaches were central to the forensic investigation.

Although whole-genome sequencing methods were eventually brought to bear on this case, the investigation began at a time when that technology was not sufficiently developed to allow it to be used with the immediacy that the circumstances demanded. Public health as well as national security considerations meant that it was critical to identify the likely source—or at least to exclude certain sources—as quickly as possible.

To that end, Paul Keim and colleagues were able to quickly perform an initial analysis of the DNA from the spores in the letters using the same MLVA system used to analyze the *B. anthracis* from the Kameido event, and with an expanded reference database. In 1991, the United Nations Special Commission had discovered weaponized anthrax spores during inspections following the Gulf War. Bacteria were recovered from ordnance, and they were identified as *B. anthracis*, but little other characterization was done at that time. After the 2001 anthrax letter attacks, there was renewed interest in the Iraqi weapon strain, given suspicions of foreign involvement from some quarters. Identifying the Iraqi weapons strain and its relationship to the strain in the anthrax letters was therefore critical.

Within just days of the hospitalization of the first victim in Florida, MLVA showed that the *B. anthracis* isolated from that victim matched the Ames strain at all eight loci. Analyses of samples from the letters also matched the Ames strain. The Ames strain is a virulent one, unlike the

Sterne strain that was deployed in the failed attack in Japan. The Ames strain was known to be used in several US government laboratories and, despite its name, it was originally isolated from Texas. The search was then on for the source of the attack strain, with three critical issues at hand. First, was the Iraqi strain also an Ames strain? Second, were other *B. anthracis* strains that could be isolated from nature similar enough to the Ames strain that they would produce a match at all eight MLVA loci? And third, what higher-resolution techniques could be employed to distinguish among sublineages within the clade that contains the Ames strain and its close relatives to trace the attack strain to a specific source?

In December 2001, the Iraqi strain was characterized using the MLVA method, and a match at all eight loci was established to another strain called Vollum. In fact, an Iraqi scientist had purchased the Vollum strain from a culture collection in 1986, indicating the likely source of that strain. Importantly, the Ames and Vollum strains differ at multiple MLVA loci. These differences meant that the *B. anthracis* strain discovered at the Iraqi bioweapons facility could be excluded, with a high degree of confidence, as the source of the spores in the letter attacks.

So, what was the source of the Ames-related material in the letters? Was the material derived from the Ames strain, which had been distributed to various laboratories? Or could it be a different isolate that just happened to match the Ames strain at all eight loci used in the MLVA testing? In fact, the database showed that an isolate obtained in 1997 from a goat in Texas also matched the Ames strain at all eight loci. The circumstances of the attacks made it clear that these anthrax cases were not a natural outbreak. In principle, someone might have reisolated a *B. anthracis* strain from nature that happened to be a close relative of the Ames strain. In any case, it became imperative to employ genetic methods that would allow maximum resolution to determine the source of the attack material.

To that end, whole-genome sequencing was employed to find genetic differences that could be analyzed using phylogenetic methods. SNPs are ideal for this purpose because reversion mutations should be rare in such young lineages as *B. anthracis* and especially the clade containing the Ames strain. Indeed, the extent of *homoplasy* in species-wide SNP data is only about 0.1 percent across the entire species. This approach identified four SNPs specific to the laboratory Ames strain, which could be used to differentiate it from natural isolates. By screening for these four SNPs, it was possible to exclude other strains, including the isolate from the Texas goat (which had matched the Ames strain at all eight loci used in the MLVA test), as well as additional isolates from the same geographic region. Thus, it became possible to determine that a strain was a member of the Ames group of lab-derived isolates with much more confidence than with the fragment-matching approach of MLVA.

Whole-genome sequencing was also employed in other lines of the investigation. With the increasingly strong evidence that the attacks had used spores derived from the Ames strain present in several laboratories, the key genetic issue became one of searching for mutations in the attack materials that might match mutations found in some laboratories but not others. Thus, one line of the investigation involved comparing the genome sequences of the *B. anthracis* isolated from the Florida victim and another Ames-derived strain, called Porton, whose virulence plasmids had been "cured" (eliminated). The Porton strain was used because it was already in the process of being sequenced and analyzed prior to the attacks, thus expediting the investigation. In fact, several mutational differences were discovered between the Florida and Porton derivatives of the Ames strain. However, these differences turned out to be useless for the investigation because all of them were unique to the Porton strain; the mutations probably arose during the mutagenic procedures employed to eliminate the plasmids.

The other line of investigation using genomic sequences proved to be more useful but also quite complicated. In the early stages of the investigation, microbiologists had allowed some of the *B. anthracis* spores taken from the letters to germinate and produce colonies. They observed subtle variation in the appearance of colonies, with one predominant type and several variants at lower frequencies. Thus the differences were heritable, which implied that the differences in colony "morphology" had resulted from mutations. If confirmed, the mutant subpopulations might then provide a signature to distinguish possible sources of the spores used in the attacks. In summary, several *clones* with variant morphologies were sequenced, and mutations were identified. These mutants had not been seen in previous sequencing attempts because they were rare in the population of spores, and the resulting sequence represented a consensus sequence from the sampled cells. Next, the sampled cells were selected specifically to include these morphological variants. Molecular assays could then be developed to screen for four of these mutations.

In the meantime, the Federal Bureau of Investigation (FBI) had created a repository of more than 1000 samples, all derived from the Ames strain, from about 20 laboratories known to have worked with that strain. These samples were then screened for the four mutations. None of the four mutations were detected in most of the samples, but eight of them gave positive results for all four mutations. (There are many complications related to the sensitivity and specificity of the assays used to detect the mutations, as well as other issues that in the interest of brevity, are not presented here but are discussed in a 2011 report prepared by a committee of experts convened by the National Research Council.) The eight samples were all apparently derived from the same source—a flask of spores identified as RMR-1029—based on information obtained by the FBI. The contents of the flask had been generated by pooling several separately grown batches of spores, to produce a single large stock of material for experiments that would be performed at different times. This manner of preparing the flask of spores might account for the diversity of variant colony types that led to this line of investigation. In any case, these results pointed toward a particular flask and samples taken from that flask as a possible source of the spores placed in the attack letters. The criminal investigation was thus also focused on those individuals who had access to the RMR-1029 flask and its derivatives.

This chapter is focused on the role of evolutionary thinking in microbial forensics; it is not the place to discuss other aspects of the criminal investigation. But for those readers who want to know, very briefly, the outcome of this investigation, the FBI identified a government scientist as the lone suspect of the anthrax letter attacks. Before the US Department of Justice could bring formal charges, that individual committed suicide.

Genomic technologies continue to advance at a rapid pace, and it is possible that spores from the attack letters and from the RMR-1029 flask could be examined even more fully by so-called deep sequencing. That approach could, in principle, expand the analysis of diversity in those samples well beyond the four mutations that were discovered based on the variation in colony morphologies.

6. FROM MOLECULAR EPIDEMIOLOGY TO MICROBIAL FORENSICS AND BACK

Over the course of several decades, increasingly powerful molecular-based methods have been used to identify the source and track the spread of infectious diseases. These methods also served as the starting point for the forensic investigation of the anthrax attacks. Nonetheless, that investigation pointed to the limitation of these methods. The urgency and resulting high levels of funding to investigate the anthrax letters enabled the application of whole-genome sequencing—an approach that molecular epidemiologists had not been able to employ previously owing to its high costs. The genomic methodologies and analytical approaches have now become much less expensive, and so they should be applied much more broadly in molecular epidemiological studies motivated by public health concerns. Both forensic and epidemiological investigations are also well served by using phylogenetic approaches to analyze genomic data for determining relationships among samples, especially as the number of key samples becomes progressively smaller when homing in on a probable source.

Thus, we predict with confidence that the use of whole-genome sequencing to understand evolutionary relationships will become common in public health. This technological change will bring with it changes in data analysis such that the full power of evolutionary theory, models, and methods can be used to determine infectious sources during natural disease outbreaks.

FURTHER READING

Budowle, B., S. E. Schutzer, R. G. Breeze, P. S. Keim, and S. A. Morse, eds. 2011. Microbial Forensics, 2nd ed. New York: Elsevier. *A comprehensive collection of approaches to microbial forensics.*
Committee on Review of the Scientific Approaches Used during the FBI's Investigation

of the 2001 *Bacillus anthracis* Mailings. 2011. Review of the scientific approaches used during the FBI's investigation of the 2001 anthrax letters. Washington, DC: National Academies Press. *A hard look at the FBI's investigative methods and results.*

Hillis, D. M. 2009. Evolution Matters. National Institutes of Health, videocast.nih.gov/launch.asp?15187. *An overview of the relevance of evolution to emerging diseases and solving certain crimes.*

Jobling, M. A., and P. Gill. 2004. Encoded evidence: DNA in forensic analysis. Nature Reviews Genetics 5: 739–751.

Keim, P., T. Pearson, and R. Okinaka. 2008. Microbial forensics: DNA fingerprinting "anthrax." Analytical Chemistry 80: 4791–4799. *An overview of DNA methods used for investigating the anthrax letter attacks.*

Keim, P., and D. M. Wagner. 2009. Humans, evolutionary and ecologic forces shaped the phylogeography of recently emerged diseases: Anthrax, plague and tularemia. Nature Reviews Microbiology 7: 813–821. *A population model for the emergence and global spread of pathogens.*

Morelli, G., Y. Song, C. J. Mazzoni, M. Eppinger, P. Roumagnac, D. M. Wagner, M. Feldkamp, et al. 2010. Phylogenetic diversity and historical patterns of pandemic spread of *Yersinia pestis*. Nature Genetics 42: 1140–1143. *A detailed evolutionary look at an important clonal pathogen that causes plague.*

Takahashi, H., P. Keim, A. F. Kaufmann, K. L. Smith, C. Keys, K. Taniguchi, S. Inouye, and T. Kurata. 2004. Epidemiological and laboratory investigation of a *Bacillus anthracis* bioterrorism incident, Kameido, Tokyo, 1993. Emerging Infectious Disease 10: 117–120. *The public health report of the Kameido incident.*

PART III

Reshaping Our World

CHAPTER 12

DOMESTICATION AND THE EVOLUTION OF AGRICULTURE

Amy Cavanaugh and Cameron R. Currie

OUTLINE

1. Domestication
2. Evolution under domestication
3. Agriculture as a mutualism
4. Agriculture in ants
5. Conclusions

Agriculture is an ancient and important factor shaping life on earth. Through the cultivation of food, populations of agriculturalists are able to greatly expand and can even develop a division of labor. This chapter explores the evolution of agriculture, including domestication and selection under domestication, along with the evolutionary events and consequences of farming. It also describes how agricultural associations are perhaps best viewed in the framework of a coevolved mutualism.

GLOSSARY

Artificial Selection. Evolutionary change caused by human breeding in populations of domesticated (or experimental) plants and animals.

Coevolution. Reciprocal evolutionary change between interacting species.

Domestication. Acquisition from the wild of one species by another and breeding it in captivity.

Domestication Syndrome. A suite of traits characteristically found in domesticated species.

Mutualism. An interaction between two species that benefits both.

The *domestication* of one species by another for food is one of the most significant evolutionary innovations in the history of life on the planet. Indeed, shifting to an agricultural lifestyle, and the concomitant expansion in numbers and range it allows, inexorably alters not only the biology of the species involved but also the ecosystems in which they occur. By establishing a reliable reserve of food, agriculturalists gain an advantage over their hunter-gatherer brethren; the ready source of calories allows the agriculturalist populations to greatly expand and ultimately facilitates the development of a division of labor. Agriculture originated among humans in the Fertile Crescent, but contrary to popular belief, humans were not the world's first farmers. That distinction belongs to a group of ants in the Amazon basin. These fungus-farming ants maintain specialized gardens of domesticated fungi that serve as the primary nutrient source for the colony. After the origin of agriculture in ants, but still millions of years before humans appeared, other groups of insects also transitioned to farming. In parallel with fungus-growing ants of the New World, some termites farm fungus for food in the Old World. The most diverse farmers are the Ambrosia beetles, represented by more than 3000 species. In all these cases, the utilization of a farmed food source has enabled these insects to expand into a new ecological niche, leading to their diversification and, in some cases, allowing them to become dominant members of their ecosystems. Other insects engage in more rudimentary forms of farming, and some ants even practice animal husbandry by tending aphids and treehoppers. Besides the insects, a marine snail cultivates fungus, and a species of damselfish farms red algae, and recently it has been found that even some amoebas practice a rudimentary form of bacterial husbandry.

The most recent origin of agriculture is in our own species, approximately 10,000 years ago, and has ultimately resulted in our domination of most of the ecosystems on the planet. Humans have cultivated around 100 different plant species, which serve primarily as a reliable and more readily stored source of nutrients. Humans have also domesticated a number of animals, obtaining a variety of benefits, including sources of nutrients (e.g., meat and milk), labor (e.g., plowing fields, transporting of goods, and protecting and herding other domesticated animals), and military advancement (e.g., cavalry). Thus, farming provided a reliable source of calories, allowing for an increase in human population size, decrease in birth intervals, and specialization of labor leading to stratified societies, while animal husbandry allowed agricultural societies to expand beyond their borders and ultimately to dominate the nonfarming populations with which they came in contact. Based on these advantages it can be argued, as Jared Diamond does in his Pulitzer Prize–winning book *Guns, Germs, and Steel*, that agriculture is the single most important force shaping human history.

Just as agriculture has shaped human society and history, it has also had an important role in the development of evolutionary theory. This influence is evident in Darwin's *On the Origin of Species*, which begins with a thorough discussion of domestication and the evolutionary changes caused by human breeding of domesticated plants and animals—an evolutionary force he termed *artificial selection*—even before introducing the tenets of natural selection:

> It is . . . of the highest importance to gain a clear insight into the means of modification and coadaptation. At the commencement of my observations it seemed to me probable that a careful study of domesticated animals and of cultivated plants would offer the best chance of making out this obscure problem. Nor have I been disappointed; in this and in all other perplexing cases I have invariably found that our knowledge, imperfect though it be, of variation under domestication, afforded the best and safest clue.

Of the different domesticated species Darwin investigated to "mak[e] out this obscure problem," the domestic pigeon was the subject of one of his most in-depth studies. After thoroughly examining all the breeds of pigeons he could acquire, he determined that more than 20 different characters varied among these breeds. Yet it was believed at the time, and confirmed by his additional studies, that all these diverse breeds had descended from a single wild species, the rock pigeon. Darwin argued that the key factors in creating all this variability among breeds were "man's power of accumulative selection" and use of the large body of literature, both modern and ancient, in which breeders and horticulturalists described in great detail the ways in which they had modified their animals and plants by selectively mating only those individuals with the desired characteristics.

In this chapter, we discuss evolutionary aspects of domestication and selection under domestication in agriculture by humans. We then argue that a useful way to conceptualize the evolution of agriculture is as a mutualism shaped by coevolution. Expanding on this argument, we end with a discussion on the evolution of agriculture in ants, drawing parallels with humans.

1. DOMESTICATION

Domestication is the practice whereby an organism is acquired from the wild and bred in captivity. The population or species that is domesticated can be referred to as the *domesticate*. Domesticates undergo genetic changes during the process of cultivation or breeding that make them more useful to the domesticator and ultimately differentiate them from their wild ancestors.

The first domestication of a plant by humans occurred about 10,000 years ago, when people living in the Middle East (parts of modern Iraq, Iran, Turkey, Syria, and Jordan) began to purposefully plant barley, peas, lentils, chickpeas, muskmelon, flax, and two species of wheat. Not long after agriculture had been established in the Middle East, it arose independently in eastern China. There the available wild species differed, so the first domesticated crops of Southeast Asia included rice, soybeans, adzuki beans, mung beans, hemp, and two species of millet. Populations within the tropical West African and Sahel regions also appear to have independently begun domesticating species including sorghum, millet, rice, cowpeas, yams, bottle gourds, and cotton. Though the dates are uncertain, people in Ethiopia domesticated coffee, and people in New Guinea domesticated sugarcane and bananas. Although populations in the Americas also independently established themselves as farmers, this transition took place later than those in Eurasia and Africa, most likely owing to the inherent differences in the available wild species. Between 9000 and 3000 years ago, humans began domesticating animals including sheep, goats, cattle, pigs, chickens, and horses in Eurasia and northern Africa. Again, populations in the Americas independently domesticated some animal species, such as the llama and the guinea pig, but they were limited in their efforts because most of the available wild species were unsuitable for domestication.

It might seem that almost any wild species could be domesticated, but history has shown that this is not the case. Although humans have domesticated a number of species, they represent an extremely small proportion of the plants and animals that occur in nature. The wild progenitors of the first crop species were already edible, grew quickly and easily, could be stored, and were self-fertilizing. This last trait is crucial in that self-fertilizing plants will directly pass traits on to their offspring largely unchanged. Species that have never been domesticated fail to meet one or more of the preceding criteria. For example, the oak tree, despite producing nutrient-rich acorns, has never been domesticated, for many reasons. First, the oak is an extremely slow-growing tree, taking more than 10 years to grow from an acorn to a fruit-bearing tree. Second, the bitterness of the acorn is under the influence of many genes, which combined with the long generation time, makes it very difficult to select for mutant, sweet acorns. Finally, acorns are a primary food source for another animal, squirrels. By burying large numbers of acorns, squirrels would undermine any human attempt to plant acorns only from oak trees with desirable traits.

Animals that have been successfully domesticated also share many traits. First, most domesticated animals are herbivores. Owing to the successive loss of energy through each trophic level, it takes much less food to support the growth of an herbivore than a carnivore; therefore, raising herbivores is far more efficient. Although we now eat carnivorous fish, we have only recently begun farming them, and whether this leads to their domestication remains to be seen. Second, as with plants, successfully domesticated animals grow quickly. Extremely large mammals, such as elephants, grow too slowly to be candidates for domestication. Third, domesticated species breed readily in captivity. As Darwin noted, this is a particularly rare trait among animals. Fourth, the animal must have a relatively pleasant disposition. While all large animals, and many small ones, are capable of killing humans, the species that have been successfully domesticated are much less prone to aggression. Fifth, they must not be prone to panic, particularly panic that results in the animals' battering themselves to death while trying to escape. This behavioral issue has been a limiting factor in the domestication of many otherwise-suitable herd species, such as gazelles. Finally, many successfully domesticated animals live in herds with well-developed hierarchies and overlapping home ranges; these animals are able to live in proximity to one another and will usually accept a human as the herd leader.

2. EVOLUTION UNDER DOMESTICATION

Although domesticates are species whose wild ancestors possess specific traits suitable for domestication, they are greatly altered by the process of artificial selection imposed by the domesticator. To Darwin, artificial selection was not merely analogous to natural selection but rather represented a clear example of natural selection under a particular set of conditions. The principles are the same, but the environmental conditions in play under artificial selection are those of the human-constructed habitat as opposed to a habitat of nature's making under natural selection. For either selective force to operate there must be variation in the trait under selection, heritability of that trait, and a tendency for individuals with some version of that trait to reproduce, or be bred, more than others with a different version. As people consciously or unconsciously selected the plants and animals that met human needs and preferentially grew and bred them, they were practicing artificial selection. At the same time, people were creating a novel environment for these plant and animal species, with natural selection increasing the frequency of traits that would lead to success in this constructed environment.

Domestication of plants and animals undoubtedly involved the conscious selection of numerous traits. In plants, early protofarmers likely preferentially collected the largest fruits or seeds to consume and to subsequently plant, and likely selected for taste, choosing the least bitter seeds and sweetest fruits. While many plants were selected for their fruit or seeds, others would have been selected for size or fleshiness of other nutritional parts of the plant (e.g., the roots or leaves), their oil content (e.g., olives and sunflowers), or length of fibers (e.g., flax and hemp). Animals likely were consciously selected on the basis of size—for those raised for meat, or reproductive physiology—for those raised for milk or eggs. Sheep and llamas would have been selected for the retention, rather than shedding, of the wool fibers in their coats, while dogs would have been selected for traits such as size, sense of smell, hunting ability, trainability, and herding ability.

Plants and animals were also subjected to a great deal of unconscious selection. For example, the wild progenitors of cereals and legumes typically drop their seeds as a dispersal mechanism. Mutant plants that did not drop seeds would die out quickly because they would leave no offspring. However, such plants would prove beneficial to humans trying to efficiently gather food, as it is much easier to collect a handful of seeds from the top of a stalk than to pick each individual seed from the ground. Once humans began cultivating plants, selection would have also favored plants with faster germination times. After planting, those plants that sprouted first were more likely to be harvested and replanted, compared with those that delayed germination. Finally, while consciously selecting for traits such as size and taste, humans were also unconsciously selecting for plants capable of self-fertilization. In plants that self-fertilize, as most crops do, favorable mutations are maintained, not diluted by recombination with their neighboring wild progenitors.

Humans attempting to breed the largest or best milk-producing variants of a species would also have inadvertently been selecting for animals with the ability to reproduce in captivity. Domestic animals reach sexual maturity earlier than wild animals and have more frequent reproductive cycles. These traits may have been both consciously and unconsciously selected for by humans—consciously by selectively breeding the animals that reached maturity earliest and breeding them as often as they were receptive to it, and unconsciously by eliminating the nutritional constraints that would have limited their reproduction in the wild.

Together, the forces of artificial and natural selection have led to changes in domesticated plants that have come to be known as the *domestication syndrome*. These traits include (1) increased size of reproductive organs (e.g., fruits and seeds); (2) increased tendency for mature seeds to remain on the plant rather than dropping to the ground; (3) faster germination as well as synchronized, predictable germination times; (4) changed allocation of biomass (e.g., larger roots, stems, leaves, or buds); and (5) reduced physical and chemical defenses. Domesticated animals also possess a suite of traits that distinguish them from their wild counterparts. Morphologically, domesticated animals typically exhibit greater variation in overall body size as well as in the size of particular body parts (e.g., length of legs in dogs), as compared with their wild ancestors. Additionally, domestic species have different coloration of fur and feathers than their wild relatives, typically an increase in white or spotted coloration. Although such colors make individual animals more visible and therefore more vulnerable to predation, humans could have inadvertently selected for such individuals because they were easy to see and recover if they wandered away.

3. AGRICULTURE AS A MUTUALISM

Agriculture can be thought of as a *mutualism*—an interaction that benefits both the agriculturalists and the domesticated species. The benefits to humans are obvious, as discussed earlier. But, to some, the benefits of being an "enslaved" plant or animal might not be so clear. However, domesticates do receive numerous benefits, broadly falling into three general categories: (1) protection, (2) increased reproduction, and (3) dispersal. Agriculturalists protect their domesticated crops and animals by significantly reducing interspecific competition, herbivory, and predation. This protection includes growing domesticates in controlled environments and actively weeding, pruning, guarding, and applying chemical treatments. Through the careful planting and cultivation of seeds, farmers increase the probability of seed germination, thus increasing the reproductive rates of domesticated crops. Similarly, domesticated animals have higher reproductive rates, typically owing to shortened interbirth and interlaying intervals. Finally, as agricultural populations spread, they bring their crops and animals with them. By altering the new habitat to be suitable for domesticated species of their homeland, people increase the range of these species. Given the tremendous efforts humans undertake to care for their domesticates and the huge expansion of some plant and animal species following their domestication, Michael Pollan argues in *The Botany of Desire* that it is worth considering the question, who is domesticating whom?

Even as humans directed the evolution of the species they domesticated, they created new selection pressures on themselves. The transition to an agricultural lifestyle led to changes in both human behavior and physiology. For example, as with domesticated animals, human agriculturalists have increased reproductive rates compared with those of hunter-gatherers. Most likely owing to the increased reliability of a higher-calorie diet, interbirth intervals are much shorter in farming societies than in hunter-gatherer societies. In addition, two enzymes, lactase and amylase, show increased expression in members of agricultural societies compared with hunter-gatherers as well as with chimpanzees, our closest nonhuman relatives. In the case of lactase, an enzyme that digests the sugar found in milk, all mammals produce the enzyme as infants but then stop producing it rapidly after weaning. However, in many human populations, a mutation allows the persistent expression of this enzyme into adulthood. The geographic distribution of this mutation is strongly correlated with pastoralism, particularly the raising of animals for milk production. In a case of parallel evolution, two different mutations have been shown to cause lactase persistence in different populations. Both these mutations occur in the promoter region of the lactase gene. Similar enzymatic evolution is seen in amylase, an enzyme that breaks down starch. In this case, it appears that populations that switched to the starchier agricultural diet evolved extra copies of the gene that produces salivary amylase. These changes in humans, in response to shifting to an agricultural lifestyle, support the view of agriculture as a mutualism. In fact, they suggest that agriculture represents a mutually beneficial association shaped by *coevolution*, given that both interactors—the farmer and the domesticate—undergo genetic modification in response to the association.

4. AGRICULTURE IN ANTS

Other than agriculture by humans, the best-studied agricultural association is that of fungus-growing ants. Agriculture in ants is ancient, having originated approximately 45 million years ago. As humans have domesticated many species of plants and animals, fungus-growing ants have domesticated multiple species of fungal crops; there are as many as seven different events of free-living fungi being domesticated. Within this agricultural mutualism the ants and their fungal cultivars have coevolved and diversified. Fungus-growing ants include more than 200 species in 13 genera. Likewise, the cultivated fungi are represented by substantial diversity of strains within specific groups of cultivated lineages. At the pinnacle of evolution of agriculture in fungus-growing ants are the charismatic leaf-cutters, which shape neotropical ecosystems through the sheer mass of leaf material that the ants harvest.

The cultivated fungus, maintained in underground garden chambers in most species, serves as the primary food source for workers, larvae, and the queen. The fungus produces specialized structures called *gongylidia*, which are rich in lipids and carbohydrates. The gongylidia appear to represent an optimized nutrient source for the ants, likely evolved under a form of artificial selection. The ants cannot survive without their fungal crops; without them they literally starve. When establishing new colonies, queens ensure the initial presence of the cultivar by bringing a small ball of fungus collected from her parent colony, effectively transferring the fungus from one generation to the next. Recent genomic studies on leaf-cutters have revealed that fungus-growing ants (like humans) have evolved genetically in response to their dependence on agriculture; in particular, they have lost the ability to synthesize an essential amino acid that they likely obtain from the fungus garden.

Leaf-cutter ants have evolved a complex set of behaviors for cultivating the fungus. Like many human-domesticated species, the ants' fungal crop is unable to survive without the ants. The ants selectively forage for leaf material that promotes the growth of the fungus garden. The garden matrix is thus composed of the fungus and the vegetative substrate that worker ants obtain from outside the nest and then integrate into the fungus garden. Once this leaf material is brought to the colony, the ants lick and chew the material into small pieces. This process breaks down the physical barriers of the leaf that would otherwise prevent the growth of the fungus on the leaf surface. Just as human farmers work manure into the soil, the ants work the leaf pulp into the top layers of the fungus garden. They then bring fungal hyphae from older parts of the garden, plant it onto the surface of the fresh leaf pulp, and continuously add fresh material to the top of the garden.

Besides adding substrate to the garden, the ants also promote the growth of their fungal crop in numerous ways. The ants open and close tunnels to the surface so they can regulate the temperature and humidity within the growth chambers. There is also evidence that the ants damage the fungus, in a manner akin to pruning, to stimulate increased fungal growth. The fungus produces enzymes that can become disadvantageously concentrated in the garden. When that happens, the ants ingest these enzymes in the areas of high concentration and then defecate them into areas of low concentration, thus creating a more even distribution of the enzymes throughout the garden.

The cultivation of monocultures of clonally propagated crops has led to increased susceptibility to disease. The ants' fungus garden is host to specialized and potentially virulent agriculture pathogens, microfungi in the genus *Escovopsis*. *Escovopsis*—known only from the fungus gardens of these ants—consumes the ants' fungal cultivar and has coevolved with the ants and their fungal crop. The ants engage in meticulous behaviors to deal with the pathogen. They groom out *Escovopsis* by pulling pieces of the fungal cultivar through their mouthparts and collecting the invading microbes in their infrabuccal pocket, a cavity and filtering device within the mouthparts of ants. The ants then deposit this material in the refuse chambers. In cases where the garden has become diseased, the ants remove the affected area in a behavior called *weeding*, which involves ripping out and discarding the infected garden material. Further paralleling human methods for dealing with agriculture pests, the ants employ chemical methods of crop protection. Whereas humans control pests by developing and then spreading chemicals on their crops, the ants form a symbiosis with antibiotic-producing bacteria. These symbionts live on the ants' cuticle and produce antifungal compounds that inhibit the garden pathogen *Escovopsis*.

In summary, agriculture in ants, much like human agriculture, has led to their dominant role in many of the ecosystems in which they occur. Further, they share many of the hallmarks of human agriculture, including multiple domestications of wild species, artificial selection of the domesticates, and cultivation including physical and chemical methods for crop protection. Finally, the recent evidence for agriculturally related genetic changes in both the domesticates and the domesticators in human and ant agriculture suggests they represent coevolved mutualisms.

5. CONCLUSIONS

The ability to cultivate and breed plants and animals represents one of the most important developments in human history, allowing for rapid and tremendous population expansion. Today domesticated plants and animals constitute an immense proportion of the global caloric intake by humans. Species that have been successfully domesticated share some important characteristics that predispose them to agriculture, and they have undergone significant genetic modification during domestication. Although the changes in domesticated plants and animals have been recognized for millennia, recent work has shown that humans, too, have undergone evolutionary changes in response to agriculture. These genetic changes in humans have occurred in response to farming and consuming specific plants or animals, and they illustrate the coevolutionary nature of agriculture. These general findings have parallels in agriculture by ants, and they show that agriculture and its evolutionary benefits and processes are not unique to humans.

FURTHER READING

Belyaev, D. K. 1979. Destabilizing selection as a factor in domestication. Journal of Heredity 70: 301–308. *An experimental study of domestication in the silver fox.*

Currie, C. R., J. A. Scott, R. C. Summerbell, and D. Malloch. 1999. Fungus-growing ants use antibiotic-producing bacteria to control garden parasites. Nature 398: 701–704.

Diamond, J. 1999. Guns, Germs, and Steel. New York: W. W. Norton.

Diamond, J. 2002. Evolution, consequences and future of plant and animal domestication. Nature 418: 700–707.

Hölldobler, B., and E. O. Wilson. 2011. The Leafcutter Ants. New York: W. W. Norton.

Pinto-Tomas, A. A., M. A. Anderson, G. Suen, D. M. Stevenson, F.S.T. Chu, W. W. Cleland, P. J. Weimer, and C. R. Currie. 2009. Symbiotic nitrogen fixation in the fungus gardens of leaf-cutter ants. Science 326: 1120–1123. *This paper describes another similarity between human crops and ant crops—the need for symbiotic, nitrogen-fixing bacteria.*

Pollan, M. 2001. The Botany of Desire. New York: Random House.

Suen, G., C. Teiling, L. Li, C. Holt, E. Abouheif, E. Bornberg-Bauer, P. Bouffard, et al. 2011. The genome sequence of the leaf-cutter ant *Atta cephalotes* reveals insights into its obligate symbiotic lifestyle. PLoS Genetics 7: e1002007.

CHAPTER 13

DIRECTED EVOLUTION

Erik M. Quandt and Andrew D. Ellington

OUTLINE

1. Directed evolution of nucleic acids
2. Directed evolution of proteins
3. Directed evolution of cells
4. The future of directed evolution

Directed evolution is a process in which scientists perform experiments that use selection to push molecular or cellular systems toward some goal or outcome of interest. The objectives of this work include the production of substances of value and improved understanding of the evolutionary process. Elucidating the precise mechanisms by which improvements occur is often of particular interest. In general, directed evolution requires a genetic system in which information is encoded, heritable, and mutable; a means for selecting among variants based on differences in their functional capacities; and the ability to amplify those molecules or organisms that have been selected.

GLOSSARY

Aptamer. A short nucleic acid molecule that binds to a specific target molecule.

Bacteriophage. A virus that infects bacteria.

Esterase. An enzyme that splits esters into an acid and an alcohol in a chemical reaction with water, also known as *hydrolysis*.

Fluorescence-Activated Cell Sorting (FACS). A method for sorting a heterogeneous mixture of cells into two or more containers, one cell at a time, based on the specific light scattering and fluorescent characteristics of each cell.

Lipase. An enzyme that catalyzes the hydrolysis of ester chemical bonds of lipid substrates.

Messenger RNA (mRNA). The product of transcription of a DNA template, which in turn encodes the sequence for the production of a protein.

Peptide. A short polymer of amino acids linked by peptide bonds.

Polymerase Chain Reaction (PCR). Enzymatic reaction in which a small number of DNA molecules can be amplified into many copies.

Protease. A protein capable of hydrolyzing (breaking) a peptide bond.

Quasispecies. A large group of related genotypes that exist in a population that experiences a high mutation rate, in which a large fraction of offspring are expected to contain one or more mutations relative to the parent.

Ribozyme. An RNA molecule with a defined tertiary structure that enables it to catalyze a chemical reaction.

Transfer RNA (tRNA). An RNA molecule linked to an amino acid that is involved in the translation of an mRNA transcript into protein.

Transcription. The process of creating a complementary RNA copy (mRNA) of a sequence of DNA.

Transcription Factor. A protein that binds to specific DNA sequences, thereby affecting the transcription of genetic information from DNA to mRNA.

Translation. The process of decoding an mRNA molecule into a polypeptide chain.

Tumor Necrosis Factor-α (TNF-α). A protein involved in the regulation of certain immune cells that induces inflammation and cell death and thereby inhibits tumorigenesis and viral replication.

Directed evolution involves guiding the natural selection of molecular or cellular systems toward some goal of interest. That goal may be either to enhance basic understanding of natural systems or to produce something of value. In either case, this approach relies on changing the frequency of genotypes over time, with concomitant changes in function and phenotype, just as natural evolution does. However, from the point of view of an "evolutionary engineer" the mechanism by which changes in frequency are obtained is often of particular interest. In general, directed evolution requires a genetic system (a system in which information is encoded, heritable, and mutable), a means for sieving the variants that are present in that system (by differences in either function or fitness), and the ability to amplify those molecules or organisms that pass through the sieve.

The directed evolution of molecules and cells has been carried out for decades, although the methods used for directed evolution have gained in technical sophistication and in the breadth of systems that can be tamed. Indeed, if one includes animal and plant husbandry, then directed evolution has been coincident with the evolution of human society. This chapter first discusses molecular evolution, focusing on the selection of nucleic acids and proteins that have novel functions. The simple rule set for directed evolution described can be satisfied in a surprisingly large variety of ways, including in molecular systems that at first glance appear to have the properties of cells but that are not actually cellular. The chapter then considers how the evolution of cells can be directed and accelerated.

1. DIRECTED EVOLUTION OF NUCLEIC ACIDS

The forefather of the directed evolution of molecules was Sol Spiegelman of the University of Illinois at Urbana-Champaign. Spiegelman and his group studied a small *bacteriophage*, called Qbeta, that has an RNA genome. In nature, this virus infects host cells of the bacterium *Escherichia coli* to replicate. However, Spiegelman's team found that the protein involved in replicating this bacteriophage, Qbeta replicase, was capable on its own of replicating RNA molecules in a test tube. The only requirements for replication were an initial RNA template, the replicase, nucleoside triphosphates, and appropriate buffer conditions. However, once freed from the confines of a cell, the Qbeta replicase tended to make multiple mutations and deletions in the RNA template, ultimately leading to smaller and smaller RNA molecules. The evolutionary fates of these so-called minimonster variants could be altered depending on the experimental conditions (Saffhill et al. 1970). Spiegelman's demonstration was highly influential and became an icon for the field. However, Qbeta replicase proved too difficult to control, since any successful variants that arose were transient and quickly mutated into a complex and ever-shifting *quasispecies*. Thus other methods and systems were required to advance the study of directed evolution.

The modern era of directed molecular evolution had to await the development of several technologies, chief among them the chemical synthesis of DNA and the advent of the *polymerase chain reaction* (PCR). Chemical DNA synthesis allowed a defined, yet random, pool of nucleic acids to be generated, providing the perfect substrate for directed evolution experiments. If constant sequence regions were included at the termini of this pool, then its members could be exponentially amplified by the PCR. All that remained to ensure the selection of functional nucleic acids was to impose some sort of selection that would differentiate the members of the pool from one another, a feat that was performed by Jack Szostak and coworkers (Ellington and Szostak 1990). Each of the different sequences in the pool can fold into a different shape; each of the different shapes therefore potentially has a different function or phenotype. One of the first selections from a random sequence pool involved identifying single-stranded DNA molecules that could bind to a particular molecular dye. The nucleic acid pool was poured down a column containing immobilized dye molecules; some variants stuck to these molecules, while most of the population flowed through. Once the bound variants were eluted (by unfolding the nucleic acids) they were amplified by the PCR, and single strands were prepared from the double-stranded product. Iterative cycles of selection and amplification resulted in the gradual accumulation of those molecules that had high affinity for a given dye. One analogy that is often used to describe these experiments is that they are akin to looking for a needle in a haystack; every time you grab a handful of hay that contains a single needle, you then convert that needle into a handful of needles. Given that the hay outnumbered the needles in the experiment described by a factor of almost 10^{10}:1, iterative selection and amplification were essential to the purification of the needle.

The selection of nucleic acids that can bind to ligands became known as *in vitro selection*, and the binding sequences that result from this approach are called *aptamers* (from the Latin *aptus*, "to fit"). In both of the experiments described so far, the visions of the experimenters were driven not so much by specific hypotheses as by the availability of the technology and a desire for unfettered exploration. Spiegelman set off into the unknown to determine whether viral replication in the test tube (outside the host cell) was even possible, whereas Szostak took the great leap that there would be at least a few, previously unknown, shapes in the haystack that could bind to a dye. However, these technological innovations also ended up providing some support for the hypothesis that life may have got its start from self-replicating nucleic acids. Along with the discovery by Tom Cech and Sidney Altman that RNA could act as a catalyst (North 1989), the finding that nucleic-acid-binding variants could be selected from random strings of information implied that nucleic acids might avoid the "chicken-and-egg problem" with respect to the origin of living systems. That is, proteins are the functional machines, while nucleic acids bear information. Proteins are needed to replicate, while nucleic acids must be replicated. It seemed unlikely that both these complex biopolymers arose simultaneously, and this problem was a huge conundrum for thinking about origins. However, once it was clear that nucleic acids were in fact "chicken-eggs" (being both functional machines and information bearing), this conundrum was deftly resolved.

Around the same time that the antidye aptamers were being generated, Larry Gold and Craig Tuerk at the University of Colorado at Boulder showed that protein-binding nucleic acids could be selected from random sequence pools (Tuerk and Gold 1990). The same technology that might contribute to explaining life's origins therefore could also be used to create new drugs. Probably the most famous aptamer produced by directed evolution is known as *Macugen*. It was selected to bind to and thereby inhibit the function of the human protein vascular endothelial growth factor (Ng et al. 2006). This experimentally evolved RNA molecule is now used clinically to combat wet macular degeneration, a common cause of blindness in the elderly.

Catalytic nucleic acids can also be selected from random sequence pools. One fascinating and important ribozyme is known as the *Class I Bartel ligase*, after David Bartel, who discovered it (Bartel and Szostak 1993). This ribozyme has been of seminal importance in understanding the origin and evolution of life. The Bartel ligase seems, on first impression, to be a miracle. The probability of selecting it from a random sequence pool can be calculated, and it turns out that it should have been found once every 10,000 times that Bartel carried out his directed evolution experiment. Although it is possible that Bartel was extraordinarily lucky, the alternative and probably better explanation is that while any particular ligase was unlikely to have evolved, the ligase function itself was much more likely to have arisen. Thus, if the experiment were to be carried out again, a molecule of similar functionality and complexity—but with an entirely different sequence—would emerge and be selected. This finding is extremely important because it means that there are many possible routes from origins to modern, complex systems. As the late paleontologist Steven Jay Gould suggested, if we were to run the tape of life again, we'd probably get a very different answer.

2. DIRECTED EVOLUTION OF PROTEINS

It has also proven possible to direct the evolution of proteins. However, in this instance genotype and phenotype are not embedded in the self-same molecule—that is, they are not chicken-eggs in the way that some nucleic acids are. Therefore, there must be some other way to connect genotype and phenotype to perform directed evolution on proteins. One of the first ways this was done was by appending a short random library to the gene that encodes a coat protein of a bacteriophage, such that the library of peptides would then be expressed on the surface of the bacteriophage. The displayed peptide variants could then be selected on the basis of their ability to bind a ligand, as with nucleic acids, and amplified not by the PCR but by passage through cells that could be infected by the selected bacteriophage (Smith and Petrenko 1997). Such phage display methods have become very popular and are the basis for selecting not just peptides but antibodies and enzymes, as well. For example, Humira, an antibody drug effective in the treatment of rheumatoid arthritis, was selected via phage display. By displaying an antibody library against a protein involved in inflammation response (TNF-α), researchers were able to select and amplify high-affinity antibodies that could bind to the protein.

The same methods were later expanded to cell surfaces, so that individual cells now are the vehicle connecting protein variants on the surface to the genes encoding those proteins. As with the phage, the proteins on the cell surface could be selected for either binding or catalysis. For example, George Georgiou and coworkers showed that libraries of peptide proteases expressed on the surface of *E. coli* could be selected for altered substrate specificity (Varadarajan et al. 2005). Cleavage of a new desired "green" substrate by a given protease variant led to the accumulation of that color on the surface of the bacteria, while there was a parallel opportunity to cleave and accumulate a parental undesired "red" substrate. Both positive and negative selections could thus be applied to tune substrate specificity. The cells expressing protease variants with altered, desired specificities (colored green, but not red) were screened from the protease libraries based on a technique known as *fluorescence-activated cell sorting* (FACS) in which individual cells are sorted based on their fluorescence. The selected variants were further amplified by bacterial growth, and multiple cycles of screening and amplification led to the winnowing of the initial population to those few proteases with the desired new specificities. The ability to tune protease specificities may someday have practical applications, such as destroying undesirable proteins, including viral proteins.

The selection of proteins inside cells is similar to cell-surface display, except that instead of selecting directly for the ability of individual proteins to bind or catalyze reactions, researchers must instead select for the impact of binding or catalysis on cellular phenotypes. For example, some antibiotic resistance genes encode enzymes that modify antibiotics in some way. New resistance functions can be selected by challenging cells with a different antibiotic. For cells to survive, the resistance element must accumulate mutations that change its function. For example, as early as 1976 Hall and Knowles showed that the beta-lactamase enzyme could be directed to evolve to cleave not its normal substrate, pencillin-like drugs known as beta-lactams, but instead a new class of drugs, cephalopsorins (Hall and Knowles 1976).

The challenge is often to make sure that selection is focused on a particular enzyme of interest and that the cell does not follow some other evolutionary pathway that leads to survival (i.e., a different cellular enzyme than the one of interest might mutate in a way that leads to antibiotic resistance). To focus selection on a particular enzyme, user-mutagenized libraries of enzymes can be generated, just as they were for the ribozyme and protease selections described earlier.

Jeremy Knowles took advantage of new technologies for the synthesis of DNA to create oligonucleotide pools that could be used to randomize DNA, thereby creating populations of enzymes that could be subjected to selection. The essential enzyme triose phosphate isomerase was mutagenized and variants that led to faster cell growth were selected (Hermes et al., 1990). Enzyme phenotypes could be screened, as well as selected, and human intervention in growth (choosing active enzymes that produced more of a colored product) proved to be broadly useful for evolving many enzymes. Frances Arnold at Caltech has greatly advanced such technologies, showing that sequential rounds of screening and mutagenesis could lead to the accumulation of mutations that greatly improve enzyme function (Kuchner and Arnold 1997). Over several decades, the Arnold lab has used these methods to optimize enzymes for a wide variety of functions, from improving the ability of proteases and esterases to function in organic solvents (Moore and Arnold 1996), to evolving cytochrome P450 to perform hydroxylation reactions that may prove important for biofuel production and the remediation of hydrocarbon pollution, to carrying out completely new organic transformations that go far beyond what natural enzymes can do (Arnold 2015).

There are various ways to both mutagenize and select a given library. In the 1990s, Pim Stemmer came up with a brilliant technique to speed the evolution of proteins by allowing for in vitro recombination (Stemmer 1994). In this method, known as *DNA shuffling*, different enzyme variants are selected from a library (or may otherwise already be present). By cutting the genes for the enzymes into pieces, and then using PCR to recombine and eventually reassemble them into the full-length gene, many mutations can be brought together in the same gene. This approach is much faster than natural recombination, which generally involves only two gene copies at a time. In either case, the interesting assumption is made that combinations (or at least some combinations) of favorable mutations will themselves be favorable. This assumption has in large measure turned out to be true, perhaps because genes have evolved to evolve. That is, the types of protein sequences and structures amenable to recombination are those that have been successful at responding to changing conditions during the long course of evolution.

The methods so far described all require living cells either to make phage or to express proteins. Other researchers have devised clever ways to carry out protein evolution even without cells, by using in vitro transcription and translation systems that contain all the components necessary to make proteins, including ribosomes and tRNAs. For example, Andreas Pluckthun and his group managed to stall ribosomes in the process of translating mRNAs and thereby could connect the mRNA information being read with the protein function being translated (Hanes and Pluckthun 1997). Similarly, Jack Szostak's group figured out how to use the antibiotic puromycin to covalently couple a protein being translated on the ribosome to the mRNA making that protein (Roberts and Szostak 1997). As with natural selection and the various schemes described earlier, the coupling of genetically encoded information with function is essential for sieving through large sequence libraries to find rare functions. One of the advantages of these ribosome-based methods is that they can be used to look through much larger sequence populations than cell-based methods.

Researchers have also begun to create cell-like bubbles to assist with directed evolution (Griffiths and Tawfik 2006). Water-in-oil emulsions can be created by simply mixing these two components and shaking them (much like making salad dressing). If in vitro transcription and translation components are added to the aqueous component, then each small aqueous bubble in the sea of oil will be capable of making proteins. If only one DNA template or mRNA molecule is captured per bubble, then only one type of protein will be made in that bubble. The problem then becomes how to capture the bubble making the protein variant of interest. One solution is to have the protein feed back on the nucleic acid that produced it, and various methods have been developed that either mark the nucleic acid (for example, by methylation), amplify the nucleic acid (via a translated polymerase), or capture the nucleic acid (via a binding protein). Once the mixture is demulsified, all the nucleic acids are remixed together, but only those that have encoded a functional protein are marked, amplified, or bound, and they can be carried into subsequent rounds of selection and amplification. Another solution to the problem of connecting the appropriate bubble with the phenotype of interest has been to develop methods to capture the bubble itself. Adding a lipid coat to the aqueous bubbles allows them to be stabilized and sorted by FACS. Thus, no direct feedback loop to nucleic acids is required, which simplifies the procedures and greatly expands what kinds of enzymes can be selected. For example, lipases that act on esterases can turn over fluorescent substrates, so that more active lipase variants will accumulate more fluorescence in their bubbles, which can in turn be sieved from a larger background population (Griffiths and Tawfik 2003). As with all the other techniques described, amplification in vitro provides a selective advantage to the functional lipases by increasing their representation in the next generation.

3. DIRECTED EVOLUTION OF CELLS

The directed evolution of whole cells can yield both the simplest and most complex products. For example, the adaptation of cells to ferment beer (by producing ethanol) and help with baking (by producing carbon dioxide) is almost as old as human society, and researchers from Louis Pasteur onward have bred cells for industrial purposes. In 1928 Alexander Fleming discovered that the antibiotic penicillin was made by the fungus *Penicillium notatum*. While this discovery would in time change the world of medicine, further engineering was necessary. Strong demand for the drug during the Second World War necessitated increased production beyond what the organism naturally produced. To direct the evolution of the organism toward greater antibiotic production, cultures were subjected to X-ray radiation, which was known to mutagenize DNA, thereby accelerating the accumulation of genetic diversity. This mutagenized population was then screened for those cells that produced the most penicillin (Backus, Stauffer, and Johnson 1946), and these selected cultures produced enough penicillin to meet the demand. This use of random mutagenesis followed by screening for a desired phenotype is commonly referred to as *strain improvement* and is a now common practice in many industries in which the production of a particular product is dependent on microbial synthesis.

Experiments that could support these sorts of applications were initiated by Barry Hall, of the University of Rochester. Hall (2003) wondered what would happen if an enzyme of *E. coli* was deleted. This enzyme, β-galactosidase, was responsible for the ability of these cells to grow on the sugar lactose. When the enzyme was deleted, the cells could not grow on lactose, at least not initially. Over time, however, the cells began to grow slowly and, eventually, evolved to grow more rapidly, because another gene in the organism had accumulated mutations that allowed it to break down lactose. By focusing selective pressure on one function, Hall turned the entire organism into a vehicle for finding and improving a suitable enzyme.

While research into the directed evolution of cells led to a better understanding of the source, rate, and type of mutations that could lead to new or modified proteins, it was still difficult to target mutations and selection within a genome, especially for complex phenotypes like antibiotic production that required the adaptation of multiple genes and enzymes in parallel. Going well beyond Pasteur and his contemporaries required the development of tools that can recombine and modify individual sites in a genome, and advances in molecular biology that allow entire genomes to be sequenced. These modern techniques have had a dramatic effect on the time and effort required to generate improved bacterial strains.

Just as protein shuffling was developed to facilitate the accumulation of favorable mutations, other methods have been developed for shuffling entire genomes. Following a single round of classical strain improvement (random mutagenesis and screening or selection), cells are stripped of their cell walls (turned into protoplasts) and then induced to fuse with one another in what is essentially a multiparent mating event. Because a cell can generally accommodate only one genome, the genetic material from fused cells must be resolved into a single unit by the cell's recombination and repair machinery. The recombination process generates mosaic genomes, thereby amplifying the population's genetic diversity. This process can be repeated multiple times, allowing the improved strain to accumulate many additive and even synergistic mutations with respect to the desired phenotype. In a striking example of the power of this method, a group improved the production of the antibiotic tylosin by *Streptomyces fradiae* by about ninefold after only two rounds of genome shuffling that took roughly one year (Zhang et al. 2002). The resulting strains were found to produce as much tylosin as strains that had been independently subjected to 20 rounds of classical strain improvement over 20 years.

Other efforts have focused on reprogramming the cell as a whole by mutating the master regulators the cell uses to control how and when proteins are made. Changes to these *transcription factors* simultaneously affect the levels of expression for many genes, thereby altering the levels of many proteins and broadly affecting how the cell operates and behaves. The altered regulatory program is presumably not optimal for the organism in its natural context but may be much more productive in an industrial setting. By screening mutant libraries of transcription factors, researchers have isolated strains with increased tolerance to industrial processes and by-products, as well as enhanced the production of small molecules.

While genome shuffling and transcription-factor engineering can accelerate evolution, these processes still rely on nondirected changes in sequence or expression as their inputs. To produce a revolution in genome engineering on the scale of that already occurring in protein engineering, it was necessary to direct mutations to particular genes, an achievement that George Church and coworkers published as *multiplex automated genome engineering*, or MAGE (Wang et al. 2009). This technique involves synthesizing single-stranded DNA oligonucleotides corresponding to particular genomic locations. The oligonucleotides can enter the cells, and mutations engineered into the oligonucleotides may then be incorporated into the genome through recombination. By using an automated system to grow cells and deliver the mutant oligonucleotides, the researchers were able to achieve a high rate of mutation at several genomic sites in only a few cycles. This ability to target mutations to specific genomic locations enables researchers to precisely manipulate cells in ways that could be used to optimize the production of proteins or other molecules for industry, or even to rewrite the genetic code at large.

These amazing advances in genomic engineering require concomitantly amazing new analytical tools. With the development of so-called next-generation sequencing platforms that generate massive amounts of sequence data, the directed evolution of a microbial population can now be observed at the level of individual sequence changes within entire genomes. In 1988 Richard Lenski established his "long-term evolution experiment" by inoculating a strain of *E. coli* into minimal media and subsequently transferring a small amount of the previous culture into fresh media each day. Over time, the cells have become adapted to these environmental conditions, and their growth rate has accelerated. The genomes of the evolved organisms from various points through 40,000 generations were sequenced to determine the many underlying mutations responsible for this adaptation (Barrick et al. 2009). As the genomic engineering techniques described earlier are increasingly melded with large-scale acquisition of sequence data, the ability to understand and shape genomes will become commonplace.

4. THE FUTURE OF DIRECTED EVOLUTION

For the pioneers of directed evolution, including Spiegelman and Hall, the experimenter was in charge of directing the selection, whether it was in a population of molecules or organisms. These researchers had to take whatever random mutations the experimental system provided and then study which of these mutations led to changes and functions of interest. This process remains characteristic of the field of directed evolution, since many researchers are still addressing basic questions such as, What outcomes can be produced? What can we make? However, the field is now also beginning to develop models that are predictive, in part based on the paired abilities to direct where mutations occur and to analyze the repertoire of phenotypes associated with vast numbers of mutations. This process will accelerate as protein-structure analysis and prediction provide an understanding of why some mutations work and some do not, and as the field of systems biology begins to more precisely define cellular states. In turn, the shift from phenomenology to quantification and prediction promises to be one of the most exciting aspects of evolutionary biology in the future, and it should ultimately yield a mature discipline of evolutionary engineering.

FURTHER READING

Arnold, F. H. 2015. The nature of chemical innovation: New enzymes by evolution. Quarterly Reviews of Biophysics 48: 404–410.

Backus, M. P., J. F. Stauffer, and M. J. Johnson.1946. Penicillin yields from new mold strains. Journal of the American Chemical Society 68: 152.

Barrick, J. E., D. S. Yu, S. H. Yoon, H. Jeong, T. K. Oh, D. Schneider, R. E. Lenski, and J. F. Kim. 2009. Genome evolution and adaptation in a long-term experiment with *Escherichia coli*. Nature 461 (7268): 1243–1247. *A single* E. coli *strain was evolved over thousands of generations. The analysis of this evolutionary path continues to illuminate how bacterial genomes can adapt to new conditions.*

Bartel, D. P., and J. W. Szostak. 1993. Isolation of new ribozymes from a large pool of random sequences. [See comment.] Science 261 (5127): 1411–1418. *The first, surprising isolation of a large and complex ribozyme from a completely random sequence pool. The ClassI Bartel ligase is to this day one of the fastest and most complex ribozymes ever discovered.*

Ellington, A. D., and J. W. Szostak. 1990. In vitro selection of RNA molecules that bind specific ligands. Nature 346 (6287): 818–822. *One of the first demonstrations that complex RNA molecules could be isolated from a completely random pool. The selected phenotype, specific dye binding, was unexpected.*

Griffiths, A. D., and D. S. Tawfik. 2003. Directed evolution of an extremely fast phosphotriesterase by in vitro compartmentalization. EMBO Journal 22 (1): 24–35.

Hall, A., and J. R. Knowles. 1976. Directed selective pressure on a beta-lactamase to analyse molecular changes involved in development of enzyme function. Nature 264: 803–804.

Hall, B. G. 2003. The EBG system of *E. coli*: Origin and evolution of a novel beta-galactosidase for the metabolism of lactose. Genetica 118 (2–3): 143–156. *While strain improvement by directed evolution was known, Hall revealed the precise molecular underpinnings of the evolution of lactose utilization. Otherwise cryptic proteins in the* E. coli *genome can evolve novel functions.*

Hanes, J., and A. Pluckthun. 1997. In vitro selection and evolution of functional proteins by using ribosome display. Proceedings of the National Academy of Sciences USA 94 (10): 4937–4942.

Hermes, J. D., S. C. Blacklow, and J. R. Knowles. 1990. Searching sequence space by definably random mutagenesis: Iimproving the catalytic potency of an enzyme. Proceedings of the National Academy of Sciences 87: 696–700.

Kuchner, O., and F. H. Arnold. 1997. Directed evolution of enzyme catalysts. Trends in Biotechnology. 15: 523–530. *An early review that reveals the power of using iterative mutagenesis and screening to accumulate mutations that gradually improve phenotype.*

Moore, J. C., and F. H. Arnold. 1996. Directed evolution of a para-nitrobenzyl esterase for aqueous-organic solvents. Nature Biotechnology 14: 458–467.

Ng, E.W.M., D. T. Shima P. Calias, E. T. Cunningham Jr., D. R. Guyer, and A. P. Adamis. 2006. Pegaptanib, a targeted anti-VEGF aptamer for ocular vascular disease. Nature Reviews Drug Discovery 5: 123–132. doi:10.1038/nrd1955.

Roberts, R. W., and J. W. Szostak. 1997. RNA-peptide fusions for the in vitro selection of peptides and proteins. Proceedings of the National Academy of Sciences USA 94 (23): 12297–12302.

Saffhill, R., H. Schneider-Bernloehr, L. E. Orgel, and S. Spiegelman. 1970. In vitro selection of bacteriophage Q-beta ribonucleic acid variants resistant to ethidium bromide. Journal of Molecular Biology 51 (3): 531–539. *Even before the advent of modern molecular biology techniques such as DNA synthesis and sequencing, Saffhill proved it possible to evolve phage RNAs in vitro for novel phenotypes.*

Smith, G. P., and V. A. Petrenko. 1997. Phage Display. Chemical Reviews 97 (2): 391–410.

Stemmer, W. P. 1994. Rapid evolution of a protein in vitro by DNA shuffling. Nature 370 (6488): 389–391. *Pim Stemmer and coworkers developed a radical new technique for in vitro recombination and consequently demonstrated remarkable enhancements in the speed of directed evolution.*

Tuerk, C., and L. Gold. 1990. Systematic evolution of ligands by exponential enrichment: RNA ligands to bacteriophage T4 DNA polymerase. Science 249 (4968): 505–510.

Varadarajan, N., J. Gam M. A. Olsen, G. Georgiou, and B. L. Iverson. 2005. Engineering of protease variants exhibiting high catalytic activity and exquisite substrate specificity Proceedings of the National Academy of Sciences USA 102: 6855–6860. doi:10.1073/pnas.0500063102.

Wang, H. H., F. J. Isaacs, P. A. Carr, Z. Z. Sun, G. Xu, C. R. Forest, and G. M. Church. 2009. Programming cells by multiplex genome engineering and accelerated evolution. Nature 460 (7257): 894–898. *Until the development of the technique known as MAGE, it was impossible to carry out multiple, site-directed alterations to a bacterial genome in parallel. This technique and others like it should continue to accelerate classic methods for the directed evolution of bacteria.*

Zhang, Y. X., K. Perry, V. A. Vinci, K. Powell, W. P. Stemmer, and S. B. del Cardayre´. 2002. Genome shuffling leads to rapid phenotypic improvement in bacteria. Nature 415 (6872): 644–646.

CHAPTER 14

EVOLUTION AND COMPUTING

Robert T. Pennock

OUTLINE

1. Unexpected links and shared principles
2. How evolutionary biology joined forces with computer science
3. How evolutionary computation is helping evolutionary biology
4. Evolutionary computation takes off
5. The future of evolution and computing

Shared principles between evolution and computing are opening up fruitful areas for research. This chapter discusses some unexpected connections between evolutionary biology and computer science, such as the core ideas of code, information, and function, and how these are leading to theoretical and practical ways in which each is benefiting the other. The chapter highlights the emerging field of evolutionary computation, giving a brief history and some examples of its utility not only in helping solve basic research problems in biology and computer science but also for generating novel designs in engineering.

GLOSSARY

Digital Evolution. The evolution of digital organisms in a system that instantiates the causal processes of the evolutionary mechanism through random variation, inheritance, and natural selection.

Digital Organism. A model organism, typically with a genome composed of simple instructions, in a computer environment.

Evolutionary Computation. The general term for research and procedures in computer science that take inspiration and utilize insights from evolutionary biology.

Evolutionary Engineering. Use of evolutionary computation approaches for solving design problems in applied engineering contexts, including robotics.

Experimental Evolution. Investigation of evolutionary processes by direct experimental methods, including replications and controls, rather than by indirect comparative methods.

Genetic Algorithm. One form of evolutionary computation; pioneered by John Holland.

What does evolution have to do with computing? What does computing have to do with evolution? At first glance, these fields almost seem to be opposites. On the one hand, evolutionary biology deals with the lush and tangled extravagance that is the living world. Living organisms grow, reproduce, and proliferate in abundant variety and complexity. It was Charles Darwin's genius that began to unravel this complexity and discovered some of the fundamental principles that produce new species and their astounding adaptations. In the century and a half since the publication of *On the Origin of Species*, evolutionary science has become a powerful explanatory framework that illuminates the entire organic world.

Computer science, on the other hand, deals not with organisms but with machines. Machines may get bigger, and computing machines have gotten more powerful, but they don't grow—they are built. The artificiality of computers stands in stark contrast to the naturalness of organisms. Computers are complex, but in quite a different way from living things. One would never confuse the specific patterns of complexity that characterize computing machines designed and built by human beings with the patterns that we find in evolved, living organisms. It is differences of this sort between the natural biological world and the technological world of objects designed and built by human beings that initially made it questionable whether it was even sensible to think that there could be, in Herbert Simon's term, a "science of the artificial" (Simon 1969).

This chapter discusses some of the ways in which evolutionary biologists and computer scientists are discovering, to their mutual benefit, that their fields actually have many concepts in common and that there are significant ways in which they may be united through deep, shared principles. After reviewing some of these principles, we will briefly look at how computer science came to recognize the applicability of evolution to computer science and began to figure out how to incorporate Darwin's findings into its own algorithmic way of thinking. We will see how the mechanism of evolution by natural selection that Darwin discovered can now be not just simulated but actually causally instantiated in a computer, and how this opens the door to surprising new ways for biologists to experimentally investigate evolutionary processes. And finally, we will look at how this new field of evolutionary computation can be applied in practical ways, such as in solving difficult design problems in engineering. As we shall see, evolutionary design has reached the point at which it can equal and sometimes surpass our own problem-solving abilities.

1. UNEXPECTED LINKS AND SHARED PRINCIPLES

The obvious contrasts between living organisms and computing machines hide significant points of commonality between the two fields of study because they focus on *products* rather than on *processes*. Once we begin to compare biology and computing in terms of processes, we find significant and fundamental linkages that were previously overlooked.

One conceptual commonality is the idea that both fields deal at the deepest level with the idea of coded functions. On the computational side, even laypersons understand that computer code runs everything, from their notebook or tablet computer to the largest mainframe. It is the coded instructions of the software loaded in one's machine that make it function. On the biological side, everyone knows that organisms similarly depend on their genetic code for their functions. A mistake in the coded program of life can make an organism unable to perform some function as surely as a mistake in a software program can cause a function error. With little exaggeration, one may say that natural organisms are biological machines that run on genetic software that codes for the myriad, complex features that make them work in their native environments—or that make them fail to work. A severe mistake in the genetic code of an organism can cause it to die, just as a serious coding mistake in some application you are running can bring up the dreaded blue screen of death.

A second, closely related conceptual commonality between both fields is the idea of information. Here, too, even the language of the two disciplines resonates with deeply shared notions about the significance of information and its flow. For instance, rather than speaking narrowly of computer science, it is becoming more common for many computer scientists to identify their work as *information science*. Again, the term *computer science* makes it seem as though their subject matter is *computers*, whereas they take their real subject matter to be *computing*, which they see as the most basic form of information processing. An information-theoretical approach has not yet been developed nearly as far on the biological side, but here, too, the language of the discipline rings with this idea. An organism's genome is said to code for "biological information," while RNA and DNA are spoken of as "informational molecules."

These and other commonalities have long hovered in the background of scientific investigations in both the biological and computing communities, but as the deep conceptual connections are becoming more appreciated, they are coming to the foreground and being recognized as providing an opportunity for cutting-edge research.

To give just one example, in 2011 the National Science Foundation (NSF) published a letter to researchers calling attention to what it called Biological and Computing Shared Principles (BCSP). Issued jointly by NSF's Biological Sciences (BIO) and Computer & Information Science & Engineering (CISE) directorates, the BCSP letter highlighted a revolutionary transition occurring in the relationship between the fields. There have always been points of mutual influence between biological and computing research, but these are no longer limited to applications of one discipline to the other; the letter highlights "the convergence of central ideas and problems requiring the theoretical, experimental, and methodological competencies of both biology and computing" (US NSF 2011). The reason for the excitement is that shared principles between biology and computing may contribute to *conceptual* advances for both fields.

The BCSP letter identifies a variety of novel areas that are ripe for the identification and investigation of shared principles. Many of these involve specific properties of common interest such as adaptation to unanticipated novel conditions; self-repair and maintenance; coevolution and defense against adaptive adversaries; and general robustness and reliability. Other topics are more abstract, such as knowledge extraction; information flow, processing, and analysis; representations and coding;

pattern recognition and pattern generation; network structure, function and dynamics; functions of stochasticity; and theory of biological computation. Although the BCSP letter speaks broadly about computing and "biology," in fact, many of these properties are connected conceptually to *evolutionary* biology.

2. HOW EVOLUTIONARY BIOLOGY JOINED FORCES WITH COMPUTER SCIENCE

Computers are as much a key instrument in biology today as the microscope was in the nineteenth century. They no longer serve just as fancy calculators that make statistical analysis of lab and field data go faster; their flexibility and power as a universal machine now allows them to also serve a fundamental role in the production of data. One important new role is to allow sophisticated simulations of complex biological entities and processes. To give just one example, recent work by Donohue and Ascoli (2008) used computers to model the morphology of neurons and the developmental processes that lead to the elongation, branching, and taper of dendrites. Starting with parameters measured from real cells, modelers can create statistical distributions that can be resampled to form virtual trees and even to simulate somatic repulsive forces thought to be responsible for shaping cells. Such simulations can reveal patterns that may point to important developmental principles. For studying evolution in particular, an even more important advance is that computers now allow scientists to model evolutionary processes directly.

The insight that led to this revolutionary approach was made independently by several researchers (most in the 1960s) who recognized that the mechanism of evolution that Darwin discovered could be instantiated not just in biological systems but in other physical systems as well, including in computers. Probably the most influential of these was University of Michigan computer scientist John Holland, who coined the term *genetic algorithm* for the idea and who implemented it at the level of binary strings of 0s and 1s that could recombine, mutate in a computer, and be subject to selection. In Germany, aeronautical engineer Ingo Rechenberg had a similar idea and developed it with Hans-

Paul Schwefel under the name *evolutionary strategies*. A third line of research, dubbed *evolutionary programming*, was begun by electrical engineer Lawrence Fogel. For their pioneering work, Holland, Rechenberg, and Fogel are credited as founders of what now goes by the general term *evolutionary computation*. This is not the place to recount the history of these and other early pioneers, but it is worth mentioning that these initial research streams proceeded separately for over a decade and a half before they discovered one another and began to interact. Today it is recognized that these and other evolutionary computation approaches share the same underlying core principles (De Jong 2006), and a community of researchers has formed around these ideas, spawning a variety of professional societies, conferences (many of which eventually joined together as GECCO, the Genetic and Evolutionary Computation Conference), and journals. *Evolutionary Computation*, the main journal in the field, was introduced in 1993.

This short history returns us to the idea of shared principles between evolution and computing. When evolution is seen as a special sort of algorithmic process, then it becomes possible for a computer to become the evolutionary biologist's lab bench. Properly understood, *digital evolution* can do more than simulate evolutionary processes, it can instantiate them (Pennock 2007). To see this we need only review the basic elements of the causal principle that Darwin discovered.

Descent with modification, as Darwin defined evolution, occurs whenever three conditions hold. The first is the random production of variations—a diversity of structure, constitution, habits, and so on. The second is that these variations be heritable, meaning that they can be passed on in the process of reproduction to the next generation. Darwin called this the "principle of inheritance." The method of inheritance is not so important as the basic causal principle of heritability itself—the key is *that* the genetic information be copied to the offspring, not the specific mechanism by which that is done. The third condition is that these heritable variations be naturally selected by the environment. If the genome of an individual happens to provide it with some slight variation that gives it any advantage over its competitors in their environment, that individual becomes more likely to survive to the point that it can reproduce, which causes the next generation to have a greater proportion of individuals with its heritable variations than those of its competitors. It is the environment, understood broadly, that naturally selects from among the extant variations, generation after generation. All these causal processes—random variation, heritability, and natural selection—can be instantiated in a computer.

Moreover, once one comes to see evolution as a universal causal law—one whose action is not limited to the familiar realm of DNA and flesh—then it becomes easier to see the possibility of sharing other biological concepts with computer science. To give just one example, the concept of a *genome* is not limited to the chromosomes of an organism; it also refers more generally to the complete information-transmitting material of any replicating entity, whether a biological organism or a self-replicating computer program. Indeed, chromosomes are best conceived of as but one possible instance of a genome. Other structures could have been found, and may yet be found, that carry heritable genetic information. Historically, it was not until many decades after the *Origin* that chromosomes were identified as the location of the genetic material. Darwin had simply spoken of "factors" in an abstract causal sense and it was not relevant to his law in *which* material they turned out to be instantiated. In this sense, it is not a metaphor or an analogy to speak of a coded sequence of instructions of a *digital organism* as its genome, for it is the genetic material of that individual in just the same sense as it is for a biological organism.

There is not the space here to lay out the full argument for this claim, but the rationale for understanding these concepts at this level of abstraction should be clear enough to see the value of bringing them and other shared concepts and principles to the foreground. Recognizing that the evolutionary causal processes can be instantiated in other physical systems, including in a computer, means that digital evolution goes beyond even the utility that a simulation can provide and provides a truly experimental system.

3. HOW EVOLUTIONARY COMPUTATION IS HELPING EVOLUTIONARY BIOLOGY

The late John Maynard Smith, a distinguished evolutionary biologist, was quick to recognize the scientific potential of marrying evolution and computing. In particular, he called attention to how digital evolution provides a way for biologists to escape from the inconvenient limits of our single planet. "So far," he wrote in a 1992 article in *Nature*, "we have been able to study only one evolving system, and we cannot wait for interstellar flight to provide us with a second. If we want to discover generalizations about evolving systems, we will have to look at artificial ones." Since then, others have opened this digital wormhole further.

To illustrate some of the advantages of digital evolution as a model system, we describe a study that used digital evolution to investigate the evolution of complex features (Lenski et al. 2003). Darwin advanced a number of hypotheses about how evolution could produce what he called "organs of extreme perfection and complication." Recognizing that features such as the eye were too complex to have arisen in a single leap, he proposed that their evolution would have involved incremental changes through intermediate forms, including changes of structures from one function to another. Darwin provided indirect evidence for his hypotheses by comparisons across different species but wrote that it would have been ideal if it were possible to precisely trace the details of a single line of descent. With an evolving digital system this is now a reality.

This study used the Avida platform, which is a well-developed model system for digital evolution research. In Avida, the genome of a digital organism (an "Avidian") is composed of simple computer instructions that do little of interest by themselves but when ordered in specific complex sequences can in principle perform any computable function. Such computational functions are the digital organism's phenotype. Among other properties, the genome of an Avidian has the potential ability to self-replicate. In its digital environment, however, the copying process is imperfect, so descendant organisms may have random mutations in their code. They also have to compete for the energy needed to execute their genetic programs. In this system, the digital organisms get energy by performing logical operations that also require specific sequences of instructions to function. Simple functions provide the organism with a small energy boost; more complex functions provide more energy, allowing them to run faster. This process provides an analogue to biological metabolism, but if a digital organism is to perform more complex metabolic functions, it must evolve them, because the ancestor could replicate but not perform any logic functions. As in nature, the digital environment naturally selects those variations that give organisms a competitive advantage. The mutations that arise in the genome are usually deleterious (and some may destroy the Avidian's ability to replicate) or neutral, but a few may improve the organism, resulting in faster replication and thus more offspring with those variations. With all the conditions of the Darwinian mechanism in place, a population of Avidians naturally evolves on its own without any outside assistance.

To investigate how complex features evolve, Lenski and collaborators ran 50 replicate populations, all under identical conditions. Ensuring identical replicates is simple in a digital evolution experiment and more precise than can be done in most natural systems, so statistical replication is simple. The digital system provided other advantages that no natural system could match, in that it allowed the experimenters to track complete lines of descent (from the ancestor across thousands of generations and millions of descendants), record every mutation along the way, and measure whether each was deleterious, neutral, or beneficial. The researchers could thus observe directly as beneficial mutations accumulated to produce one or another logic function, which themselves were later lost or modified for some other, more complex function. And although a given complex function first emerged in a line of descent as the result of just one or two mutations, systematic knockout experiments (removing individual instructions of the genome one at a time) showed that the function always depended on some specific sequence of many instructions that had previously evolved as parts of other functions and that their removal would eliminate the new feature. In an evolving digital system one has the ability to monitor such changes and to analyze them with an extraordinary degree of precision.

Nor are the results of such experiments always predictable. Because this is a real evolving system rather than a numerical simulation, the dynamics of the evolutionary mechanism can yield surprising results, just as in a biological evolving system. One unexpected finding in this particular study was that occasional deleterious mutations were in the line of descent leading to the most complex function in some populations. Some were only slightly deleterious, but a couple reduced fitness by more than 50 percent. Further tests ruled out the possibility that these mutations were just accidental hitchhikers; although the mutations were highly deleterious when they occurred, they became highly beneficial in combination with subsequent mutations.

This study is but one of many that used digital organisms to examine basic questions about evolutionary processes that would have been exceedingly difficult or impossible to perform with biological organisms. In a digital evolution system, one may "replay the tape of life," as Stephen Jay Gould put it, and directly observe, for example, the role that historically contingent events such as mass extinctions can play in the course of evolution. One can devise appropriate controls to test how altruistic behaviors evolve under different selective pressures. One also may investigate the effect of natural selection on different methods of phylogeny reconstruction. Digital-evolution researchers have done all this and more.

This is not to say that digital evolution works as a model system for all the kinds of questions evolutionary biologists want to ask. While digital evolution instantiates the core causal processes of the evolutionary mechanism, it does not model, for instance, the defining features of any particular species or the unique properties of particular molecular structures. Thus it will be of little use to someone investigating questions for which such physical structures are salient. But for the biologist interested in questions about the cause-effect relationships of Darwin's law or seeking generalizations that will apply to any evolving system, *experimental evolution* with digital organisms is a revelation.

Evolution, broadly understood, requires neither DNA nor even living organisms. Evolutionary computation can help biologists understand shared properties such as robustness and nonexpressed code (Foster 2001). The power and flexibility of digital evolution gives researchers unprecedented opportunities to test evolutionary hypotheses, especially those requiring manipulations that are impractical in biological systems or numbers of generations that cannot directly be observed. Equally exciting are the practical applications of these shared principles embodied in evolutionary computation to such fields as engineering.

4. EVOLUTIONARY COMPUTATION TAKES OFF

Asked in 2010 for his judgment about the future course of his field, the president of the National Academy of Engineering, Charles M. Vest, wrote in the *New York Times* that "we're going to see in surprisingly short order that biological inspiration and biological processes will become central to engineering real systems. It's going to lead to a new era in engineering." Vest was no doubt thinking of a range of ways that engineering is beginning to make what we might call "the biological turn," including biomimicry and other uses of the products of evolution, but the use of evolutionary computation is certainly one of the most compelling ways that biological processes are being applied in engineering. As before, let us look at just one example in a bit of detail to show just how far evolutionary engineering has gone—in this case literally into outer space.

In 2004 NASA was preparing for the launch of its Space Technology 5 mission, whose aim was to test technology for measuring the effect of solar activity on the earth's magnetosphere. One of NASA's needs for the mission was a specialized antenna that had to meet a variety of precise specifications. Given certain transmit-and-receive frequencies, it had to operate within specified ranges for important functional properties. Moreover, the antenna had to fit within a 6 in. cylinder. Members of the Evolvable Systems Group at the NASA Ames Research Center decided to see whether an evolutionary approach could solve the problem.

The research team began by setting up a virtual world with a genetic encoding scheme that could represent the construction of three-dimensional wire forms—the space of possible antenna shapes. They used a tree-structured encoding in which, for example, a branch in the genotype would represent a branch in the wire form. The genotype was allowed to vary at random so as to produce diverse forms with different numbers, lengths, and angles of branches. In an initial population, 200 individuals were evaluated for their fitness for the task that NASA had set—think of this as a competition in which the virtual antennas were tested against each other. The best individuals in a given generation were automatically selected, and most of these were again randomly mutated or recombined to form new variations for the next generation. As this process was iterated over many generations, the shapes of the individuals in the population evolved little by little to better match the functional requirements that NASA had set. The antenna shapes that evolved in these runs were unlike anything that antenna engineers would have come up with themselves. They were very small—not much bigger around than a quarter—and they looked rather like a bunch of randomly twisted paper clips. But at the end of the run, the team built the device that had evolved in the virtual world and found that this misshapen bunch of wires met the required specifications.

The evolved antenna had additional technical benefits, including requiring less power, having more uniform coverage, and not requiring a matching network or a phasing circuit, thus simplifying the design and fabrication. Moreover, all these benefits were achieved with a shorter design cycle: the prototype for the antenna took three person-months to design and fabricate, compared with five person-months for a conventionally designed antenna (Hornby, Lohn, and Linden 2011).

What is especially impressive about this evolved design is that it succeeded where human engineers had failed. Prior to the Evolvable System Group's tackling the problem, NASA had contracted with an antenna engineering group that had produced a prototype design using conventional techniques, after a bidding process among several competing groups. However, the conventional design they produced did not meet the exacting mission requirements, while the evolved design did.

On March 22, 2006, the evolved antenna was launched into space. This was the first time that evolved hardware had reached such heights, both metaphorically and physically, but it was not an isolated instance of the power of evolutionary engineering. Indeed, evolutionary approaches have advanced to the degree that they now routinely equal or surpass human engineers in a variety of design tasks (Koza 2003).

This might seem to be a bold claim. By what measures can one say that evolutionary designs can equal or surpass those of human beings? Since 2004, GECCO has held a contest for human-competitive results, and it judges entries using a variety of criteria. To qualify for the competition, an evolved solution must meet at least one of several standards, such as producing a patentable design or a result that is publishable in its own right in a peer-reviewed journal, independent of the fact that it was mechanically created. The evolved antenna shared the Gold Award in 2004. Since then winners have been recognized for human-competitive results in areas as disparate as photonic crystal design, automated software repair, and protein structure prediction. These awards for human-competitive evolved design are appropriately called the "Humies."

5. THE FUTURE OF EVOLUTION AND COMPUTING

One promise of evolutionary approaches to computation and engineering is that they will solve real-world problems. This promise is already being fulfilled: evolutionary computation harnesses the power of evolution, allowing evolutionary processes to work in a digital world just as they work in nature. A further promise of evolutionary computation is that it will help reveal the deeply shared principles between what initially appeared to be quite distinct fields of research. Think of what it means to recognize that it is not a mere metaphor to say that living organisms are running a genetic program. Think of what it means to understand that the functional properties of life—those astounding adaptations—that are coded in the genome were *programmed by evolution*. If this lesson can be learned, the marriage of evolution and computing will have been profound indeed.

See also chapter 13.

FURTHER READING

Avida-ED Project: Technology for teaching evolution and the nature of science using digital organisms. avida-ed. msu.edu/. *With free downloadable digital evolution software as well as background materials and model exercises for undergraduate and AP biology courses, this award- winning project gives students everything they need to perform evolutionary experiments on their own computers.*

Clune, J., H. Goldsby, C. Ofria, and R. T. Pennock. 2010. Selective pressures for accurate altruism targeting: Evidence from digital evolution for difficult-to-test aspects of inclusive fitness theory. Proceedings of the Royal Society B 278: 666–674. *One example of the kind of basic science research that experimental evolution with digital organisms makes possible, this study tested hypotheses about inclusive fitness and the evolution of altruistic behavior.*

De Jong, K. A. 2006. Evolutionary Computation: A Unified Approach. Cambridge, MA: MIT Press. *An authoritative account of the common theoretical underpinnings of different varieties of evolutionary computation.*

Donohue, D. E., and G. A. Ascoli. 2008. A comparative computer simulation of dendritic morphology. PLoS Computational Biology. 4(5): e1000089. doi:10.1371/journal.pcbi.1000089.

Foster, J. A. 2001. Evolutionary computation. Nature Reviews Genetics 2: 428–36. *An excellent review article giving an overview of the field of evolutionary computation for biologists.*

Holland, John H. 1992. Adaptation in Natural and Artificial Systems. Cambridge, MA:

MIT Press. *Holland's pioneering book, originally published in 1975, gave a detailed account of how what he called genetic algorithms could model the process of evolutionary adaptation in a computer system.*

Hornby, G. S., J. D. Lohn, and D. S. Linden. 2011. Computer-automated evolution of an X-band antenna for NASA's Space Technology 5 Mission. Evolutionary Computation 19 (1): 1–23. *Scientific account of how a NASA space antenna was produced by evolution engineering techniques.*

Koza, J. R., M. A. Keane, M. J. Streeter, and W. Midlowec. 2003. Genetic Programming IV: Routine Human-Competitive Machine Intelligence. Norwell, MA: Kluwer Academic. *The fourth in a series by John Koza and colleagues about genetic programming, this book lays out the case for how evolutionary techniques have advanced to be able to routinely match human intelligence for a wide variety of problems.*

Lenski, R., C. Ofria, R. T. Pennock, and C. Adami. 2003. The evolutionary origin of complex features. Nature 423:139–144. *This pioneering study used digital evolution to perform a direct experimental test of some of Darwin's hypotheses about the evolutionary mechanisms that pro- duce complex features.*

Pennock, R. T. 2007. Models, simulations, instantiations and evidence: The case of digital evolution. Journal of Experimental and Theoretical Artificial Intelligence 19 (1): 29–42. *This paper sorts out common confusions about model-based reasoning using digital evolution, explaining why it is a mistake to think of these models as just simulations of evolution and why they are real instances of the causal mechanism that Darwin discovered.*

Simon, H. A. 1969. The Sciences of the Artificial. Cambridge, MA: MIT Press. *Herbert Simon, who was later to win a Nobel Prize, wrote this prescient book about why artificial systems, including computers, could properly be treated as objects for scientific study.*

Smith, John Maynard. 1992. Byte-sized evolution. Nature 355: 772–773. doi:10.1038 /355772a0.

CHAPTER 15

EVOLUTION AND CONSERVATION

H. Bradley Shaffer

OUTLINE

1. Evolution, genetics, and conservation
2. Process versus pattern and why both matter
3. The enemies to watch out for
4. What genomics brings to the table
5. Concluding thoughts and prospectus

Traditionally, evolutionary biology has had a distant relationship to conservation compared with ecology and field-based natural history. However, this situation has changed dramatically in the last two decades, particularly as abundant molecular data have become available for at-risk species of conservation concern. As the availability of genome-level data for these species increases, the role of evolutionary biology in conservation management continues to grow to a far greater extent. The combination of these new data from microevolutionary analyses with more traditional input from phylogenetics and systematics has elevated evolutionary biology to a position of primary importance in conservation science.

GLOSSARY

Ecotone. A transition area where two distinct ecological communities meet and integrate.

Endangered Species Act (ESA). The US law that protects critically at-risk species from extinction due to human activities. It was passed into law in 1973 under President Richard Nixon and remains one of the most powerful conservation laws in existence.

Landscape Genetics/Genomics. The fields that integrate population genetics (or genomics) data with features of specific landscapes to study how those features influence the movements of genes and individual. This is a computationally intensive discipline that has become a major part of many conservation programs.

Nongovernmental Organization (NGO). An organization that is independent of any government, and generally has an important advocacy role. Several leading NGOs play a critical role, both locally and globally, in biological conservation.

Phylogenetic Diversity (PD). The amount of character change that evolves along a branch of a phylogenetic tree. PD may evolve along internal or tip branches and may be nonsymmetrical along two branches derived from a common ancestor.

Phylogenetics. The discipline that reconstructs the genealogical relationships of species and lineages.

Systematics. The discipline that names, describes, and infers the evolutionary history of species and lineages.

1. EVOLUTION, GENETICS, AND CONSERVATION

Suppose that you control environmental policy, and you have a choice: you can save either the New Zealand tuatara (*Sphenodon punctatus*) or the western fence lizard (*Sceloporus occidentalis*) from extinction. Whichever you choose, the other will go extinct. The tuatara is a lizard-like animal that is the sole surviving member of a once-diverse but now nearly extinct lineage of vertebrate life. Although that lineage was widespread and globally common 200 million years ago, it is currently down to one (or possibly two, virtually identical) species that occupies a handful of islands off the coasts of both main islands of New Zealand. If any lineage deserves the name "living fossil," it may be the tuatara. The western fence lizard is probably the most common lizard in North America. It is a widespread, abundant, and somewhat-unremarkable member of one of the most diverse and adaptable genera of lizards on earth. As of this writing, 92 species in the fence lizard genus *Sceloporus* are recognized, and new ones are constantly being described and characterized. So, how do you decide?

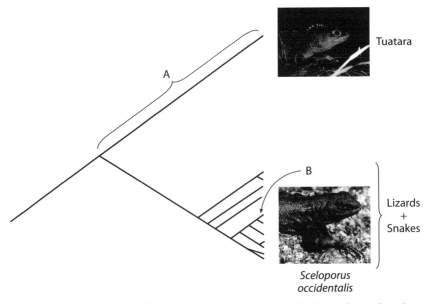

Figure 15-1. A phylogeny showing the relationships between the tuatara and its closest relatives, the snakes plus lizards. If the tuatara goes extinct, that species plus all the evolutionary history that occurred along the branch leading to it (labeled A in the figure) will be lost forever. If the western fence lizard (*Sceloporus occidentalis*) goes extinct, only that species plus the unique evolution on the much shorter branch B will be lost.

This kind of "conservation triage" is one of the arenas where evolutionary biology plays a critical role in conservation decision making. Evolutionary biology cannot tell a manager which species is more important, but it can frame the question and provide quantitative insights that can help guide the decision-making process. In this particular case, virtually all policy makers would choose the tuatara. The question is, why? Conserving evolutionary history—that is, long branches of the tree of life that provide a record of the changes that have occurred during the history of life on earth—is a universally recognized component of conservation biology. The logic is that any lineage that took 200 million years to evolve is, in some real sense, more precious than another lineage that has many close relatives with which it shares most aspects of its morphology, ecology, and natural history. The phylogenetic uniqueness of the tuatara, its lack of close relatives, and the incredible length of its branch on the tree of life (figure 15-1) are all insights that come directly from understanding its evolutionary history and are the primary reasons why it is a global conservation icon. The same is true for many other important conservation targets, including the duck-billed platypus, the remaining rhinoceros species, and the California and Chinese redwoods.

At least three different components of evolutionary biology speak directly and forcefully to problems in conservation biology. The first is *systematics* and the related discipline of *phylogenetics*, and the tuatara is one of the classic examples. Both disciplines now rely heavily on molecular—usually DNA-level—data to make inferences about organisms, and both seek to describe the diversity and interrelationships of life on earth. Given that probably the single most important tenet of conservation biology is that "you cannot protect what you don't recognize," and that one goal of systematics is the delimitation of species and lineages, it seems clear that we require a catalog of life on earth before we can realistically plan for protecting it. For example, until fairly recently it was widely considered that the living tuatara consisted of a single species. However, in 1990, Daugherty and colleagues evaluated the variation found among remnant tuatara populations and hypothesized that the animals on Brothers Island actually constituted a different species, for which they used the name *Sphenodon guntheri*. In so doing, they simultaneously presented the world with one of the rarest species of vertebrates on earth and removed one population from the small catalog of known breeding populations of the critically endangered northern tuatara, *S. punctatus*. However, more recent work, based on additional data and sampling, reversed that decision, instead concluding that the tuatara "is best described as a single species that contains distinctive and important geographic variants." By studying one endangered species in ever-greater detail, this research team has continued to refine our understanding of the evolutionary history of tuataras and thus the populations and potential species in need of conservation actions.

A second, related area in which modern evolutionary biology informs conservation and management is *phylogeography*. Originally introduced by evolutionary geneticist John Avise in 1987, phylogeography uses genetic data to understand lineage formation and evolutionary diversification within, rather than among, species of organisms. As the name implies, a key goal of phylogeographic research is determining the relationship between the geographic location of populations and genetic differentiation among those same populations. In many cases, the recognition of deeply separated lineages within species has led to their independent protection and conservation. For example, recent phylogeographic work from our laboratory on the California tiger salamander (*Ambystoma californiense*) demonstrated that the species consists of at least three genetically independent lineages; two of these are geographic isolates in the south (Santa Barbara County) and the north (Sonoma County), while the third is the larger central group from the Great Central Valley. When the species was protected under the US *Endangered Species Act*, the combination of different levels and types of threats and the phylogeographic recognition of three lineages led to the independent protection of salamanders from Santa Barbara and Sonoma counties as endangered, while the rest of the species' range was separately listed as threatened under the ESA. These different listing levels (threatened versus endangered) actually do matter and could not have been proposed or implemented without this phylogeographic research.

Finally, *population genetics* has traditionally been the cornerstone of evolutionary biology's contribution to conservation, and this tradition has grown in the last few years. Three decades ago Frankel and Soulé (1981) emphasized the close connections between population genetics and conservation in conceptual areas ranging from minimum viable population sizes to the relationship between inbreeding depression and genetic drift. Frankel and Soulé's book, the first to use the words "conservation" and "evolution" in a single title, was also among the first to explicitly point out the expected relationship between small effective population size (N_e) and population health that is predicted from population genetics theory. Because inbreeding is generally detrimental to most outbreeding populations, Frankel and Soulé argued that small populations would be particularly vulnerable to inbreeding depression and coined their "basic rule of conservation genetics," which relates the change in inbreeding coefficient, ΔF, to the likelihood that a population will survive into future generations. In particular, they suggested that ΔF greater than about 1 percent constitutes "a threshold rate of inbreeding, above which fitness relentlessly declines" and populations go extinct. A related concept is Frankel's 50/500 rule, which states that on average, populations with a persistent N_e less than 50 may be in immediate danger of extinction, whereas over long time periods, populations with N_e less than 500 may not contain enough genetic variation to adapt to changing conditions (Braude and Low 2010). Although controversial, these "rules" emphasize a key point—when populations become too small and isolated, genetic drift can overcome natural selection, and low-fitness genotypes can rise in frequency by chance alone. If this happens too often, or for too long, extinction may follow. Based on this fundamental insight, conservation geneticists have recently invested a considerable amount of both theoretical and empirical research into the related concepts of evolutionary and genetic rescue as an important, often underutilized component of modern conservation biology.

2. PROCESS VERSUS PATTERN AND WHY BOTH MATTER

A key question in conservation biology is deciding what to conserve and why. The resolution of this question depends on many factors. The country where the action is taking place may have strong conservation laws, like the US ESA, or it may have virtually no history or capacity for even the weakest protection of taxa or landscapes. *Nongovernmental organizations* (NGOs) may be prominent partners that have their own opinions and agendas, local jurisdictions may interact in a positive or negative way with national governments and NGOs, and international organizations like the United Nations or World Bank may enter into the conversation.

Regardless of the organization, its politics, or its agenda, anyone who considers human-mediated extinction to be an outcome that should be avoided is really trying to conserve an aspect of evolutionary biology. At the broadest level, one can think about this problem in two ways. First, one can focus on conserving evolutionary history that has already occurred. Alternatively, or in addition, one can attempt to conserve the potential for future evolutionary change. Interestingly, these two approaches sometimes lead to very similar actions, and sometimes to radically different conservation priorities.

Conserving Evolutionary History

The US ESA is one of the most powerful pieces of conservation law in the world. Essentially, it simply states that species (including subspecies) should not be allowed to go extinct and that actions that lead to the further decline of listed species require special permission from the federal government. For evolutionary biologists, this means that the ESA seeks to protect one of the key products of the evolutionary process—species, including incipient species and subspecies. This theme of conserving the outcome of the evolutionary process is at the core of most conservation efforts. It is a very retrospective view of what to conserve—it requires that evolutionary biologists provide a clear picture of how many species, subspecies, and distinct population segments exist in a region, an indication of whether those populations are increasing or decreasing, and a measure of how distinct they are from one another. It is then up to the conservation community to take those results and use them to prioritize species and landscapes and try to preserve as much evolutionary history as possible.

Several specific approaches to conserving evolutionary history above and beyond the basic tenet that species extinction should be avoided are worth discussing in a bit more detail. First is the issue of retaining as much of a phylogenetic tree as possible—the tuatara problem that opened this essay. Two basic approaches have dominated the thinking on this topic. First, one can attempt to conserve the phylogenetic branch length (i.e., the sum of the branches of a phylogeny)—in units of time—that might be lost if an extinction event occurs. In the case of the tuatara, different opinions exist as to when it last shared a common ancestor with its closest relatives, but 271.5 million years seems to be a reasonable estimate. That is, if you lose the tuatara, you lose not only that species but also the 271.5 million years of evolutionary history that it uniquely represents among living organisms.

An alternative approach proposed by Faith (1992) is to conserve *phylogenetic diversity*, or PD. Faith proposed PD as an explicitly character-based approach to identifying taxa to conserve—those that have evolved lots of unique features (characters) contain more important evolutionary history than those that have changed relatively little and therefore remain relatively similar to other taxa. An example might be the human-chimp-gorilla trio of species. Although they all shared a common ancestor about 6 to 8 million years ago, the human lineage has changed considerably more, in a wide variety of biologically important ways, than has the chimp or gorilla from its most recent common ancestor. Thus, in this case, humans would have a far greater PD than the chimpanzee and would be a higher conservation priority if the two species were equally threatened. To the extent that unique features accumulate over time, branch lengths and PD will prioritize species for conservation in the same order. However, evolution does not always proceed in a tidy, time-dependent manner, as the human example points out. In those cases, one must decide what matters most—time or evolutionary novelties—as a target for conservation.

A very different approach is to recognize that the primary reason for human-mediated extinction is habitat alteration and destruction, and that certain landscapes or regions tend to accumulate a great number of unique organisms. Certain regions of the world, like New Zealand or the Appalachian Mountains of eastern North America, are rich in endemic taxa found nowhere else on earth. The same can be said, of course, for many parts of the world—the Amazon basin is also very species rich, and most of the species found there are restricted to the Amazon. However, regions like New Zealand contain a disproportionately large number of old, unique lineages—like tuatara, flightless kiwis, and southern beech forests—while the Amazon abounds in species that are often members of widespread tropical genera. Many factors can contribute to these patterns, including the geological age and stability of a landscape, its isolation from other parts of the world, and the extent to which humans have disrupted ecological processes that naturally occur in the area. One emerging area of conservation phylogenetics is to mathematically combine phylogenetic distinctiveness and level of endangerment into a single *Evolutionarily Distinct and Globally Endangered* (EDGE) statistic to prioritize species. As phylogenies, and particularly as time-calibrated evolutionary trees, accumulate for the world's fauna and flora, conservation biologists can prioritize those regions that harbor the greatest depth and breadth of the tree of life by their EDGE scores and

protect them into the future. In so doing they are saving the most endangered species, but they are also saving the longest branches in the
tree of life.

Conserving the Potential for Future Evolutionary Change

In 1997 Tom Smith and colleagues promoted a very different approach
to using insights from evolutionary biology to conserve biodiversity.
Smith argued, based on an analysis of a dozen populations of an African
bird species, that there is a tremendous level of morphological differentiation between birds in the forest and the same species in the *ecotone*
between forest and savanna habitats, and that this variation persists in
the face of ongoing movement of individuals and gene flow. While the
argument is fascinating in its own right, Smith and colleagues took it
one step further, arguing that the ecotone habitat selects for morphologically very different birds from those in the rain forest. Their conclusion was quite radical: if you want to preserve evolutionary processes
that generate diversity, you should preserve ecotones in addition to pure
rain forest.

The importance of preserving habitats critical for the functioning of
normal evolutionary processes within species has gained considerable
traction in the last decade. The entire discipline of *landscape genetics*,
including the emerging subdiscipline of *landscape genomics*, focuses on
exactly this issue, and it represents one of the major growth areas in research on the genetics of natural populations. Here, the goal is to use
standard population genetics data, in combination with geographic information system (GIS) data layers, to quantify the ways that organisms
move across the landscapes they occupy. The approach is directly relevant to landscape management and conservation planning because it
takes genetic data from organisms on landscapes and asks whether potential migration corridors and barriers to gene flow function to promote or to disrupt population connectivity. The results can provide unexpected insights into the ways organisms use their environment and
can identify those habitat patches and corridors that are most important
for maintaining normal evolutionary processes. Such data can take years
to collect with traditional mark-recapture methods, but only weeks or

months using the insights gained from landscape genetics. Particularly for threatened species, for which decision makers must act quickly, landscape genetics is a powerful conservation tool.

In a related area of conservation biology, researchers have begun to focus on evolution itself as a management tool to enhance a population's ability to respond to human-induced environmental change. Such population *rescue* comes in at least three forms. *Demographic rescue* has been the most commonly used to date, and simply involves adding new individuals to a declining population to bolster its numbers. However, it is not explicitly evolutionary, and therefore is not discussed further in this chapter.

Genetic rescue, whereby individuals are added to a population to introduce new genetic variation and counter inbreeding depression, has been successfully implemented a number of times in populations with extremely small effective population sizes. Probably the best known example of genetic rescue involves the stunning recovery of the Florida panther, *Puma concolor coryi*. Designated the state animal of Florida in 1982, this isolated subspecies of panther (also known as puma, mountain lion, or cougar in other parts of the species' vast range) was reduced to about 20 individuals in the 1970s, at which time it was showing clear signs of genetic inbreeding depression and was sliding toward extinction. In response, the population was intentionally supplemented with panthers from the adjacent, much larger, and more outbred population in east Texas from the subspecies *Puma concolor stanleyana*. The two panther subspecies interbred, the telltale signs of inbreeding depression disappeared, and Florida's state animal now appears to be in a strong phase of population growth and genetic recovery. This case also had its downside (see later), but it is without question an example of the importance of adequate genetic variation for population viability.

Evolutionary rescue involves the genetic response of a population to a novel selective environment that otherwise would have caused it to go extinct. Evolutionary rescue may well be the most far-reaching form of rescue, since its success rests on an evolutionary change leading to a self-sustaining, demographically healthy population or species. Although the population genetic theory of evolutionary rescue is well developed, actual empirical examples for conservation are still rare. Given the impact of people on natural landscapes, human activities ranging from habitat modification and climate change to introduction of invasive species clearly exert strong selection pressures on populations in the wild, and many populations and species will have to either evolve or be lost. One fascinating example comes from the intertwined fates of human fisheries, conservation actions, and evolutionary rescue. Fisheries inevitably remove the largest individuals first, which are generally old females with very high reproductive outputs. This intense selection on the largest, most fecund females leads to a well-documented evolutionary decrease in body size, an earlier age at first reproduction (since having a few babies at a small body size is a better strategy than waiting and being caught by a fisherman before reproducing at all), and a lower total biomass and reproductive output for individuals and the species overall.

To counteract this evolutionary response, marine protected areas (MPAs) have been established in many parts of the world. The idea is that these no-fishing zones will offset human fishing pressure, keeping stocks of large-sized females, and their large-sized genes, available to restock depleted fishing grounds that have evolved smaller, less fecund females. Recent models on the effects of MPAs in preventing these conservation disasters have found that some MPAs ameliorate evolutionary responses to human fishing, but others do not. If there is extensive gene flow between fishing grounds and MPAs, then smaller females will migrate to, and breed in the MPA, and larger, protected females will migrate outside MPAs, where they will be caught and killed. In this case, little is gained genetically or evolutionarily with the MPA. Alternatively, if the MPA is very large, or if the adult fish tend not to migrate extensively, then the larger, protected females will remain in the MPA and provide a constant source of young, genetically unmodified individuals, some of which may migrate out to the fishing grounds. In that case, the MPA has provided the needed genetic variation to evolutionarily rescue the exploited fishery.

Clearly, evolutionary rescue can and does occur. The question for conservation biology is how to better apply the power of evolutionary adaptation to novel environments for critical conservation targets. Humans now constitute a potent force leading to rapid evolutionary change, and future conservation efforts that introduce specific alleles that allow, for example, threatened species to adapt to climate change or environmental pollutants should become a part of the evolutionary conservation toolkit. We clearly need to add ourselves to the list of important processes at the intersection of evolution and conservation.

3. THE ENEMIES TO WATCH OUT FOR

Conservation biology is a complex business that involves equal parts of biology, politics, and economics if real progress is to be made and sustained. An essential element of conservation is to think clearly about what one wants to protect and what one wants to avoid in terms of conservation outcomes. Evolutionary genetics in particular has made very substantial contributions to the identification of these problems and their solutions. For example, the Florida panther is widely touted as one of the best examples of genetic rescue in the wild. Except—is the subspecies really recovering? There are definitely healthy panthers back in the Florida Everglades, but are they *Puma concolor coryi*? Or has that subspecies been driven extinct by an invasive hybrid panther that was intentionally created by well-meaning wildlife managers? The example raises the critical question, what does one want to conserve, and why? Is one protecting native genes, naturally evolved lineages, ecological roles, or something else entirely? Is it better to keep some native Florida panther genes on the Florida landscape than none at all (the hybrid panthers definitely have a lot of native Florida genes), or is hybrid "impurity" worse than extinction? Evolutionary genetics can provide the data, but not the answers, to the moral dilemmas that these questions pose.

Hybridization

Hybridization happens all the time, both because of natural processes and because humans meddle with species and landscapes. Evolutionary genetics can bring great clarity to the status of populations of plants and animals, including precise estimates of the fraction of the genome that is native versus derived from a different species. Most practitioners naturally assume that the primary goal of conservation biology is to preserve pure genetic lineages on the landscapes where they evolved. Thus, in a case that our lab has worked on for several years, human-transported, nonnative barred tiger salamanders (*Ambystoma tigrinum mavortium*) from Texas and New Mexico have successfully hybridized with native, endangered California tiger salamanders (*A. californiense*) across much of central California, and at least 20 percent of the range of the California tiger salamander is now occupied by hybrids (Fitzpatrick et al. 2010). In this and many other cases, hybrids are viewed as a conservation threat, to be identified, eliminated, and replaced with pure natives if at all possible. Trout, salmon, escaped genes from agricultural plants, and domestic dog genes infiltrating wolf and coyote populations are a few of the better-studied examples of this phenomenon, and the evolutionary analysis of hybridization constitutes the key data on which conservation actions have been based. The contrasting view is that genetic rescue for populations that have dipped below a genetically sustainable size remains a viable and occasionally used strategy to augment populations that would otherwise go extinct without genetic intervention. These are cases in which evolutionary biology can provide the key insights on expected and realized inbreeding depression, can track the fate of nonnative genes as they move through populations, and can measure the fitness consequences of hybridization. What evolutionary biology cannot do is tell us, as managers of the fate of species, whether and when we should bring in foreign genes as a last-ditch conservation effort.

Population Bottlenecks, Population Isolation, and Effective Population Size

More books and papers discuss the interface of population genetics and conservation biology than any other aspect of evolutionary conservation biology. There are many reasons for this, but the primary one is that the connection between classical problems in population genetics and conservation biology is both direct and clear. Population geneticists tend to worry about the relationships between genetic drift caused by small effective population sizes, the efficacy of natural selection in shaping variation in the field, and the interplay among mutation, selection, and drift. Conservation biologists spend a great deal of time and energy trying to understand the health of populations in nature, including the effects of small population sizes. Both groups recognize that small populations have a higher chance of becoming inbred and that high levels of standing genetic variation are critical for the current and future health and adaptability of populations. Any good field ecologist knows that populations fluctuate over time and that low numbers are sometimes unavoidable. However, when populations become completely isolated, then migrants from larger populations cannot rescue those populations demographically or genetically. The result is small, isolated populations that lose genetic variation over time, become inbred, express deleterious mutations at a higher frequency, and have limited resilience to bounce back from unavoidable population crashes. And given their isolation, when they go locally extinct, they cannot be repopulated—they stay extinct.

One of the holy grails of both population genetics and conservation biology has been to infer past and current demographic parameters using the standing genetic variation that exists in natural populations. At least for genetic markers that are unaffected by strong natural selection (so-called neutral genetic variation), there is a long, rich history of using patterns of variation within and among populations to estimate the amount of migration (or gene flow) among populations. Here, the idea is straightforward—if a mutation arises in one population, and no migrants successfully leave that population and reproduce in a new population, then the mutation will remain exclusively in its site of origin. Such "private" variants will build up over time, such that the longer a population remains in isolation, the greater will be its genetic distinctiveness from other populations. Sewall Wright, one of the pioneers in the field of population genetics, developed a series of statistical methods to quantify this kind of genetic differentiation, and his F-statistics remain the primary way in which such realized gene flow is measured in nature. High values imply little or no gene flow among populations, whereas low values suggest that successful migration and breeding occur regularly. Newer methods can measure the movement of individuals by recognizing that if an occasional migrant moves between somewhat-differentiated populations, that individual can be "assigned" to its population of origin based on its multigene identity. Conducting such assignment tests with confidence requires a great deal of genetic (or genomic) data, but it represents a powerful addition to the conservation biologist's toolkit for measuring how organisms successfully traverse landscapes in nature.

Some of the most compelling and exciting new developments in population genetics allow conservationists to study, with far greater precision, the actual size of a breeding population in nature. Population biologists recognize two different ways to measure population size—N_e, or the effective population size, and N_c, the census population size. The difference is straightforward and absolutely critical: N_c is the number of individuals in a population, whereas N_e reflects the number of individuals who breed and contribute to the genetic variation in the species (note that this represents a simplification of a mathematically complicated concept). For example, if a population has 50 males and 50 females, its census size will be 100. However, if only one of those males and 10 of those females actually breed, its effective size will be close to 10; this latter population will suffer much greater genetic drift and potential inbreeding depression than one in which all 100 individuals breed. Both N_c and N_e are important, and they measure different aspects of the health of a population. Recent advances in molecular population genetics have provided new tools to measure N_e in nature, sometimes from only a single individual. The math is complex, and the requirements for both the number of genetic markers and the proportion of the population sampled may be large, but the results indicate that the effective population size can be estimated, often relatively easily and quickly.

4. WHAT GENOMICS BRINGS TO THE TABLE

Genomics means many things to many people, ranging from data on the full DNA sequence of an organism assembled into complete chromosomes to having "a lot of sequence data." A truly complete genome, in which every base pair has been sequenced and assembled into contiguous chromosomes, has yet to be completed for any vertebrate, although several species, including humans, have essentially complete genomes. However, an increasingly large number of species have had many thousands of genes sequenced, sometimes for multiple individuals and populations. In either case, genomics always means having lots of data for each study organism—it may mean billions of nucleotides of sequence data (many vertebrate genomes are around 2–3 billion nucleotides in length), or it may mean thousands, but it is always a lot.

It seems clear that in the next few years, genomic data will dominate population genetics, phylogenetics, and conservation genetics research. In 2008, sequencing a complete tortoise genome would have been inconceivable; in 2013, our lab fully sequenced 270 endangered Mojave desert tortoises (*Gopherus agassizii*) to quantify the effects of solar development on the species' genetic connectivity. The genomic future, even for endangered species, has arrived.

Aside from the general truism that more data are always better than fewer data, this onslaught of new information should open several critical avenues of research at the interface of evolutionary and conservation biology. First, genomic data allow one to study the genetics of functionally important genes as well as neutral ones not affected by natural selection. Presumably, conservation geneticists will increasingly focus on genetic variation in the functionally important genes, since they are most important to survival and the ability to adapt to future change. For neutral loci, genetic variation per se is not important to population health, but the standing levels of variation at those loci reflect the past and current effective population size and levels of gene flow or genetic isolation. Both are important to evolutionists and conservationists, but the two are very different. When a large part of the genome is subject to study, the neutral and selected loci can be neatly separated, leading to important insights from both genomic components.

In a similar vein, genomic data help the evolution and conservation community focus much more clearly on exactly what needs to be conserved. For example, the major histocompatibility complex (MHC) is a set of genes involved in the immune response to disease of many vertebrates. Certain diseases, including the fibropapilloma tumors in green sea turtles or the devil facial tumor disease in Tasmanian devils of Australia, may be involved in bringing these endangered taxa to the brink of extinction, and conserving and managing populations for MHC variation may be a way to increase the likelihood of evolutionary rescue. Genes that allow cold-adapted plants and animals to better cope with human-induced climate change in the next decade are another key class of functional genes that genomics may bring to the conservation table.

The impact of genomic data on evolutionary biology in general, and conservation in particular, is huge, multifaceted, and largely unexplored. It stands as perhaps the most important frontier at the intersection of evolution and conservation biology.

5. CONCLUDING THOUGHTS AND PROSPECTUS

Evolution is all about change—changes in allele frequencies over time, in population size and distributions, and in species composition due to extinction and speciation. Conservation is about managing for change— climate change, invasive species, hybridization, human habitat modifications, and a host of others. As large-scale genetic analyses become increasing available for nonmodel organisms, it seems inevitable that evolutionary genetic analyses will move to center stage in the conservation and management of declining species. Consider, for example, being able to track the reproductive output of captive-reared organisms that are repatriated into the wild, allowing resource managers to measure the impact of their conservation efforts, in the wild, in real time. Or imagine having the data to be able to determine, with very high accuracy, exactly how many individuals have moved between habitat patches historically, and using that information to mimic those patterns with human-assisted migration in fragmented habitats. Or being able to quantify, for any newly proposed protected park, exactly how much of the phylogenetic tree of life is contained in that park—not for specific taxa and a handful of genes, but for all life. These are heady ideas, but as genomics, metagenomics, and phylogenomics become affordable and easier to accomplish, they are also very realistic. And they just might help conserve a bit more of our declining biosphere.

FURTHER READING

Allendorf, F. W., and G. Luikart. 2007. Conservation and the genetics of populations. Malden, MA: Blackwell. *A wonderful reference, particularly for the more mathematical* aspects of conservation genetics. This book is particularly strong on the interface of population genetics and conservation.

Braude, S., and B. S. Low, eds. 2010. An Introduction to Methods and Models in Ecology, Evolution, and Conservation Biology. Princeton, NJ: Princeton University Press.

Carlson, S. M., C. J. Cunningham, and P.A.H. Westley. 2014. Evolutionary rescue in a changing world. Trends in Ecology and Evolution 29: 521–530. A clear, concise review of the differences between demographic, genetic, and evolutionary rescue, and their applications in conservation biology.

DeWoody, J. A., J. W. Bickham, C. H. Michler, K. M. Nichols, O. E. Rhodes Jr., and K. E. Woeste, eds. 2010. Molecular Approaches in Natural Resource Conservation and Management. New York: Cambridge University Press. An edited volume, with some very specific chapters that may be of limited general interest, but others with quite broad appeal. A great source of recent examples and case studies using a wide range of molecular genetic tools to inform conservation.

Faith, D. P. 1992. Conservation evaluation and phylogenetic diversity. Biological Conservation 61: 1010. The paper that introduced the phylogeny-based concept of conserving character evolution into the mainstream of conservation thinking.

Fitzpatrick, B. M., J. R. Johnson, D. K. Kump, J. J. Smith, S. R. Voss, and H. B. Shaffer. 2010. Rapid spread of invasive genes into a threatened native species. Proceedings of the National Academy of Sciences USA 107: 3606–3610. Following a well-documented, human-mediated introduction, this paper shows that some invasive genes can sweep across landscapes at incredibly rates, while other genes are much slower.

Frankel, O. H., and M. E. Soulé. 1981. Conservation and Evolution. Cambridge: Cambridge University Press. The original book that brought together the fields of conservation biology and evolutionary genetics—a "must-read."

Frankham, R., J. D. Ballou, and D. A. Briscoe. 2004. A Primer of Conservation Genetics. Cambridge: Cambridge University Press.

Höglund, J. 2009. Evolutionary Conservation Genetics. Oxford: Oxford University Press.

Schonewald, C. M., S. M. Chambers, B. MacBryde, and W. L. Thomas, eds. 2003. Genetics and Conservation. Caldwell, NJ: Blackburn.

CHAPTER 16

ADAPTATION TO A CHANGING WORLD: EVOLUTIONARY RESILIENCE TO CLIMATE CHANGE

Martha M. Muñoz and Craig Moritz

OUTLINE

1. Introduction
2. Plastic and genetic responses to climate change
3. Behavioral shifts in response to climate change
4. Species' range shifts in response to climate change
5. Assisted translocation: If you build it, will they evolve?
6. Predicting evolutionary response to climate change

The rate and magnitude of global climate change is imposing unprecedented levels of selection on the world's organisms such that the earth may be at the cusp of the sixth major extinction crisis. Over the past 75 years, the field of evolutionary biology has transformed from a traditionally historical science into a more predictive one. By examining past patterns and observing the process of evolution in real time, we can infer how organisms can adaptively respond to selection imposed by global climate change. Our knowledge of the evolutionary process can inform better conservation efforts. What we have discovered over the past decades is that (i) traits can shift remarkably fast in response to climate change, but the ability to rapidly evolve differs considerably among species and (ii) many species are tracking their preferred conditions up mountain slopes or to higher latitudes, a phenomenon also seen in the fossil record during previous warming episodes. When organisms cannot adapt or track habitats, they are highly vulnerable to extinction. Connectivity among populations facilitates dispersal of adaptive alleles and habitat tracking, but increasing habitat fragmentation poses a great threat to both these processes. In cases where habitat fragmentation is severe, human-mediated mitigation efforts such as assisted translocation will be required to enable species' survival.

GLOSSARY

Acclimation. Adjustment to the environment through plastic shifts in trait values.

Adaptation. A genetically determined trait that confers higher fitness in an environment.

Assisted Migration. The human-mediated translocation of individuals across space to enhance population (and species) viability and resilience.

Habitat Tracking. Changes in species' geographic ranges over time to track preferred habitat conditions.

Outbreeding Depression. A decline in fitness for crossed individuals from increasing distance of relatedness.

Phenology. The timing of a major life history event, such as flowering, mating behavior, or metamorphosis.

Phenotypic Plasticity. The ability of a genotype to express different phenotypes as a result of the environment it experiences.

1. INTRODUCTION

Throughout earth's 4.5-billion-year history, its climate has continually been in flux, with periodic major changes that have contributed to pulses and dips in the diversity of life on the planet, sometimes including high rates of extinction. Dramatic increases in temperature and greenhouse gases marked the end of the Permian (~252 Ma), which, along with other factors, precipitated an extinction event that eventually led to the death of nearly all the world's species. Today's organisms are all descended from those few that survived the crisis. After a large asteroid hit earth at the end of the Cretaceous (62.5 Ma) and spewed steam, dust, and ash into the atmosphere and blanketed the planet in debris, the ensuing rapid climate change led to the extinction of nonavian dinosaurs and many other organisms.

If the world's climate has continually been changing, and species "naturally" come and go, then we must reasonably ask whether the current episode of climate change presents a particularly pernicious risk to global biodiversity. The answer is a resounding yes. The pace and potential magnitude of contemporary global warming could be unprecedented in the geological record. Over the last century land and sea-surface temperatures have risen 1 °C overall, and the rate of warming has doubled. The frequency and ferocity of droughts and heat waves have also dramatically increased. Even if the strictest emission mitigation efforts are employed, global temperatures are predicted to rise an additional 1.5 °C over the next century. Past episodes of climate change were typically thousands of years long; today, comparable changes are being recorded in decades. What is particularly alarming for scientists is the rate at which large changes in climate are accruing. Can species evolve fast enough to meet the pace of climate change?

In *On the Origin of Species*, Charles Darwin laid the foundation for our current understanding of the evolutionary process. Although Darwin has been proven right about many aspects of evolution, he was wrong about the pace at which evolution can occur, which he believed occurred over long time spans. However, many studies in the past half century, such as those on the Galápagos finches that bear his name, have demonstrated convincingly that evolutionary change can occur rapidly—in many cases over only a few generations. That evolution can be extraordinarily rapid may be promising in terms of species' survivability, but one question prevails: Can *adaptation* to climate change meet the swift pace of environmental change predicted over the coming century?

A basic tenet of evolutionary theory/biology is that species need time to respond to selection; the amount of time depends on certain properties of the organisms, as will be discussed in this chapter. But it is clear that the current pace of global climate change is so dramatic that life on earth may be at the brink of a major extinction crisis. Hence, determining how organisms can respond to climate change is perhaps the most important biological question of our generation (figure 16-1). What we've learned so far is that it isn't all gloom and doom: some species are responding to climate change and, some of these, by rapid adaptation (see table 16-1). The major questions that remain are how best to assist

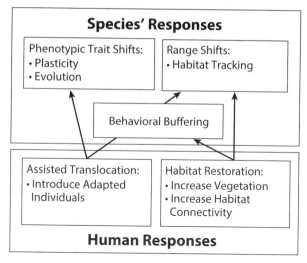

Figure 16-1. Response of species to climate change (top) and how humans can help them survive (bottom). See the text for specifics.

species through mitigation efforts, particularly those species that cannot adapt as quickly as others, and whether our efforts will be sufficient to avert a biodiversity crisis.

2. PLASTIC AND GENETIC RESPONSES TO CLIMATE CHANGE

Over the past several decades, many studies have documented phenotypic shifts in life history, morphology, and physiology linked to global climate change. Shifts in *phenology*—the timing of life history events such as reproduction—account for a large proportion of observed phenotypic changes. Many types of organisms are initiating springtime activity earlier in the year: on land, plants are leafing or flowering earlier; and in the oceans, phytoplankton blooms have significantly advanced. Fish and birds are initiating seasonal migrations earlier, arthropods and amphibians are significantly advancing their first emergence, and mammals are coming out of hibernation sooner. Many types of organisms—insects, amphibians, mammals, and birds, to name a few—are exhibiting earlier breeding and nesting behavior. In some cases, organisms are also delaying winter dormancy or autumn migrations.

Although phenological changes are certainly the most readily observed, they are not the only phenotypic shifts detected. For instance, some corals and seagrasses have adjusted their physiology to withstand higher water temperatures that would otherwise kill them. And life history traits are changing in numerous species. For example, some amphibians and mammals have smaller clutch sizes and litter sizes, respectively, than those of a few decades ago. Body size at sexual maturity is now considerably smaller in various amphibians, mammals, and freshwater invertebrates, among others. Body size of birds is shrinking, and plumage is darkening for those that live in regions with less winter snow cover. In some reptiles that exhibit temperature-dependent sex determination, rising temperatures are biasing sex ratios toward females.

In short, it is clear that many species are exhibiting phenotypic shifts in response to climate change. Nonetheless, a central question prevails: Are the observed shifts *evolutionary* adaptations to new conditions? Adaptive evolution occurs when natural selection leads to changes in the genetic makeup of a population. For example, following a major drought on the Galápagos island of Daphne Major, larger and thicker-billed finches (*Geospiza fortis*) were better able than smaller individuals with narrower beaks to forage the hard woody fruits that remained. Size and beak shape are highly heritable: by the next year the frequency of large, thick-billed birds had increased substantially, indicating that adaptive evolution had occurred.

However, not all the observed responses to climate change are necessarily the result of genetic change due to natural selection. In many cases, organisms with the same genotype can produce different phenotypes in response to changes in their environment, a phenomenon known as *phenotypic plasticity*. For example, when *Daphnia* water fleas are grown in the presence of predators, they develop defensive head shields and tail spines.

Distinguishing between these two mechanisms—phenotypic plasticity and adaptive evolution—is challenging because many, if not most, of the genes underlying phenotypic trait shifts still remain poorly understood. Nonetheless, there are many examples in the literature of trait shifts confidently ascribed to genetic changes, plastic variation, or both (table 16-1).

Table 16-1. A summary of examples of adaptive evolution in response to climate change

Organism	Location	Observed Trait Shift
Plants		
Field mustard (*Brassica rapa*)	Southern California	Earlier flowering time
Mediterranean thyme (*Thymus vulgaris*)	Southern France	Increase in phenolic anti-herbivory compounds
Wild cereal (*Triticum dicoccoides* & *Hordeum spontaneum*)	Israel	Earlier flowering time
Invertebrates		
Pitcher plant mosquito (*Wyeomyia smithii*)	Eastern North America	Decrease in photoperiod response
Asian tiger mosquito (*Aedes albopictus*)	United States	Decrease in photoperiod response
Water strider (*Aquarius paludum*)	Japan	Decrease in photoperiod response
Tabletop coral (*Acropora hyacinthus*)	American Samoa	Increased heat resistance
Two-spotted lady beetle (*Adalia bipunctata*)	Netherlands	Reduced melanism
Vertebrates		
Pink salmon (*Oncorhynchus gorbuscha*)	Alaska, USA	Earlier migration timing
Sockeye salmon (*Oncorhynchus nerka*)	Northwestern USA	Earlier migration timing
Tawny owl (*Strix aluco*)	Southern Finland	Increase in darker plumage coloration

One key finding is that different mechanisms can create the same phenotypic pattern. For example, in response to rising temperatures, both great tits (*Parus major*) from the United Kingdom and field mustard plants (*Brassica rapa*) from North America have advanced the timing of their reproduction, but through different mechanisms. British great tits have advanced their egg-laying date by approximately two weeks over the past half century to tightly match shifts in the abundance booms of their caterpillar prey. The shift in egg-laying behavior is not the result of genetic changes caused by natural selection; rather, the timing of egg-laying behavior is plastic, and the birds take their timing cues from the environment. Although the overall trend is that spring temperatures are rising, year-to-year temperatures still fluctuate a bit, and the activities of both prey and birds are tightly linked with these fluctuations: during slightly cooler years, they delay emergence and reproduction.

Conversely, field mustard plants have genetically advanced their reproduction in Southern California, where strong droughts occurred from 2000 to 2004. When a drought curtails the wet season, plants that fail to bloom in time will not be able to reproduce, and so there is selection for earlier reproduction during dry years. Thus, postdrought field mustard plants flowered up to 9 days earlier than predrought plants. In contrast with the great tits, field mustard plants have advanced their flowering date via genetic changes. Even when postdrought plants are grown under wet conditions, they still flower early, indicating that the shift is genetically controlled. Through these genetic shifts, mustard plants advanced their first flowering date by approximately one week to successfully reproduce in spite of droughts.

Another key finding is that a single phenotypic shift can sometimes be driven by both plasticity and genetic adaptation, as in the case of resistance to heat stress in corals. Across the globe, corals—and the diverse communities they support—are quite literally "in hot water" owing to bleaching events initiated by spikes in ocean temperatures. Bleaching occurs when heat stress forces corals to eject the photosynthetic algae that live within them and provide them with nutrition. Some Pacific *Acropora* corals, however, are highly resistant to heat and acidification. Such resistant corals have acclimated to higher temperatures and are more efficient at retaining chlorophyll, the food produced by their symbiotic algae. Resistant corals increase their production of proteins that maintain normal cellular functioning during heat stress (heat-shock proteins), which helps them persist in waters with temperatures that exceed their expected thermal tolerances. Furthermore, resistant corals exhibit fixed differences—that is, unrelated to environmental conditions—in gene expression relative to vulnerable corals, suggesting that adaptive evolution is also influencing their heat resistance.

Although climate change is a pernicious threat that stresses most animals, in some cases organisms have evolved to capitalize on the potential benefits of higher environmental temperatures. Many temperate organisms go dormant over the winter, when resources become scarce, but the milder autumn and winter conditions of the changing climate are prolonging growing seasons. Pitcher plant mosquitoes (*Wyeomyia smithii*) have genetically delayed the onset of their winter dormancy (*diapause*) to exploit the longer growing seasons. Mosquitoes initiate diapause as days shorten, when hours of daylight reach a critical threshold, termed *critical photoperiod*. But through genetic shifts, these mosquitoes have extended their critical photoperiod to maintain activity and capitalize on food resources even as day length shortens. Remarkably, the shift in critical photoperiod occurred over only 5 years, indicating rapid adaptation to changing seasonal conditions.

Like the pitcher plant mosquitoes, Mediterranean thyme (*Thymus vulgaris*) plants have also evolved to capitalize on rising temperatures, but in this case to bolster their defenses against herbivores. Among the compounds that thyme plants produce to deter herbivores, phenols are the most effective, but they make the plant vulnerable to freezing. Release from selection—fewer freezing nights during the winter—enabled thyme plants to genetically increase phenol production to deter herbivores. As winter temperatures have increased in southern France, so, too, has the prevalence of phenolic compounds in populations of thyme. Since the 1970s, phenols are now found in populations where they were historically absent, and in greater frequency where previously they were rare.

A few decades ago almost nothing was known about how organisms were responding to climate change. Now, there is abundant evidence for phenotypic shifts and, in several cases, concrete evidence for adaptive evolution (table 16-1). Not all organisms, however, are equally evolvable. Evidence for adaptive evolution in mammals, for example, is exceedingly scant. Only one study has claimed to find adaptive evolution in response to climate change: in American red squirrels (*Tamiasciurus hudsonicus*) first parturition date advanced by several weeks in only a decade. However, subsequent work on this species was unable to confidently link the shift in phenology to climate change.

In general, and given suitable genetic variation, the ability to evolve rapidly appears to be a function of population size and population growth rate. All else being equal, some populations with large numbers of breeding individuals, short generation times, and a high potential growth rate have been shown to adapt within a few years. Owing to their relatively long generation time and small population sizes, mammals should be particularly unsuited to evolve on the timescale necessary to match the pace of climate change. Mammals are not unique in this regard: many long-lived organisms or those with low population densities—which include many vertebrates, invertebrates, and plants—should also exhibit low evolutionary potential.

3. BEHAVIORAL SHIFTS IN RESPONSE TO CLIMATE CHANGE

Phenotypic trait shifts are just one way in which organisms can respond to climate warming, and the ability to do so, either through plasticity or adaptation, varies greatly among species (figure 16-1). In many cases, however, species do not need to shift their phenotypes to match new conditions if they can alter their activity patterns and habitat use to continue to experience their historic environmental conditions. Through behavior, organisms can buffer the effects of environmental change. For example, if species cannot adapt to physiologically tolerate more heat, they can avoid being active during the hotter times of day.

Abundant evidence indicates that organisms can behaviorally respond to rising environmental temperatures. For example, in warm habitats water dragons (*Intellegama lesueurii*), a type of Australian lizard, dig deeper nests to access cooler soils, which allows them to maintain relatively constant incubation temperatures for their clutches in different thermal environments. In vegetated habitats, banded grove snails (*Cepaea nemoralis*) seek shade to avoid stressfully hot temperatures. In less vegetated habitats such as sand dunes, however, the snails lack the access to shade necessary to thermoregulate; instead, they have evolved lighter-colored shells to absorb less heat.

As with the banded snails, access to shade will be critical for various types of diurnal organisms to behaviorally regulate their body temperature, especially in the tropics where temperatures are already quite high and animals appear to be living near their physiological limits. However, in many tropical areas there have also been dramatic decreases in vegetation cover due to habitat conversion and increasing fire intensity, restricting access to cool microhabitats. We underscore that mitigation efforts for climate change and those for habitat restoration are intrinsically linked: Conserving and restoring the structural features of the habitat, particularly vegetation complexity, will enhance behavioral resilience to climate change.

Often, however, behavioral changes are not enough to buffer the effects of environmental warming. For instance, the three-lined skink (*Bassiana duperreyi*) of Australia has adjusted to rising temperatures by digging deeper burrows. This species has also shifted its phenology by breeding several weeks earlier. Nonetheless, the combination of behavioral compensation and advanced breeding phenology has not been enough to fully accommodate environmental change; *B. duperreyi* exhibits temperature-dependent sex determination, and incubation temperatures are now high enough in this species that most offspring are likely to be female.

Higher environmental temperatures also appear to be leading to a greater frequency of females in the bearded dragon (*Pogona vitticeps*) of Australia. In this lizard, sex is determined both genetically (i.e., via chromosomes) and environmentally; females result from higher incubation temperatures. In recent years, however, scientists have been detecting high numbers of females with male chromosomes, indicating that high incubation temperatures have shifted control of sex determination from genetics to the environment. Because females with male chromosomes can successfully mate with male lizards, climate change could lead to the loss of the female-specific chromosome (i.e., the W chromosome).

Even when behavioral shifts can occur, it is likely that they incur potential fitness costs. For example, North American elk (*Cervus elaphus*) from montane deserts consistently foraged in areas where they could avoid thermal stress, although those environments had poorer-quality food. Similarly, to avoid being active during the hottest times of day, some animals forage more at night. In turn, this behavior may increase their exposure to nocturnal predators or parasites. When organisms retreat to burrows during the hottest times of the day, they miss out on opportunities for foraging and reproduction. As these costs mount, populations will be increasingly unlikely to maintain positive growth and will be at a greater risk of extinction.

4. SPECIES' RANGE SHIFTS IN RESPONSE TO CLIMATE CHANGE

When behavioral adjustments and phenotypic trait shifts, whether genetic or plastic, cannot match the pace of environmental warming, the only option for populations to avoid extinction is to disperse to suitable habitat, a phenomenon termed *habitat tracking*. The fossil record indicates that such geographic range shifts were a dominant response to prior climate change episodes, including the warming event that occurred during the Pleistocene-Holocene transition. Likewise, contemporary climate change is precipitating dramatic changes in the distribution of species in both the marine and terrestrial realms. Over the past several decades, for example, terrestrial species have been shifting poleward in latitude and uphill in elevation. Shrubs are now present in Arctic habitats from which they were previously absent; similarly, lichens and mosses have newly expanded into more southerly latitudes on Antarctica. Many different types of animals—birds, butterflies, amphibians, and reptiles, to name only a few—have moved to cooler habitats. Armadillos have recently expanded into the northern half of the United States. On average, land plant and animal ranges have shifted 10 m higher in elevation and a staggering 17 km higher in latitude per decade.

In the marine realm, species have also been moving poleward, often at even more dramatic rates. On average, bony fish and phytoplankton have shifted 72 km poleward per decade. Zooplankton communities along the North Atlantic coasts are increasingly including more tropical species. When latitudinal shifts are not possible, such as in the Gulf of Mexico, which is oriented east to west, species are moving to deeper waters.

Nevertheless, whereas many species are tracking their climatic requirements uphill or poleward, many are not, and the reasons for this are not yet clear. Perhaps the latter are adapting in situ, or their movement across landscapes is inhibited. Understanding this variation in responses is now a key research question.

How fast a species' range changes depends on its dispersal capacity and habitat specialization. Highly mobile generalists, for example, can shift their ranges more rapidly than can sessile habitat specialists. Because there is such variability in species' range shifts, the composition of communities is changing. Species that used to not coexist are being brought together, and others that used to co-occur are being separated. These novel community associations are leading to new selective pressures as species adapt to the presence of novel predators, prey, mutualists, and pathogens.

The ability of species to respond to climate change by shifting their range is limited by continuity of suitable habitat. Gaps between suitable habitats larger than a species' dispersal capability may prevent it from tracking suitable environments. Consequently, the long-term survival of many species will depend on their ability to respond to climate change.

5. ASSISTED TRANSLOCATION: IF YOU BUILD IT, WILL THEY EVOLVE?

As discussed previously, most species have at least some capacity to respond to climate change phenotypically or behaviorally, or to disperse to suitable habitats. However, the rate of environmental change is so rapid that many species may not be able to keep pace because of life history or genetic constraints, or fragmentation of habitats that reduces dispersal potential, or both. For ecologically important or threatened species subject to these constraints (such as large mammals), many scientists and conservation managers are actively considering *assisted migration*—the deliberate movement of individuals (or their gametes) from more- to less-adapted populations, that is, from localities that will become unsuitable to those where suitable conditions will be created as warming continues (e.g., from lower to higher altitude or latitude).

Genetic translocations have been used in ecological restoration projects to introduce individuals with adapted alleles to new habitats. Conservation managers in British Colombia have already acted on the risk of climate change and translocated the seeds from various species of trees to habitats predicted to be suitable with continued warming. This practice is in sharp contrast to the usual emphasis on using locally sourced seeds or individuals in restoration ecology and wildlife translocations to retain "local adaptation." Longer-distance translocations are sometimes viewed as risky, because mixing widely separated populations might result in less-fit hybrids or loss of local adaptation. But in practice, this more conservative view may not meet the challenge of rapid climate change. The risk of *outbreeding depression*—lower fitness for individuals from genetically distant parents—could be outweighed by the benefit of introducing genes that are preadapted to future climate conditions, especially for long-lived organisms.

Based on simple genetic theory, the optimal solution would be to introduce a small proportion (e.g., 5%–10%) of nonlocal individuals from preadapted populations—for example, from hotter or drier locations if that is the direction of climate change. The introduction of these individuals will lead to a more adapted local population. For species with reasonably short generation times, adapted alleles should quickly increase in frequency. To reduce the risk of outbreeding depression, such translocations should also mimic the historical extent of gene flow, meaning that the distances over which individuals are translocated should not exceed the distance over which gene flow occurred previously.

All this methodology relies on identifying species for genetic management, giving priority to those thought to be most exposed and least able to adjust in situ. Once target species have been identified, it then needs to be determined (i) how climates currently vary across a species' range and how they are projected to change in the future, (ii) how the relevant genetic diversity is stored across that range, and (iii) what capacity the traits and populations have to respond to selection. Addressing these challenges is currently an active field of research. In short, identifying suitable source populations for assisted translocation should focus on introducing adaptive variation and mimicking ancestral levels of gene flow while minimizing the effects of deleterious alleles and inbreeding depression.

6. PREDICTING EVOLUTIONARY RESPONSE TO CLIMATE CHANGE

From both theory and evidence, we now have firm evidence that rapid adaptation to climate change is occurring, but conditions strongly favor taxa with short generation times and high growth rates. As described above, similar adaptive responses, such as shifts in reproductive timing, can occur through plasticity (as in great tits) or through genetic evolution (as in field mustard). To make better forecasts of how organisms will respond to climate warming, we need a better understanding of the conditions under which each of these mechanisms will occur. Further, we need to address the disconnect between theory and evidence relating to the evolution of plasticity, and how adaptation, behavior, and dispersal interact to determine population viability.

What is clear is that given accelerating rates of climate change, long-lived organisms are less likely to adapt in situ. This means that many of the earth's keystone species, including top predators, are particularly at risk for extinction. Without these organisms, ecosystems can collapse. For these taxa, management to maximize access to cool microhabitats in existing populations and to enhance dispersal potential is the best option. Specifically, protecting and expanding forested areas, particularly in the tropical regions of the world, will go a long way toward protecting these organisms, and the ecosystems that rely on them, from extinction. For species that are intrinsically vulnerable and with reasonably short generation times, assisted migration is also an important option to consider.

FURTHER READING

Bradshaw, W. E., and C. M. Holzapfel. 2001. Genetic shift in photoperiodic response correlated with global warming. Proceedings of the National Academy of Sciences USA 98: 14509–14511. *An empirical study showing rapid evolution in photoperiod response in pitcher plant mosquitoes.*

Chen, I. C., J. K. Hill, R. Ohlemuller, D. B. Roy, and C. D. Thomas. 2011. Rapid range shifts of species associated with high levels of climate warming. Science 333: 1024–1026. *A study documenting extensive range shifts among terrestrial species in response to climate change.*

Franks, S. J., S. Sim, and A. E. Weis. 2007. Rapid evolution of flowering time by an annual plant in response to a climate fluctuation. Proceedings of the National Academy of Sciences USA 104: 1278–1282. *An empirical study demonstrating rapid evolution in flowering time in field mustard in response to increased droughts.*

Hampe, A., and R. J. Petit. 2005. Conserving biodiversity under climate change: The rear edge matters. Ecology Letters 8: 461–467. *A review examining methodological practices for successful assisted migration efforts.*

Hoffmann, A. A., and C. M. Sgró. 2011. Climate change and evolutionary adaptation.

Nature 470: 479–485. *A review examining the features that underlie evolutionary potential, and how we can test for it in nature.*

Huey, R. B., M. R. Kearney, A. Krockenberger, J.A.M. Holtum, M. Jess, and S. E. Williams. 2012. Predicting organismal vulnerability to climate warming: Roles of behavior, physiology and adaptation. Philosophical Transactions of the Royal Society B 367: 1665–1679. *A review discussing how plasticity, adaptive evolution, and behavior interact to predict climate change vulnerability.*

Kearney, M., R. Shine, and W. P. Porter. 2009. The potential for behavioral thermoregulation to buffer "cold-blood" animals against climate warming. Proceedings of the National Academy of Sciences USA 106: 3835–3840. *A modeling-based study demonstrating that effective behavioral thermoregulation against climate change will hinge on the availability of shade, particularly at lower latitudes.*

Long, R. A., R. T. Bowyer, W. P. Porter, P. Mathewson, K. L. Monteith, and J. G. Kie. 2014. Behavior and nutritional condition buffer a large-bodied endotherm against direct and indirect effects of climate. Ecological Monographs 84: 513–532. *An empirical study showing that there are trade-offs between thermoregulatory behavior and foraging behavior in the North American elk.*

Merilä, J., and A. P. Hendry, eds. 2014. Special Issue: Climate change, adaptation and phenotypic plasticity. Evolutionary Applications 7:1–191. *Reviews on evidence for trait shifts due to plasticity and adaptation in a wide variety of plants and animals.*

Moritz, C., and R. Agudo. 2013. The future of species under climate change: Resilience or decline? Science 341: 504–508. *A review discussing the responses of organisms to past climate change episodes through range shifts and their relation to species' survivability in the current climate change event.*

Palumbi, S. R., D. J. Barshis, N. Taylor-Knowles, and R. A. Bay. 2014. Mechanisms of reef coral resistance to future climate change. Science 344: 895–898. *A synthesis of various empirical studies by the authors detailing plastic and genetic mechanisms underlying increased heat resistance in some Pacific corals.*

Seebacher, F., C. R. White, and C. E. Franklin. 2015. Physiological plasticity increases resilience of ectothermic animals to climate change. Nature Climate Change 5: 61–66. *A review discussing how phenotypic plasticity can help species survive climate change, and how plasticity differs between terrestrial and marine biomes.*

PART IV

....

Evolution in the Public Sphere

CHAPTER 17

EVOLUTION AND RELIGION: CONFLICT AND DIALOGUE

Francisco J. Ayala

OUTLINE

1. The argument from design
2. Natural theology and the Bridgewater Treatises
3. Evolutionary glimpses
4. Darwin's revolution
5. Evolution and the Bible
6. The problem of evil
7. Evolution: Imperfect design, not intelligent design
8. Evolution and the New Atheists
9. Evolution and religious beliefs

Theologians and other religious authors have over centuries sought to demonstrate the existence of God by the argument from design, which asserts that organisms have been designed and that only God could account for the design. Its most extensive formulation is William Paley's *Natural Theology* (1802). Darwin's (1859) theory of evolution by natural selection disposed of Paley's arguments: the adaptations of organisms are outcomes of a natural process that causes the gradual accumulation of features beneficial to organisms and accounts for the evolution of new species. There is "design" in the living world, but the design is not intelligent, as expected from an engineer, but imperfect and worse: defects, dysfunctions, oddities, waste, and cruelty pervade the living world. Science and religious faith need not be in contradiction. Science concerns processes that account for the natural world. *Religion* concerns the meaning and purpose of the world and of human life, the proper relation of humans to their Creator and to each other, and the moral values that inspire and govern people's lives.

GLOSSARY

Evolution. Hereditary change and diversification of organisms through the generations.

Intelligent Design. The idea that adaptations of organisms are designed by an intelligent author (= God), rather than resulting from natural processes.

Natural Selection. Differential reproduction of alternative genetic variants.

Problem of Evil. The challenge of explaining the presence of physical evil (e.g., earthquakes that kill millions of people) and biological evil (e.g., the cruelty of predators) if they are designed outcomes of an omnipotent and benevolent Creator.

Religion. Faith in and worship of God or the supernatural.

1. THE ARGUMENT FROM DESIGN

The Abrahamic religions—Christianity, Islam, and Judaism—proclaim the existence of a Supreme Being, a Creator, who accounts for the origin of the world and presides over everything that happens in it. Total world membership in these three religions amounts to more than 4 billion people, or more than half the human population. Other major religions with numerous members, accounting perhaps for about one-third of the human population, are the oriental religions. Among them Hinduism is the dominant religion in south Asia, particularly in India and Nepal, with about one billion adherents worldwide. Hinduism is the synthesis of several religious traditions, without a single God or Creator, although a majority of Hindus recognize Vishnu, Shiva, and Devi as different aspects of a Supreme Being or Brahman. Variations of Hinduism are Confucianism and Taoism, particularly in China. Shintoism is Japan's most distinctive religion, although about one-third of Japan's population is Buddhist. Buddhism is a nontheistic religion that includes a variety of beliefs and spiritual practices, mostly attributed to the teachings of Gautama Buddha, who lived around the fifth century BCE. There are, in addition, numerous tribal religions, particularly in

Africa and the Americas, but also in Australia, Asia, and Indonesia, each with idiosyncratic myths, small membership, and limited geographic presence.

Most religions, other than Judaism, Christianity, and Islam, do not profess faith in a single Supreme Being, or Creator, and are primarily concerned with moral behavior, rituals, and social customs and activities. The concept of a god, or gods, existed in Greek and Roman antiquity—the likes of Zeus, Athena, Poseidon, and many others. Although they were not necessarily associated with the creation of the world, the gods had influence on human affairs, though it was usually limited to particular concerns or human groups.

Other creation myths from antiquity are the Babylonian Epic of Gilgamesh, which may date to the third millennium BCE and includes a universal flood, in which all humans perish by turning to clay, except those in a boat built by instruction of the god Ea. Also from Mesopotamia is the Enuma Elish, dating to around 2000 BCE, which asserts that the world is not eternal but was created in time. The Enuma Elish recognizes multiple gods.

Even older are the first creation myths of Egypt, although these changed locally and over time. In one Egyptian account, in the beginning the sun-god Atum spat out Shu, the god of air, and Tefnut, the goddess of moisture, who in turn gave birth to the earth-god Geb and the sky-god Nut, who mated and produced Osiris and his consort Isis. The Egyptian myths include multiple gods. However, the two largest divisions, Upper and Lower Egypt, each had a special god with a distinctive crown (Moore 2002).

The religious concerns about the origin of the world and of living organisms are largely concentrated in the three Abrahamic religions. Christian authors over the centuries have explored God's actions concerning the world. The perceived conflicts between science and religious faith, in particular, gave rise to the metaphor of a war between religion and science which, according to the historian James R. Moore (1979), was actually introduced by Darwin's "bulldog," Thomas H. Huxley, and his followers "as part of their campaign to erect science as the new source of influence in modern society" (Bowler 2007, 4). The iconic manifesto of the conflict is *History of the Conflict between Religion and Science*, published in 1875 by Huxley's American disciple J. W. Draper.

Christian authors have over the centuries argued that the order, harmony, and design of the universe are incontrovertible evidence that the universe was created by an omniscient and omnipotent Creator. Notable Christian authors include Augustine (353–430 CE), who wrote in *The City of God* that the "world itself, by the perfect order of its changes and motions and by the great beauty of all things visible, proclaims . . . that it has been created, and also that it could not have been made other than by a God ineffable and invisible in greatness, and . . . in beauty." Thomas Aquinas (1224–1274), considered by many to be the greatest Christian theologian, advances in his *Summa Theologiae* five ways to demonstrate, by natural reason, that God exists. The fifth way derives from the orderliness and designed purposefulness of the universe, which are evidence that it has been created by a Supreme Intelligence: "Some intelligent being exists by which all natural things are directed to their end; and this being we call God."

This manner of seeking a natural demonstration of God's existence became later known as the "argument from design," which is a two-pronged argument. The first prong asserts that the universe evinces that it has been designed. The second prong affirms that only God could account for the complexity and perfection of the design. A forceful and elaborate formulation of the argument from design was *The Wisdom of God Manifested in the Works of Creation* (1691) by the English clergyman and naturalist John Ray (1627–1705). Ray regarded as incontrovertible evidence of God's wisdom that all components of the universe— the stars and the planets, as well as all organisms—are so wisely contrived from the beginning and perfect in their operation. The "most convincing argument of the Existence of a Deity," writes Ray, "is the admirable Art and Wisdom that discovers itself in the Make of the Constitution, the Order and Disposition, the Ends and uses of all the parts and members of this stately fabric of Heaven and Earth."

The design argument was advanced, in greater or lesser detail, by a number of authors in the seventeenth and eighteenth centuries. John Ray's contemporary Henry More (1614–1687) saw evidence of God's design in the succession of day and night and of the seasons: "I say that the Phenomena of Day and Night, Winter and Summer, Spring-time and Harvest . . . are signs and tokens unto us that there is a God . . . things are so framed that they naturally imply a Principle of Wisdom and Counsel in the Author of them. And if there be such an Author of external Nature, there is a God." Robert Hooke (1635–1703), a physicist and eventual Secretary of the Royal Society, formulated the watchmaker analogy: God had furnished each plant and animal "with all kinds of contrivances necessary for its own existence and propagation . . . as a Clock-maker might make a Set of Chimes to be a part of a Clock" (Hooke 1665, 124). The clock analogy, among other analogies such as temples, palaces, and ships, was also used by Thomas Burnet (1635–1703) in his *Sacred Theory of the Earth*, and it would become common among natural theologians of the time. The Dutch philosopher and theologian Bernard Nieuwenfijdt (1654–1718) developed, at length, the argument from design in his three-volume treatise, *The Religious Philosopher*, where, in the preface, he introduces the watchmaker analogy. Voltaire (1694–1778), like other philosophers of the Enlightenment, accepted the argument from design. Voltaire asserted that in the same way as the existence of a watch proves the existence of a watchmaker, the design and purpose evident in nature prove that the universe was created by a Supreme Intelligence. The Irish archbishop of Armagh James Ussher, in his *Annales Veteris et Novi Testamenti* (1650–1654, Annals of the Old and New Testament), estimated that the earth was created in 4004 BCE at midday on Sunday, October 23.

Biblical skeptics included the French naturalist Comte de Buffon, head of the king's botanical gardens in Paris, who suggested in 1749 that the earth might be 70,000 years old, 10 times older than calculated by Ussher. The editor of the *Encyclopédie*, the atheist Denis Diderot, thought that there were not fixed species or divinely ordained creatures but rather that nature was subject to constant flux, producing organisms at random, without plan or purpose. The Scottish philosopher David Hume (1711–1776), in his posthumously published *Dialogues on Natural Religion*, written in the 1750s, criticized the analogy between human artifacts and organisms, denying that the element of design was involved in living species.

2. NATURAL THEOLOGY AND THE BRIDGEWATER TREATISES

William Paley (1743–1805), one of the most influential English authors of his time, formulated at length in his *Natural Theology* (1802) the argument from design, based on the complex and precise design of organisms. Paley was an influential writer of works on Christian philosophy, ethics, and theology, such as *The Principles of Moral and Political Philosophy* (1785) and *A View of the Evidences of Christianity* (1794). With *Natural Theology*, Paley sought to update Ray's *Wisdom of God* (1691). But Paley could now carry the argument much further than Ray, by taking advantage of one century of additional biological knowledge. Darwin, while he was an undergraduate student at the University of Cambridge between 1827 and 1831, read Paley's *Natural Theology*, which was part of the university's canon for nearly half a century after Paley's death. Darwin writes in his autobiography of the "much delight" and profit that he derived from reading Paley. "I was charmed and convinced [at that time] by the long line of argumentation."

Paley's keystone claim is that there "cannot be design without a designer; contrivance, without a contriver; order, without choice; . . . means suitable to an end, and executing their office in accomplishing that end, without the end ever having been contemplated." *Natural Theology* is a sustained argument for the existence of God based on the obvious design of humans and their organs, as well as the design of all sorts of organisms, considered by themselves and in their relations to one another and to their environment. Paley's first analogical example in *Natural Theology* is the human eye. He points out that the eye and the telescope "are made upon the same principles; both being adjusted to the laws by which the transmission and refraction of rays of light are regulated." Specifically, there is a precise resemblance between the lenses of a telescope and "the humors of the eye" in their figure, their position, and the ability of converging the rays of light at a precise distance from the lens—on the retina, in the case of the eye.

Paley makes two remarkable observations about the complex and precise design of the eye. The first is that rays of light should be refracted by a more convex surface when transmitted through water than when passing out of air into the eye. Accordingly, "the eye of a fish, in that part of it called the crystalline lens, is much rounder than the eye of terrestrial animals. What plainer manifestation of design can there be than this difference? What could a mathematical instrument maker have done more to show his knowledge of [t]his principle [. . .]?" The second remarkable observation made by Paley in support of his argument is dioptric distortion: "Pencils of light, in passing through glass lenses, are separated into different colors, thereby tinging the object, especially the edges of it, as if it were viewed through a prism. To correct this inconvenience [. . .] a sagacious optician . . . [observed] that in the eye the evil was cured by combining lenses composed of different substances, that is, of substances which possessed different refracting powers." The telescope maker, accordingly, corrected the dioptric distortion "by imitating, in glasses made from different materials, the effects of the different humors through which the rays of light pass before they reach the bottom of the eye. Could this be in the eye without purpose, which suggested to the optician the only effectual means of attaining that purpose?" (Paley 1802, 23).

Could the eye have come about without design or preconceived purpose as a result of chance? Paley had set the argument against chance in the very first paragraph of *Natural Theology*, arguing rhetorically by analogy: "In crossing a heath, suppose I pitched my foot against a *stone*, and were asked how the stone came to be there, I might possibly answer, that for any thing I knew to the contrary it had lain there forever. . . . But suppose I had found a *watch* upon the ground, and it should be inquired how the watch happened to be in that place, I should hardly think of the answer which I had before given, . . . Yet why should not this answer serve for the watch as well as for the stone; why is it not as admissible in the second case as in the first? For this reason, and for no other, namely, that when we come to inspect the watch, we perceive . . . that its several parts are framed and put together for a purpose, . . . that if the different parts had been differently shaped from what they are, or placed after any other manner or in any other order than that in which they are placed, either no motion at all would have been carried on in the machine, or none which would have answered the use that is now served by it."

The strength of the argument against chance derives, Paley tells us, from what he names "relation," a notion akin to what some contemporary antievolutionist writers have named "irreducible complexity" (and that some of them have given themselves credit for its discovery): "When several different parts contribute to one effect . . . , the fitness of such parts . . . to one another for the purpose of producing, by their united action, the effect, is what I call *relation*; and wherever this is observed in the works of nature or of man, it appears to me to carry along with it decisive evidence of understanding, intention, art." The outcomes of chance do not exhibit relation among the parts or, as we might say, organized complexity.

Natural Theology has chapters dedicated to the human frame, which displays a precise mechanical arrangement of bones, cartilage, and joints; to the circulation of the blood and the disposition of blood vessels; to the comparative anatomy of humans and animals; to the digestive tract, kidneys, urethra, and bladder; to the wings of birds and the fins of fish; and much more. After detailing the precise organization and exquisite functionality of each biological entity, relationship, or process, Paley draws again and again the same conclusion: only an omniscient and omnipotent Deity could account for these marvels of mechanical perfection, purpose, and functionality, and for the enormous diversity of inventions that they entail.

Paley's *Natural Theology* fails, even in his time, when seeking an account of imperfections, defects, pain, and cruelty that would be consistent with his notion of the Creator. This is his general explanation for nature's imperfections: "Irregularities and imperfections are of little or no weight [. . .], but they are to be taken in conjunction with the unexceptionable evidences which we possess of skill, power, and benevolence displayed in other instances" (p. 46). But if functional design is a manifestation of an intelligent designer, why should not deficiencies indicate that the designer is less than omniscient or less than omnipotent or less than benevolent? We know that some deficiencies are not just imperfections but are outright dysfunctional, jeopardizing the very function the organ or part is supposed to serve. We now know, of course, that the explanation for dysfunction and imperfection is *natural selection*, which can account for design and functionality but does not achieve any sort of perfection—nor is it omniscient or omnipotent. And not only are organisms and their parts less than perfect, but deficiencies and dysfunctions are pervasive, evidencing defective "design."

Francis Henry Egerton (1756–1829), the eighth Earl of Bridgewater, bequeathed in 1829 the sum of 8000 pounds sterling with instructions to the Royal Society that it commission eight treatises that would promote natural theology by setting forth "The Power, Wisdom and Goodness of God as manifested in the Creation." Eight treatises were published in the 1830s, several of which artfully incorporated the best science of the time and had considerable influence on the public and among scientists. *The Hand, Its Mechanisms and Vital Endowments as Evincing Design* (1833), by Sir Charles Bell, a distinguished anatomist and surgeon famous for his neurological discoveries, examines in considerable detail the wondrously useful design of the human hand but also the perfection of design of the forelimb used for different purposes in different animals, serving in each case the particular needs and habits of its owner: the human's arm for handling objects, the dog's leg for running, and the bird's wing for flying. He concludes that "Nothing less than the Power, which originally created, is equal to the effecting of those changes on animals, which are to adapt them to their conditions." William Buckland, professor of geology at Oxford University, notes in *Geology and Mineralogy* (1836) the world distribution of coal and mineral ores and proceeds to point out that they had been deposited in remote parts, yet obviously with the forethought of serving the larger human populations that would come about much later. Later, another geologist, Hugh Miller in *The Testimony of the Rocks* (1858), would formulate what may be called the *argument from beauty*, which allows that it is not only the perfection of design but also the beauty of natural structures found in rock formations and in mountains and rivers that manifests the intervention of the Creator. One additional treatise, never completed, was authored by the notable mathematician and pioneer in the field of calculating machines Charles Babbage (1791–1871). In the *Ninth Bridgewater Treatise: A Fragment*, published in 1838, he seeks to show how mathematics may be used to bolster religious belief.

In the 1990s, a new version of the design argument was formulated in the United States, named *intelligent design* (ID), which refers to an unidentified Designer who accounts for the order and complexity of the universe, or who intervenes from time to time in the universe so as to design organisms and their parts. The complexity of organisms, it is claimed, cannot be accounted for by natural processes. According to ID proponents, this intelligent designer could be, but need not be, God. The intelligent designer could be an alien from outer space or some other creature, such as a "time-traveling cell biologist," with amazing powers to account for the universe's design. Explicit reference to God is avoided, so that the "theory" of ID can be taught in the public schools as an alternative to the theory of evolution without incurring conflict with the US

Constitution, which forbids the endorsement of any religious beliefs in public institutions.

3. EVOLUTIONARY GLIMPSES

Traditional Judaism, Christianity, and Islam account for the origin of living beings and their adaptations to life in their environments—wings, gills, hands, flowers—in their accounts of the origin of the universe, as the handiwork of an omnipotent and omniscient God. Yet, we can already see among early Christian authors gleams of evolutionary ideas, that is, acceptance of the possibility that some living species may have come about through natural processes. Gregory of Nyssa (335–394) and Augustine (354–430) maintained that not all of creation was initially created by God; rather, some species of plants and animals had evolved in historical times from God's creations.

According to Gregory of Nyssa, the world has come about in two successive stages. The first stage, the creative step, is instantaneous; the second stage, the formative step, is gradual and develops through time. According to Augustine, many plant and animal species were not directly created by God but only indirectly, in their potentiality (in their *rationes seminales*), so that they would come about by natural processes later in the world. Gregory's and Augustine's motivation was not scientific but theological. For example, Augustine was concerned that it would have been impossible to hold representatives of all animal species in a single vessel, such as Noah's Ark; some species must have come into existence only after the Flood.

The notion that organisms may change by natural processes was not investigated as a biological subject by Christian theologians of the Middle Ages, but it was, usually incidentally, considered as a possibility by many, including Albertus Magnus (1200–1280) and his student Thomas Aquinas (1224–1274). Aquinas concluded, after consideration of the arguments, that the development of living creatures, such as maggots and flies, from nonliving matter, such as decaying meat, was not incompatible with Christian faith or philosophy, but he left it to others (to scientists, in current parlance) to determine whether this actually happened.

The issue of whether living organisms could spontaneously arise from dead matter was not settled until four centuries later, by the Italian Francesco Redi (1626–1697), one of the first scientists to conduct biological experiments with proper controls. Redi set up flasks with various kinds of fresh meat; some were sealed, others covered with gauze so that air but not flies could enter, and others left uncovered. The meat putrefied in all flasks, but maggots appeared only in the uncovered flasks, which flies had entered freely. Redi was a poet as well as a physician, chiefly known for his *Bacco in Toscana* (1685, Bacchus in Tuscany).

The cause of putrefaction would be discovered two centuries later by Darwin's younger contemporary, the French chemist Louis Pasteur (1822–1895), one of the greatest scientists of all time. Pasteur demonstrated that fermentation and putrefaction were caused by minute organisms that could be destroyed by heat. Food decomposes when placed in contact with germs present in the air. The germs do not arise spontaneously within the food. We owe to Pasteur the process of *pasteurization*, the destruction by heat of microorganisms in milk, wine, and beer, which can thus be preserved if kept out of contact with the microorganisms in the air. Pasteur also demonstrated that cholera and rabies are caused by microorganisms, and he invented *vaccination*, treatment with attenuated (or killed) infective agents that stimulate the immune system of animals and humans, thus protecting them against infection.

The first broad theory of evolution was proposed by the French naturalist Jean-Baptiste de Monet, Chevalier de Lamarck (1744–1829). In his *Philosophie zoologique* (1809, Zoological Philosophy), Lamarck held the enlightened view, shared by the intellectuals of his age, that living organisms represent a progression, with humans as the highest form. Lamarck's theory of evolution asserts that organisms evolve through eons of time from lower to higher forms, an ongoing process, always culminating in human beings. The remote ancestors of humans were worms and other inferior creatures, which gradually evolved into more and more advanced organisms, and ultimately humans.

The *inheritance of acquired characters* is the theory most often associated with Lamarck's name. Yet, this theory was actually a subsidiary construct of his theory of evolution: that evolution is a continuous process and that today's worms will yield humans as their remote descendants. The theory stated that as animals become adapted to their environments through their habits, modifications in their body plan occur by "use and disuse." Use of an organ or structure reinforces it; disuse leads to obliteration. Lamarck's theory further asserted that the characteristics acquired by use and disuse would be inherited. Lamarckism was disproved in the twentieth century.

Lamarck's evolutionary theory was metaphysical rather than scientific. He postulated that life possesses the innate property to improve over time, so that progression from lower to higher organisms would continually occur, always following the same path of transformation from lower organisms to increasingly higher and more complex organisms. A somewhat similar evolutionary theory was formulated a century later by another Frenchman, the philosopher Henri Bergson (1859–1940) in his *L'Évolution créatrice* (1907, Creative evolution).

Erasmus Darwin (1731–1802), a physician and poet, and the grandfather of Charles Darwin, proposed, in poetic rather than scientific language, a theory of the transmutation of life forms through eons of time (*Zoonomia, or the Laws of Organic Life*, 1794–1796). More significant for Charles Darwin was the influence of his older contemporary and friend the eminent geologist Sir Charles Lyell (1797–1875). In his *Principles of Geology* (1830–1833), Lyell proposed that earth's physical features were the outcome of major geologic processes acting over immense periods of time, incomparably greater than the few thousand years since Creation commonly believed at the time.

4. DARWIN'S REVOLUTION

Darwin occupies an exalted place in the history of Western thought, deservedly receiving credit for the theory of evolution. In *On the Origin of Species*, he lays out the evidence demonstrating the evolution of organisms. However, Darwin accomplished something much more important for intellectual history than demonstrating evolution. Darwin's *Origin* is, first and foremost, a sustained effort to solve the problem of how to account scientifically for the design of organisms. Darwin explains the design of organisms, their complexity, diversity, and marvelous contrivances as the result of natural processes.

One version of the history of the ideas sees a parallel between the Copernican and the Darwinian Revolutions. In this view, the Copernican Revolution consisted in displacing the earth from its previously accepted locus as the center of the universe, moving it to a subordinate place as just one more planet revolving around the sun. In congruous manner, the Darwinian Revolution is viewed as consisting of the displacement of humans from their exalted position as the center of life on earth, with all other species created for the service of humankind. According to this version of intellectual history, Copernicus accomplished his revolution with the heliocentric theory of the solar system. Darwin's achievement emerged from his theory of organic evolution.

Although this version of these two intellectual revolutions is correct, it misses what is most important about them, namely, that they ushered in the beginning of science in the modern sense of the word. These two revolutions may jointly be seen as the one Scientific Revolution, with two stages—the Copernican and the Darwinian. The Copernican Revolution was launched with the publication in 1543, the year of Nicolaus Copernicus's death, of his *De revolutionibus orbium celestium* (On the revolutions of the celestial spheres) and bloomed with the publication in 1687 of Isaac Newton's *Philosophiae naturalis principia mathematica* (The mathematical principles of natural philosophy). The discoveries by Copernicus, Kepler, Galileo, Newton, and others, in the sixteenth and seventeenth centuries had gradually ushered in a conception of the universe as matter in motion governed by natural laws. It was shown that earth is not the center of the universe but a small planet rotating around an average star; that the universe is immense in space and in time; and that the motions of the planets around the sun can be explained by the same simple laws that account for the motion of physical objects on our planet. These and other discoveries greatly expanded human knowledge. The conceptual revolution they brought about was more fundamental yet: a commitment to the postulate that the universe obeys immanent laws that account for natural phenomena. The workings of the universe were brought into the realm of science: explanation through natural laws.

The advances of physical science brought about by the Copernican Revolution drove mankind's conception of the universe to a split-personality state of affairs. Scientific explanations derived from natural laws dominated the world of nonliving matter, on the earth as well as in the heavens. However, supernatural explanations, which depended on the unfathomable deeds of the Creator, were accepted as explanations of the origin and configuration of living creatures. Authors such as William Paley argued that the complex design of organisms could not have come about by chance, or by the mechanical laws of physics, chemistry, and astronomy, but rather was accomplished by an omniscient and omnipotent Deity, just as the complexity of a watch, designed to tell time, was accomplished by an intelligent watchmaker. It was Darwin's genius that resolved this conceptual schizophrenia. Darwin completed the Copernican Revolution by propounding for biology the notion of nature as a lawful system of matter in motion that human reason can explain without recourse to supernatural agencies.

The conundrum faced by Darwin can hardly be overestimated. The strength of the argument from design to demonstrate the role of the Creator had been forcefully set forth by philosophers and theologians: wherever there is function or design, we look for its author. It was Darwin's greatest accomplishment to show that the complex organization and functionality of living beings can be explained as the result of a natural process—natural selection—without any need to resort to a Creator or other external agent. The origin and adaptations of organisms in their profusion and wondrous variations were thus brought into the realm of science.

Organisms exhibit complex design, but it is not, in current language, "irreducible complexity," emerging all of a sudden in full bloom. Rather, according to Darwin's theory of natural selection, the design has arisen gradually and cumulatively, step by step, promoted by the reproductive success of individuals with incrementally more adaptive elaborations.

Natural selection accounts for the "design" of organisms, because adaptive variations tend to increase the probability of survival and reproduction of their carriers at the expense of maladaptive, or less adaptive, variations. The arguments of Paley against the incredible improbability of chance accounts of the adaptations of organisms are well taken as far as they go. But neither Paley, nor any other author before Darwin, was able to discern that there is a natural process (namely, natural selection) that is not random but rather is oriented and able to generate order or "create." The traits that organisms acquire in their evolutionary histories are not fortuitous but determined by their functional utility to the organisms, "designed" as it were to serve their life needs.

5. EVOLUTION AND THE BIBLE

To some Christians and other people of faith, the theory of evolution seems to be incompatible with their religious beliefs, because it is inconsistent with the Bible's narrative of creation. The first chapters of the biblical book of Genesis describe God's creation of the world, plants, animals, and human beings. A literal interpretation of Genesis seems incompatible with the gradual evolution of humans and other organisms by natural processes. Even independent of the biblical narrative, the Christian beliefs in the immortality of the soul and in humans as "created in the image of God" have appeared to many as contrary to the evolutionary origin of humans from nonhuman animals.

In 1874, Charles Hodge, an American Protestant theologian, published *What Is Darwinism?*—one of the most articulate assaults on evolutionary theory. Hodge perceived Darwin's theory as "the most thoroughly naturalistic that can be imagined and far more atheistic than that of his predecessor Lamarck." Echoing Paley, Hodge argued that the design of the human eye reveals that "it has been planned by the Creator, like the design of a watch evinces a watchmaker." He concluded that "the denial of design in nature is actually the denial of God."

Some Protestant theologians saw a solution to the apparent contradiction between evolution and creation in the argument that God operates through intermediate causes. The origin and motion of the planets could be explained by the law of gravity and other natural processes without denying God's creation and providence. Similarly, evolution could be seen as the natural process through which God brought living beings into existence and developed them according to his plan. Thus, A. H. Strong, the president of Rochester Theological Seminary in New York State, wrote in his *Systematic Theology* (1885): "We grant the principle of evolution, but we regard it as only the method of divine intelligence." He explains that the brutish ancestry of human beings was not incompatible with their excelling status as creatures in the image of God. Strong drew an analogy with Christ's miraculous conversion of water into wine: "The wine in the miracle was not water because water had been used in the making of it, nor is man a brute because the brute has made some contributions to its creation." Arguments for and against Darwin's theory came from Roman Catholic theologians as well.

Gradually, well into the twentieth century, evolution by natural selection came to be accepted by a majority of Christian writers. Pope Pius XII in his encyclical *Humani generis* (1950, Of the human race) acknowledged that biological evolution was compatible with the Christian faith, although he argued that God's intervention was necessary for the creation of the human soul. Pope John Paul II, in an address to the Pontifical Academy of Sciences on October 22, 1996, deplored interpreting the Bible's texts as scientific statements rather than religious teachings. He added: "New scientific knowledge has led us to realize that the theory of evolution is no longer a mere hypothesis. It is indeed remarkable that this theory has been progressively accepted by researchers, following a series of discoveries in various fields of knowledge. The convergence, neither sought nor fabricated, of the results of work that was conducted independently is in itself a significant argument in favor of this theory."

Similar views have been expressed by other mainstream Christian denominations. The General Assembly of the United Presbyterian Church in 1982 adopted a resolution stating that "Biblical scholars and theological schools . . . find that the scientific theory of evolution does not conflict with their interpretation of the origins of life found in Biblical literature" (National Academy of Sciences 2008). The Lutheran World Federation in 1965 affirmed that "evolution's assumptions are as much around us as the air we breathe and no more escapable. At the same time theology's affirmations are being made as responsibly as ever. In this sense both science and religion are here to stay, and . . . need to remain in a healthful tension of respect toward one another" (National Academy of Sciences 2008).

Equally explicit statements have been advanced by Jewish authorities and leaders of other major religions. In 1984, the 95th Annual Convention of the Central Conference of American Rabbis adopted a resolution stating: "Whereas the principles and concepts of biological evolution are basic to understanding science . . . we call upon science teachers and local school authorities in all states to demand quality textbooks that are based on modern, scientific knowledge and that exclude 'scientific' creationism" (National Academy of Sciences 2008).

Christian denominations that hold a literal interpretation of the Bible have opposed these views. A succinct expression of this opposition is found in the Statement of Belief of the Creation Research Society, founded in 1963 as a "professional organization of trained scientists and interested laypersons who are firmly committed to scientific special creation": "The Bible is the Written Word of God, and because it is inspired throughout, all of its assertions are historically and scientifically true in the original autographs. To the student of nature this means that the account of origins in Genesis is a factual presentation of simple historical truths."

Many Bible scholars and theologians have long rejected a literal interpretation as untenable, however, because the Bible contains mutually incompatible statements. The very beginning of the book of Genesis presents two different creation narratives. Extending through chapter 1 and the first verses of chapter 2 is the familiar six-day narrative, in which God creates human beings—both "male and female"—in His own image on the sixth day, after creating light, earth, firmament, fish, fowl, and cattle. In verse 4 of chapter 2, a different narrative starts, in which God creates a male human, then plants a garden and creates the animals, and only then proceeds to take a rib from the man to make a woman.

Which one of the two narratives is correct and which one is in error? Neither one contradicts the other if we understand the two narratives as conveying the same message, that the world was created by God and that humans are His creatures. But both narratives cannot be "historically and scientifically true" as postulated in the Statement of Belief of the Creation Research Society.

There are numerous inconsistencies and contradictions in different parts of the Bible, for example, in the description of the return from Egypt to the Promised Land by the chosen people of Israel, not to mention erroneous factual statements about the sun's circling around the earth, and the like. Biblical scholars point out that the Bible should be held inerrant with respect to religious truth, not in matters that are of no significance to salvation. Augustine wrote in his *De Genesi ad litteram* (Literal commentary on Genesis): "It is also frequently asked what our belief must be about the form and shape of heaven, according to Sacred

Scripture. . . . Such subjects are of no profit for those who seek beatitude. . . . What concern is it of mine whether heaven is like a sphere and earth is enclosed by it and suspended in the middle of the universe, or whether heaven is like a disk and the Earth is above it and hovering to one side." He adds: "In the matter of the shape of heaven, the sacred writers did not wish to teach men facts that could be of no avail for their salvation." Augustine is saying that the book of Genesis is not an elementary book of astronomy. The Bible is about religion, and it is not the purpose of the Bible's religious authors to settle questions about the shape of the universe that are of no relevance whatsoever to how to seek salvation.

In the same vein, Pope John Paul II said in 1981 that the Bible itself "speaks to us of the origins of the universe and its makeup, not in order to provide us with a scientific treatise but in order to state the correct relationships of man with God and with the universe. Sacred Scripture . . . in order to teach this truth, it expresses itself in the terms of the cosmology in use at the time of the writer."

6. THE PROBLEM OF EVIL

Christian scholars for centuries struggled with the *problem of evil* in the world. The Scottish philosopher David Hume (1711–1776) set the problem succinctly with brutal directness: "Is he [God] willing to prevent evil, but not able? Then he is impotent. Is he able, but not willing? Then, he is malevolent. Is he both able and willing? Whence then evil?" If the reasoning is valid, it would follow that God is not all-powerful or all-good. Christian theology accepts that evil exists but denies the validity of the argument.

Traditional theology distinguishes three kinds of evil: (1) moral evil or sin, the evil originated by human beings; (2) pain and suffering as experienced by human beings; (3) physical evil, such as floods, tornados, earthquakes, and the imperfections of all creatures. Theology has a ready answer for the first two kinds of evil. Sin is a consequence of free will; the flip side of sin is virtue, also a consequence of free will. Christian theologians have expounded that if humans are to enter into a genuinely personal relationship with their maker, they must first experience some degree of freedom and autonomy. The eternal reward of heaven calls for a virtuous life as many Christians see it. Christian theology also provides a good accounting of human pain and suffering. To the extent that pain and suffering are caused by war, injustice, and other forms of human wrongdoing, they are also a consequence of free will; people choose to inflict harm on one another. On the flip side are good deeds by which people choose to alleviate human suffering.

What about earthquakes, storms, floods, droughts, and other physical catastrophes? Enter modern science into the theologian's reasoning. Physical events are built into the structure of the world itself. Since the seventeenth century, humans have known that the processes by which galaxies and stars come into existence, the planets are formed, the continents move, and floods and earthquakes occur, as well as the weather and the change of seasons are natural processes, not events specifically designed by God for punishing or rewarding humans. The extreme violence of supernova explosions and the chaotic frenzy at galactic centers are outcomes of the laws of physics, not the design of a fearsome deity. Before Darwin, theologians still encountered a seemingly insurmountable difficulty. If God is the designer of life, whence the lion's cruelty, the snake's poison, and the parasites that secure their existence only by destroying their hosts? Evolution came to the rescue.

Jack Haught (1998), a contemporary Roman Catholic theologian, has written of "Darwin's gift to theology." The Protestant theologian Arthur Peacocke has referred to Darwin as the "disguised friend," by quoting the earlier theologian Aubrey Moore, who in 1891 wrote that "Darwinism appeared, and, under the guise of a foe, did the work of a friend" (Peacocke 1998). Haught and Peacocke are acknowledging the irony that the theory of evolution, which at first had seemed to remove the need for God in the world, now has convincingly removed the need to explain the world's imperfections as failed outcomes of God's design.

Indeed, a major burden was removed from the shoulders of believers when convincing evidence was advanced that the design of organisms need not be attributed to the immediate agency of the Creator but rather is an outcome of natural processes. If we claim that organisms and their parts have been specifically designed by God, we have to account for the incompetent design of the human jaw, the narrowness of the birth canal, and our poorly designed backbone, less than fittingly suited for walking upright. Imperfections and defects pervade the living world. Consider once again the human eye. The visual nerve fibers in the eye converge to form the optic nerve, which crosses the retina (to reach the brain) and thus creates a blind spot, a minor imperfection, but an imperfection of design, nevertheless; squids and octopuses do not have this defect. Did the Designer have greater love for squids than for humans and, thus, exhibit greater care in designing their eyes than ours? Not only are organisms and their parts less than perfect but also deficiencies and dysfunctions are pervasive, evidencing incompetent rather than intelligent design. Consider the human jaw. We have too many teeth for the jaw's size, so that wisdom teeth need to be removed and orthodontists can make a decent living straightening the others. Would we want to blame God for this blunder? A human engineer would have done better.

7. EVOLUTION: IMPERFECT DESIGN, NOT INTELLIGENT DESIGN

Evolution gives a good account of this imperfection. Brain size increased over time in our ancestors; the remodeling of the skull to fit the larger brain entailed a reduction of the jaw so that the head of the newborn would not be too large to pass through the mother's birth canal. The birth canal of women is much too narrow for easy passage of the infant's head, so thousands upon thousands of babies and many mothers die during delivery. Surely we don't want to blame God for this dysfunctional design or for the children's deaths. The theory of evolution makes the situation understandable as a consequence of the evolutionary enlargement of our brain. Females of other primates do not experience this difficulty. Theologians in the past struggled with the issue of dysfunction because they thought it had to be attributed to God's design. Science, much to the relief of theologians, provides an explanation that convincingly attributes defects, deformities, and dysfunctions to natural causes.

Consider the following. About 20 percent of all recognized human pregnancies end in spontaneous miscarriage during the first two months of pregnancy. This misfortune amounts at present to more than 20 million spontaneous abortions worldwide every year. Do we want to blame

God for the deficiencies in the pregnancy process? Many people of faith would rather attribute this monumental mishap to the clumsy ways of the evolutionary process than to the incompetence or deviousness of an intelligent designer.

Evolution makes it possible to attribute these mishaps to natural processes rather than to the direct creation or specific design of the Creator. The response of some critics is that the process of evolution by natural selection does not discharge God's responsibility for the dysfunctions, cruelties, and sadism of the living world, because for people of faith God is the Creator of the universe and thus accountable for its consequences, direct or indirect, immediate or mediated. If God is omnipotent, the argument would say, He could have created a world where such things as cruelty, parasitism, and human miscarriages would not occur.

One possible religious explanation goes along the following lines of reasoning. Consider, first, human beings, who perpetrate all sorts of misdeeds and sins, even perjury, adultery, and murder. People of faith believe that each human being is a creation of God, but this does not entail that God is responsible for human crimes and misdemeanors. Sin is a consequence of free will; the flip side of sin is virtue. The critics might say that this account does not excuse God, because God could have created humans without free will (whatever these "humans" may have been called and been like). But one could reasonably argue that "humans" without free will would be a very different kind of creature, being much less interesting and creative than humans are. Robots are not a good replacement for humans; robots do not perform virtuous deeds.

This line of argumentation can be extended to the catastrophes and other events of the physical world and to the dysfunctions of organisms and the harms caused to them by other organisms and environmental mishaps. However, some authors do not find this extension fully satisfactory as an explanation that would exonerate God from moral responsibility. The point made again is that the world was created by God, so God is ultimately responsible. God could have created a world without parasites or dysfunctionalities. But a world of life with evolution is much more exciting; it is a creative world where new species arise, complex ecosystems come about, and humans have evolved. These considerations may provide the beginning of an explanation for many people of faith, as well as for theologians.

The Anglican theologian Keith Ward (2008) has stated the case even in stronger terms, arguing that the creation of a world without suffering and moral evil is not an option even for God: "Could [God] not actualize a world wherein suffering is not a possibility? He could not, if any world complex and diverse enough to include rational and moral agents must necessarily include the possibility of suffering. . . . A world with the sorts of success and happiness in it that we occasionally experience is a world that necessarily contains the possibility of failure and misery." The physicist and theologian Robert J. Russell (2007) goes even further, making the case for the existence of natural (physical and biological) evil in the world, "including the pain, suffering, disease, death, and extinction that characterize the evolution of life."

An additional point is that physical or biological (other than human) events that cause harm are not moral evil actions, because they are not caused by moral agents but are the result of natural processes. If a terrorist blows up a bus with school children, that is moral evil. If an earthquake kills several thousand people in China and destroys their homes and livelihood, no subject is morally responsible, because the event was not committed by a moral agent but was the result of a natural process. Similarly, if a mugger uses a vicious dog to brutalize a person, the mugger is morally responsible. But if a coyote attacks a person, there is no moral evil that needs to be accounted for. In the world of physical and biological nature (again, excluding human deeds), no morality is involved. This claim, of course, may or may not satisfy everyone, but it deserves to be explored by theologians and people of faith.

8. EVOLUTION AND THE NEW ATHEISTS

Some people of faith assert that evolution is not compatible with religion because the theory of evolution makes assertions about the origin of humans and of the universe that contradict the Bible narrative and other religious beliefs. On the other side of the argument are authors who assert that religion is incompatible with evolution, and indeed with science in general, because religious beliefs are outright false as well as toxic.

These authors include distinguished scientists, philosophers, and others, particularly some who profess what has become known as "New Atheism," asserting that "the tenets of many religions are *hypotheses* that can, at least in principle, be examined by science and reason. If religious claims can't be substantiated with reliable evidence, the argument goes, they should like dubious scientific claims, be rejected" (Coyne 2015, xii–xiii; italics in the original). The New Atheism includes distinguished scientists, philosophers, and other authors, such as Richard Dawkins, Sam Harris, Daniel Dennett, the late Christopher Hitchens, and, more recently, Jerry Coyne, just quoted (Pinker 2015).

Richard Dawkins has written that Darwinism makes it possible for someone to become "an intellectually fulfilled atheist": "The universe we observe has precisely the properties we should expect if there is, at bottom, no design, no purpose, no evil and no good, nothing but blind, pitiless indifference" (Dawkins 1995, 133). Increasingly, in addition to being an outspoken proponent of atheism, Dawkins has become a severe critic of organized religion, arguing that the world religions are a positive danger to humanity (Dawkins 2006). Daniel Dennett asserts that natural selection, "Darwin's idea," bears "an unmistakable likeness to universal acid: it eats through just about every traditional concept, and leaves in its wake a revolutionized world-view . . . transformed in fundamental ways" (Dennett 1995, 63). "Those evolutionists who see no conflict between evolution and their religious beliefs have been careful not to look as closely as we have been looking" (310). "[T]here is no Special Creation of language, and neither art nor religion has a literally divine inspiration" (144). The historian of science William Provine not only affirms that there are no absolute principles of any sort but draws the ultimate conclusion from a materialistic line of thinking that even free will is an illusion: "Modern science directly implies that there are no inherent moral or ethical laws, no absolute principles for human society . . . [F]ree will as it is traditionally conceived—the freedom to make uncoerced and unpredictable choices among alternative courses of action—simply does not exist" (Provine 1988). The subtitle of Christopher Hitchens's book *God is Not Great* (2007) is *How Religion Poisons Everything*. Other critics of religion include the Nobel Laureate physicist Steven Weinberg: "With or without religion, good people can behave well and bad people can do evil; but for good people to do evil—that takes religion" (Weinberg 1999).

The New Atheists assert that valid knowledge about the natural world can be acquired through the scientific process of observation and experimentation, that is, through science. Thus, religion is rejected because it does not pass the test of science. The evolutionary biologist Richard Lewontin adds an interesting twist: scientists accept science and reject religion because of their uncompromising commitment to a materialistic philosophy. "It is not that the methods and institutions of science somehow compel us to accept a material explanation of the phenomenal world, but, on the contrary that we are forced by our *a priori* adherence to material causes to create an apparatus of investigation and a set of concepts that produce material explanations. . . . [M]aterialism is absolute, for we cannot allow a Divine foot in the door" (Lewontin 1997).

The philosopher of science Michael Ruse disagrees: "Does this mean that science is just faith-based like religion? Not at all. The point about the assumptions of science is that they work. They justify themselves pragmatically. . . . Theories that correctly predict the unexpected do turn out to be powerful words of understanding" (Ruse 2015, 243). Ruse, like other self-proclaimed atheists and agnostics, has been an articulate player, defending evolutionism against the attacks of creationists and intelligent designers, yet he has disagreed openly with Dawkins and Dennett. In the United States and other countries where many hold religious beliefs, it is unwise to link the theory of evolution too strongly to materialism. As Ruse has pointed out, there are, indeed, many people of faith who are evolutionists (see Ruse 2001, 2008). The historian of science Peter J. Bowler agrees. "As a historian who has spent decades studying the response to Darwin, and as an observer of modern debates in America and Europe, I too believe that the best defense of evolutionism is to show the complexity of the religious approach to science. There are many scientists who still have deeply held religious beliefs, and many religious thinkers who are happy to accept evolution" (Bowler 2007, 3–4).

Most famously, Stephen Jay Gould, who proclaimed to be an agnostic, viewed science and religion as non-overlapping magisteria (NOMA): science and religion deal with different subjects. Science is concerned with the facts of observed reality, whereas religion is concerned with morality and other human values. "Science covers the empirical realm: what is the universe made of (fact) and why does it work this way (theory). The magisterium of religion extends over questions of ultimate meaning and moral value. These two magisteria do not overlap, nor do they encompass all inquiry (consider, for example, the magisterium of art and the meaning of beauty)" (Gould 1999, 6). "NOMA is no wimpish, wallpapering, superficial device, acting as a mere diplomatic fiction and smoke screen to make life more convenient by compromise in a world of divine and contradictory passions. NOMA is a proper and principled solution—based on sound philosophy—to an issue of great historical and emotional weight" (92).

9. EVOLUTION AND RELIGIOUS BELIEFS

Evolution and religious beliefs need not be in contradiction. Indeed, if science and religion are properly understood, they *cannot* be in contradiction because they concern different matters. Science and religion are like two different windows through which to view the world. The two windows look at the same world, but they show different aspects of that world. Science concerns the processes that account for the natural world: the movement of the planets, the composition of matter and the atmosphere, the origin and adaptations of organisms. Religion concerns the meaning and purpose of the world and of human life, the proper relation of people to the Creator and to one another, the moral values that inspire and govern people's lives. Apparent contradictions emerge only when either the science or the beliefs, or often both, trespass their own boundaries and wrongfully encroach upon each other's subject matter.

The scope of science is the world of nature, the reality that is observed, directly or indirectly, by our senses. Science advances explanations concerning the natural world, explanations that are subject to the possibility of corroboration or rejection by observation and experiment. Outside that world, science has no authority, no statements to make, no business whatsoever taking one position or another. Science has nothing decisive to say about values, whether economic, aesthetic, or moral;

nothing to say about the meaning of life or its purpose; nothing to say about religious beliefs (except in the case of beliefs that transcend the proper scope of religion and make assertions about the natural world that contradict scientific knowledge; such statements cannot be true).

Science is *methodologically* materialistic or, better, methodologically *naturalistic.* I prefer the second expression because "materialism" often refers to a metaphysical conception of the world, a philosophy that asserts that nothing exists beyond the world of matter, that nothing exists beyond what our senses can experience. The question of whether science is inherently materialistic depends on whether one is referring to the methods and scope of science, which remain within the world of nature, or to the metaphysical implications of materialistic philosophy asserting that nothing exists beyond the world of matter. Science does not imply metaphysical materialism.

Scientists and philosophers who assert that science excludes the validity of any knowledge outside science make a "categorical mistake"; that is, they confuse the method and scope of science with its metaphysical implications. Methodological naturalism asserts the boundaries of scientific knowledge, not its universality. Science transcends cultural, political, and religious differences because it has no assertions to make about these subjects (except, again, to the extent that scientific knowledge is negated). That science is not constrained by cultural or religious differences is one of its great virtues. Science does not transcend those differences by denying them or by taking one position rather than another. It transcends cultural, political, and religious differences because these matters are none of its business (Scott 2009).

Science is a way of knowing, but it is not the only way. Knowledge also derives from other sources. Common experience, imaginative literature, art, and history provide valid knowledge about the world; and so do revelation and religion for people of faith. The significance of the world and human life, as well as matters concerning moral or religious values, transcends science. Yet, these matters are important; for most people, including scientists, they are at least as important as scientific knowledge per se.

For people of faith, the proper relationship between science and religion can be mutually motivating and inspiring. Science may inspire religious beliefs and religious behavior as individuals respond with awe to the immensity of the universe, the glorious diversity and wondrous adaptations of organisms, and the marvels of the human brain and the human mind. Religion promotes reverence for creation, for humankind, as well as for the world of life and the environment. Religion often is, for scientists and others, a motivating force and source of inspiration for investigating the marvelous world of the creation and solving the puzzles with which it confronts humankind.

FURTHER READING

Aquinas, T. 1905. Of God and His creatures. In J. Rickaby, ed., Summa contra gentiles, 241–368. London: Burns & Oates. *Aquinas is often considered the greatest Christian theologian of all time.*

Aquinas, T. 1964. Existence and Nature of God. In Summa Theologiae, Blackfriars ed., 10–249. New York: McGraw-Hill.

Augustine. 1998. The City of God, R. Dyson, ed. Cambridge: Cambridge University Press. *An early classic of Christian theology.*

Augustine. 2002. Work of St. Augustine. Vol. 13. On Genesis. J. E. Rotelle, ed. Hyde Park, NY: New City Press.

Ayala, F. J. 2007. Darwin's Gift to Science and Religion. Washington, DC: Joseph Henry Press. *Develops at greater length and in greater depth the ideas of this chapter.*

Ayala, F. J. 2010. Am I a Monkey? Six Big Questions about Evolution. Baltimore: Johns Hopkins University Press. *An easy read about science and religion.*

Babbage, C. 1838. The Ninth Bridgewater Treatise: A Fragment. London: John Murray.

Bell, C. 1833. The Hand, Its Mechanisms and Vital Endowments as Evincing Design. London: William Pickering.

Bergson, H. 1907. Creative Evolution. Paris: Alcan.

Bowler, P. J. 2007. Monkey Trials & Gorilla Sermons: Evolution and Christianity from Darwin to Intelligent Design. Cambridge, MA: Harvard University Press.

Buckland, W. 1836. Geology and Mineralogy Considered with Reference to Natural Theology. London: William Pickering.

Burnet, T. 1691. Sacred Theory of the Earth. London: R. Norton.

Coyne, J. A. 2015. Faith vs. Fact: Why Science and Religion Are Incompatible. New York: Viking.

Darwin, C. 1958. The Autobiography of Charles Darwin, 1809–1882. N. Barlow, ed. London: Collins.

Darwin, C. 1859. On the Origin of Species by Means of Natural Selection. London: John Murray.

Darwin, E. 1794–1796. Zoonomia, or the Laws of Organic Life. London: J. Johnson, in St. Paul's Church-Yard.

Dawkins, R. 1995. River Out of Eden. New York: Harper Collins.

Dawkins, R. 2006. The God Delusion. London: Bantam Press.

Dennett, D. C. 1995. Darwin's Dangerous Idea. New York: Simon and Schuster.

Draper, J. W. 1875. History of the Conflict between Religion and Science. London: Henry S. King & Co.

Gould, S. J. 1999. Rock of Ages: Science and Religion in the Fullness of Life. New York: Ballantine.

Haught, J. F. 1998. Darwin's gift to theology. In R. J. Russell, et al., eds., Evolutionary and Molecular Biology: Scientific Perspectives on Divine Action, 393–418. Vatican City State: Vatican Observatory Press; and Berkeley, CA: Center for Theology and the Natural Sciences. *A modern theologian's view of how modern biology favorably affects Christian faith.*

Hitchens, C. 2007. God is Not Great: How Religion Poisons Everything. New York: Hachette Book Group.

Hodge, C. 1874. What Is Darwinism? New York: Scribner, Armstrong & Co.

Hooke, R. 1665. Micrographia; or, Some Physiological Descriptions of Minute Bodies Made by Magnifying Glasses, with Observations and Inquiries Thereupon. London: Martyn & Allestry.

Hume, D., and N. K. Smith. 1935. Dialogues Concerning Natural Religion. Oxford: Clarendon Press.

John Paul II (pope). 1996. The address of Pope John Paul II to the members of the Pontifical Academy of Sciences appeared in L'Osservatore Romano on October 23, 1996, in its French original, and on October 30, 1996, in English. Both texts are reproduced in R. J. Russell, W. R. Stoeger, and F. J. Ayala, eds.1998. Evolutionary and Molecular Biology: Scientific Perspectives on Divine Action, 2–9. Vatican City State: Vatican Observatory Press; and Berkeley, CA: Center for Theology and the Natural Sciences.

Lamarck, J. B. 1809. Philosophie Zoologique (Zoological philosophy). 1914, London: Macmillan.

Lyell, C. 1830–1833. Principles of Geology. London: John Murray.

Miller, H. 1858. The Testimony of the Rocks; or, Geology in Its Bearings on the Two Theologies, Natural and Revealed. Boston: Gould and Lincoln.

Miller, K. R. 1999. Finding Darwin's God: A Scientist's Search for Common Ground. New York: Harper-Collins. *A scientist's extended argument conciliating Darwinian evolution and Christian theology.*

Moore, J. A. 2002. From Genesis to Genetics: The Case of Evolution and Creationism. Berkeley: University of California Press.

Moore, J. R. 1979. The Post-Darwinian Controversies: A Study of the Protestant Struggle to Come to Terms with Darwin in Great Britain and America 1870–1900. Cambridge: Cambridge University Press.

National Academy of Sciences and Institute of Medicine. 2008. Science, Evolution, and Creationism. New York: National Academy of Sciences Press. *A concise, forceful argument by the most distinguished scientific institution affirming the compatibility of science and religion.*

Nieuwenfijdt, B. 1718/2007. The Religious Philosopher. Whitefish, MT: Kessinger Publishing.

Paley, W. 1785. The Principles of Moral and Political Philosophy. Collected Works, 1819. London: Rivington.

Paley, W. 1794. A View of the Evidences of Christianity. Collected Works, 1819. London: Rivington.

Paley, W. 1802. Natural Theology; or, Evidences of the Existence and Attributes of the Deity Collected from the Appearances of Nature. New York: American Tract Society. *A classical treatise expounding the traditional view that the design of the world manifests the existence of the Creator.*

Peacocke, A. R. 1998. Biological evolution: A positive appraisal. In R. J. Russell, et al., eds., Evolutionary and Molecular Biology: Scientific Perspectives on Divine Action, 357–376. Vatican City State: Vatican Observatory Press; and Berkeley, CA: Center for Theology and the Natural Sciences. *A distinguished Anglican minister and theologian asserts that modern biology provides an enlightened view of creation.*

Pinker, S. 2015. The Untenability of Faitheism. Current Biology 25: R635–653.

Provine, W. 1988. Evolution and the foundation of ethics. MBL Science 3(1): 25–29.

Ray, J. 1691. The Wisdom of God Manifested in the Works of Creation. London: Wernerian Club.

Redi, F. 1685. Bacco in Toscana (Bacchus in Tuscany). Florence.

Ruse, M. 2001. Can a Darwinian Be a Christian? : The Relationship between Science and Religion. Cambridge: Cambridge University Press.

Ruse, M. 2008. Evolution and Religion: A Dialogue. Lanham, MD: Rowman & Littlefield.

Ruse, M. 2015. Atheism: What Everyone Needs to Know. Oxford: Oxford University Press.

Russell, R. J. 2007. Physics, cosmology, and the challenge to consequentialist natural theology. In N. Murphy, et al., eds., Physics and Cosmology: Scientific Perspectives on the Problem of Natural Evil, 109–130 Vatican City State: Vatican Observatory Press; and Berkeley, CA: Center for Theology and the Natural Sciences. *Physics and astronomy are shown to be compatible with Christian faith.*

Scott, E. C. 2009. Evolution vs. Creationism, 2nd ed. Berkeley: University of California Press. *An extensive and thoughtful discussion of the controversy.*

Slack, G. 2008. The Battle over the Meaning of Everything. New York: Wiley. *A narrative of the controversies between science and religion.*

Strong, A. H. 1885. Systematic Theology, 3 vols., 1907. Westwood, NJ: Fleming Revell.

Voltaire (François-Marie Arouet). 1967. Encyclopedia of Philosophy. Vol. 8, 262–270 London: Macmillan.

Ward, K. 2008. The Big Questions in Science and Religion, 79–80 West Conshohocken, PA: Templeton Foundation. *A theologian explores a variety of scientific issues often seen as contrary to religious faith and asserts that there is no necessary opposition.*

CHAPTER 18

CREATIONISM AND INTELLIGENT DESIGN

Eugenie C. Scott

OUTLINE

1. What kind of creationist?
2. The creation-evolution continuum
3. Intelligent design
4. What does the future hold?

Many are unaware that there are several kinds of creationisms, even within the tradition of Christianity. In that tradition, the various creationisms are a function of how the Bible is interpreted, and the differences reflect how much of modern science is accepted. Intelligent design is a more recent form of creationism, but in its particulars it reflects themes similar to other forms of Christian creationism. New forms of creationism may develop in the future, but it is likely that they will reflect the same ideas as their ancestors.

GLOSSARY

Creationism. As used in this essay, the Christian doctrine of special creationism: that God created the universe, earth, and living things in essentially their present form. Varieties of Christian creationism differ on *how* and *when* God creates (thousands or millions of years ago, whether over six 24-hour days, or consecutively over time, etc.) but all forms are incompatible with evolution.

Evolution. The scientific theory that the universe, earth, and living things have cumulatively changed over time. Regarding biological evolution, evolution refers specifically to the inference of common ancestry of living things.

Geocentrism. The belief, based on a literal interpretation of the Bible, that the earth is the center of the solar system, and the sun and other planets revolve around it. In its extreme form, geocentrism claims that the earth is the center of the universe.

Heliocentrism. The scientific theory that the sun (Gr: *Helios*) is the center of the solar system, and earth and other planets revolve around it.

Intelligent Design. A minimalist special creationist view in which very complex or very improbable natural phenomena are believed to be unexplainable through natural causes. Such phenomena are believed to be explainable only through the action of an intelligence (God).

Materialism. Sometimes known as *naturalism,* the philosophical view that only material (matter and energy) phenomena exist in the universe. Materialism thus rejects the idea that there is a God or gods or any supernatural forces.

Theistic Evolution. A mainstream view in Christianity in which God uses the processes of evolution to bring about His purpose for the universe and humans in it. There are many forms, varying (among other ways) in the amount of intervention attributed to God.

1. WHAT KIND OF CREATIONIST?

There has been a long-standing tension between some religious groups and evolutionary biology, and that tension plays out in schools throughout the United States. At the National Center for Science Education, we monitor the *creationism* and *evolution* controversy, and we help parents, teachers, and others cope with challenges to evolution education. All the challenges emanate from people who call themselves—or can be called—*creationists.* Creationists believe that God has specially created the universe, earth, and living things in essentially their present forms. It is a view incompatible with the science of evolution, which explains a changing universe. But all creationisms are not alike. Often, a student will tell a teacher, "I don't believe in evolution, I'm a creationist." We recommend asking in reply, "What kind of creationist?"

It is a teachable moment: the student has probably never considered that there might be more than one type of creationism, and the teacher has the opportunity not only to expand the student's horizons but also, with luck, to reduce barriers to learning evolution. And yes, it is also an opportunity to help the student understand that scientists don't "believe" in evolution, they accept common ancestry as the best explanation for the patterned differences and similarities among living things. That is, the word *belief* evokes positions held with or without evidence; hence, *belief* is at best an ambiguous word to use in the context of science. Scientists don't "believe" in evolution any more—or less—than they "believe" in thermodynamics.

"Belief" in evolution, as it is too frequently termed, occurs at a lower frequency in the United States than in almost any other developed country. The percentage of Americans accepting that humans developed from earlier species of animals hovers near 50 percent, far less than in Western Europe and Japan, where the percentages are above 70 percent and even 80 percent, respectively. Survey research shows a major disconnect between the US public's acceptance of evolution and that of scientists. In one survey of members of the American Association for the Advancement of Science, the world's largest association of scientists, 97 percent accepted the statement that "humans and other living things have evolved over time." High school teachers are more likely than the general public to accept evolution as a scientific concept, but only about 30 percent report that they teach it extensively. Fully 60 percent admit that they either omit evolution or give it short shrift. The reasons include the teachers' apprehension that evolution is a controversial issue, personal religious beliefs, and the feeling that their education did not prepare them to teach the subject well.

Teachers in the United States can expect that their students who describe themselves as creationists will usually base their creationism on some form of Christianity, the religion of most Americans, and almost always on Christianity, Islam, or Judaism. But there are exceptions: some teachers in communities where Native Americans are numerous have also reported pushback on the teaching of evolution. Other forms of creationism based on Hindu and various New Age religious beliefs also occasionally surface in the classroom.

We should therefore speak of creationisms in the plural. This point reveals as problematic the long-standing plea of antievolutionists that teachers should "teach both" evolution and creationism. How should a teacher choose which creationist version to contrast with evolution?

Even supposing that there were some reason to privilege Christianity over other religions, there are several distinct versions of Christian creationism, corresponding to the different ways in which scripture is identified and interpreted by various denominations. Mormons revere the Book of Mormon, and Seventh-Day Adventists regard the writings of Ellen Gould White as inspired; which, if either, should a teacher present? Even if only the Bible is considered, whose Bible?—the King James Version, the New Jerusalem Bible (favored by Catholics), the New International Version (favored by Evangelicals), the New Revised Standard Version (favored by mainline Protestants), or one of the scores of texts available? And given a particular version of the Bible, who is to decide which verses are relevant and how they are to be understood? In fact, these complications only scratch the surface, and the following discussion of the varieties of Christian creationism is necessarily abbreviated.

2. THE CREATION-EVOLUTION CONTINUUM

Creationism, as usually encountered in the United States, is based on a "plain" reading of the Bible. Taking the creation narrative in Genesis 1 as authoritative, creationists hold that God specially created the universe, the planet earth, and the living things on it. There are many ways to read the Bible, and the varieties of Christian creationism can be viewed on a continuum reflecting how literally they interpret the words of Genesis and how far their interpretation lies from mainstream science (figure 18-1).

Flat-Earthism

It is almost comical to believe that *flat-earthers* can exist in the twenty-first century. Nonetheless, until his death in 2001, Charles K. Johnson was president of the International Flat Earth Research Society, a small organization whose interpretation of the Bible is so extreme that passages referring to the "circle of the Earth" (circles are two-dimensional, while spheres are three-dimensional) and the "pillars of heaven" (supports for a metal dome or "firmament" arching over a horizontal planet)

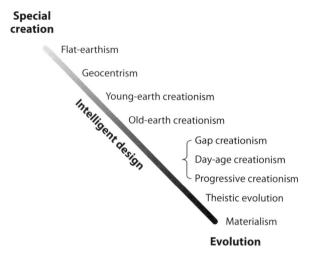

Figure 18-1. The creation/evolution continuum.

are interpreted as stating that the earth is flat. Few Christians take the Bible so literally, but geocentrists are only slightly more liberal in their exegesis.

Geocentrism

Geocentrists believe that the Bible presents earth as the center of the solar system. Passages cited include Joshua 10:12–13, in which God complies with Joshua's plea to stop the sun over the Valley of Ajalon, which requires a stationary earth. Both flat-earthers and geocentrists contend that one cannot pick and choose passages of the Bible to "interpret," so the entire Bible must be accepted as received. If Genesis is not true, they argue, how can we be sure of any of the rest of the Bible, including the New Testament and Revelations, which promise salvation? As Gerardus Bouw, a modern geocentrist author, asked, "If we cannot take God's word as to the rising of the Sun, how can we believe him as to the rising of the Son?"

Geocentrism fought it out with *heliocentrism*, the idea of a sun-centered solar system, during the sixteenth and seventeenth centuries. Heliocentrism eventually would have won because the science was right, but its acceptance was helped by a shift in church doctrine at that time away from the strict biblical literalism of the Middle Ages and early Renaissance periods. As a result, even most creationists accept heliocentrism.

Young-Earth Creationism

Young-earth creationists (YECs) agree on the importance of a plain reading of the Bible, although they do not think that requires belief in a flat earth or geocentrism. YECs are currently the most numerous creationists in the United States. YECs understand Genesis as stating that creation took place over six 24-hour days, during which God created the universe in essentially its present form. Following Archbishop James Ussher, a seventeenth-century Irish cleric, YECs hold that earth was formed only thousands, not billions, of years ago. Animals and plants were created as separate and independent "kinds" and do not share common ancestry. Humans in particular are independent creations, made in God's image.

Most YECs also support the movement known as *creation science*, which in its modern form began to be promoted in the 1960s by Henry M. Morris, founder of the Institute for Creation Research (ICR) and its leader until his death in 2006. Morris was highly respected among conservative Christians, and his influence can hardly be overestimated. The movement Morris originated contends that the data and theory of science support the claims of the Bible in all its details. The special creation of all living things by God and the existence of a worldwide Noachian flood (a literal interpretation of Genesis 6–9) are held to be supported not only by faith but also by science.

The science of creation science, however, is decidedly lacking in quality. The logic of creation science is clearly stated by Morris and his followers: evidence *against* evolution is evidence *for* creationism. This approach solves their problem of finding scientific evidence for the sudden appearance of living things in essentially their present form, which special creationism requires. But it also means focusing only on anomalies purporting to disprove evolution and ignoring the massive evidence supporting it. There is ample literature wherein scientists have examined the claims of creation science and found them both factually wrong and theoretically empty. But proponents loudly, if ineffectually, defend their claims that creationism can be made scientific.

Morris firmly believed that a universe measured in thousands of years is foundational to a proper interpretation of the Creation story in Genesis. The only acceptable understanding of creation, then, is young-earth creationism. It is true, of course, that if the earth were young, there would not have been time for much astronomical, geological, or biological evolution. For this reason, YEC institutions, including the ICR and Answers in Genesis, adhere to Morris's original vision.

YECs insist that just as a young earth is foundational to creationism, so also is the Creation story foundational to Christianity. Christianity's central pillar is the sacrifice of Jesus on the cross to redeem humankind's sins. YECs believe that if Adam and Eve had not been specially created by God, had not sinned, and had not been punished by being driven from the Garden of Eden, as described in Genesis, then there would not have been a need for a Savior to redeem the sin of Adam. If there was no need for Christ's life, death, and resurrection, as described in the New Testament, then there is no reason to believe the promise of eternal life in the Book of Revelation. The credibility of the entire Bible is thus contingent on the credibility of the special creation of earth and of Adam and Eve. Evolution is therefore unacceptable to YECs, because it is incompatible with special creationism.

In Morris's version of young-earth creationism, all sedimentary geological features are the result of Noah's flood, and scientific evidence is sought to support this conclusion. The geological column, it is claimed, only appears to present a succession of fossils showing the gradual emergence of present-day forms from earlier forms. Accordingly, the geological succession of fossils resulted from the "hydrodynamic sorting" of the remains of organisms that died in Noah's flood. It is claimed that spherical and smooth organisms such as clams would more likely be found at the bottom of the column because such shapes fall through water more readily than irregular shapes. Jointed organisms with irregular shapes, such as dinosaurs, would be found higher up. And the smarter, more mobile organisms such as mammals would likely have sought higher ground to avoid the floodwaters, explaining their occurrence higher in the geological column. These views are supported by carefully chosen examples—and by ignoring the copious data that refute them.

Old-Earth Creationism

Old-earth creationists (OECs) have perhaps the most variable positions on the continuum. OECs are special creationists, believing that God specially created living things as identifiable "kinds," and thus they reject biological evolution. But they accept the evidence from physical science that our planet and the universe are ancient. Many OECs even accept an earth that is billions of years old. A common view among OECs is to identify the Big Bang as a creative event of Genesis 1. OECs thus consider themselves true to the Bible while accepting the evidence of planetary and cosmic deep time. But among OECs, adherents "interpret" the holy text in various ways to make it compatible with an old earth.

Gap creationism requires the least tinkering with Genesis. Sometimes called "ruin and restoration" theology, gap creationism sees the possibility of two creations in Genesis, with a long period of time between them. The first creation was of a world before Adam and is referenced in the familiar words of Genesis 1:1— "In the beginning, God created the Heaven and the Earth." God then destroyed that creation, a great deal of time passed ("the Earth became without form and void, and darkness was upon the face of the deep"), and then, as stated in Genesis 1:2–31, He created the present world and its inhabitants in six 24-hour days. Gap creationists thus interpret the Bible very literally, though with room for an old earth. It is not surprising that enthusiasm for gap creationism grew in parallel with the rise of modern geological sciences in the late eighteenth and nineteenth centuries.

Day-age creationists, by contrast, believe that the six days of creation were not 24-hour days but instead long periods of indeterminate duration—perhaps hundreds of thousands or even millions of years. They retain reference to a literal Genesis six days but they, too, allow for an old earth. They cite biblical passages such as Psalm 90:4 ("For a thousand years are in your sight like a day") to suggest that the days of creation need not be 24 hours long. Yet another group of OECs downplays the idea of six days, believing instead in an interventionist God who sequentially—and specially—created living things over immense amounts of time. These *progressive creationists* thus accept the geological column as reflecting an accurate history of life on earth but do not believe that the sequences of organisms reflect evolutionary continuity.

All the positions on the continuum discussed thus far are forms of special creationism. But the continuum can be extended to include additional positions on the relationship between the Bible and science. The positions discussed next all accept the mainstream scientific findings of astronomy, geology, and biology—hence, none of them are creationist positions—but they differ from one another on theological or philosophical grounds.

Theistic Evolution

The abandonment of special creationism is clear in the next position on the continuum. *Theistic evolution* (TE) can be described as the belief that evolution has occurred but that God uses evolution and other natural processes to bring about the universe, earth, and living things. Unbeknown to most Americans, TE is mainstream Christian theology, routinely taught in Catholic and Protestant parochial schools. It is considered uncontroversial in many Protestant denominations such as Episcopalians, Presbyterians, United Church of Christ, and in the less conservative branches of Lutheranism and Methodism. Thus, when teachers hear students say "I don't believe in evolution, I'm a Catholic," it should be evident that the student is unclear on both science and theology. Embedded in the floor of the hallway of the Jordan Hall of Science at Catholic Notre Dame University is a large mosaic, 5 ft in diameter, that quotes an aphorism by a famous twentieth-century geneticist, Theodosius Dobzhansky: "Nothing in biology makes sense except in the light of evolution."

In TE, God did not have to create organisms as we see them today: organisms can descend with modification from earlier forms. Believers in TE scorn biblical literalism, being critical of the literalists not only for views incompatible with modern science but also for theological views they consider outmoded and inconsistent. The TE view is that Christian theology must reflect what we know of the world from science if it is to be coherent. Unlike YEC and OEC, the TE view accepts standard scientific interpretations of evidence from geology, physics, chemistry, and biology that indicate that the universe has a long history and that organisms have evolved.

Like all theists, adherents to TE believe that the universe was created for a purpose, which science cannot address, although science can address and explain the processes involved in the creation of that universe. TE believers range along a continuum of their own, varying in how much and in what ways God intervenes over time. Divine intervention is usually conceived as miraculous: with miracles, God violates His created laws, such as by raising Jesus from the dead. However, it is also important in TE for there to be minimal intervention. This "economy of miracles" reflects theological issues not germane to this discussion, such as free will, and the consequences of God "breaking" his own laws. So varieties of TE differ in the degree to which God was "hands-off" in the creation—from one in which God set forth the laws of the universe and allows it to evolve without intervention, to another interpretation in which God also created the first replicating organism (after which evolution proceeded naturally), to yet another in which God intervened also to bring about the evolution of humankind.

The amount of divine action in the creation of the universe is not the only criterion shaping TE views. Like other Christians, believers in TE are also concerned with the degree to which the Deity is personal: an entity who is involved in a meaningful way with the self. One extreme is again a God who created the laws of the universe and is thereafter uninvolved. At the other extreme is the interventionist God to whom one might pray and hope to receive an answer.

The continuum thus far has expressed a greater or lesser reliance on biblical literalism. It has also reflected an inverse acceptance of modern science, with the flat-earthers and geocentrists rejecting some of the most basic facts of modern science, YECs rejecting less familiar but core principles of physics and geology (such as radioisotopic dating) and biology, OECs more or less accepting the physical sciences but rejecting modern biology, and TEs accepting the conclusions of all modern science. All the positions discussed so far have been theistic ones: God exists and is in some way involved in creating the universe in which we live. Next on the continuum are materialists, who reject the concept of a God or higher power.

Materialism

Because this chapter deals with creationism, the nuances and variations of *materialism* will not be discussed in detail. Briefly, *materialists* believe that matter and energy not only are sufficient to explain the physical universe, as with science, but also are sufficient in a metaphysical sense: there are no gods or supernatural forces or powers. Among materialists there are agnostics, who agree with Thomas Henry Huxley (who coined the term *agnostic*), that one can never know for certain whether there is a God. Agnostics suspend belief. Atheists deny belief in God or gods, and there is a debate among them whether atheism is a philosophical system or merely the denial of the supernatural. Humanism is a nontheistic philosophical system with deep historical roots.

3. INTELLIGENT DESIGN

What about the *intelligent design* (ID) movement? On the diagram of the continuum (figure 18-1), it is shown straddling OEC and YEC, because ID is, at heart, special creationism, but carefully formulated not to take a stance on the issues that separate OEC and YEC. (In the words of one of the early collections of ID writings, ID espouses "mere creation.") While ID is sometimes erroneously conflated with TE, in practice the ID movement has consistently been antievolutionary in its focus. Also, leaders of the ID community have strongly rejected TE, and the rejection is mutual. Nonetheless, ID has also been criticized by proponents of YEC and, despite some initial enthusiasm, by some leaders in the OEC community.

The reasons for this apparent contradiction lie in the history and content of ID and the strategy its leaders have used to promote their view to the public. The history of ID shows it emerged from a group of OECs (and some YECs) in the mid-1980s. These conservative Christians were dissatisfied with the lack of progress of the YECs in convincing the public to reject evolution or, at least, to accompany its teaching with some form of creationism (such as creation science). At the time, laws promoting equal time for creation science were being tested in the courts, and after a thorough defeat in an Arkansas federal district court, creationists realized that creation science was too obviously tied to Christian religion to survive the Establishment Clause of the United States Constitution. That clause requires public institutions to be religiously neutral. Teaching creation science was judged to be the promotion of religion and thus unconstitutional.

ID emerged as a stripped-down form of creationism out of a series of private meetings (attended by both YECs and OECs) and from the production of a supplemental high school textbook, *Of Pandas and People*, intended to "balance" standard evolution-based textbooks. It ignored creation science favorites, such as the age of the earth and Noah's flood, in favor of the core creationist principle of special creation, although the term *creationism* was (and is) carefully avoided. The ID movement reflected the "argument from design" of William Paley's 1802 book, *Natural Theology*, which compared highly complex biological structures to human-made artifacts. Paley contended that just as a pocket watch could not have assembled itself but required a watchmaker, so, too, a complex biological structure such as the human eye also required a designer and artificer—God. Modern ID examples tend to focus on the complexity of molecular structures. The flagellum of a bacterium is a favorite example of a biological "engine" that is "irreducibly complex" (supposedly too complex to have been produced through natural selection) and thus, it is argued, the product of design by an intelligent agent. Such irreducibly complex structures are called forth in abundance: DNA, the first cell, the body plans of invertebrate phyla of the Cambrian explosion, and so on. Whenever such irreducibly complex structures are discovered, an intelligent designer is invoked, because great complexity is assumed to be unattainable through natural causes.

Who is the intelligent designer responsible for such structures? Proponents of ID are often coy, suggesting that it could be extraterrestrial aliens or time-traveling cell biologists from the far future. However, the more candid among them will acknowledge that they believe the designer to be God, even while agreeing that that is a conclusion unwarranted by science. But when an intelligent agent is invoked at every appearance of an irreducibly complex structure, what is being proposed is actually a form of progressive special creationism. At the grassroots level, ID is understood to be about creationism, with God as the designing agent—even if the leadership of the movement attempts to obscure these identifications to avoid running afoul of the Establishment Clause.

In 1987 the Supreme Court declared in *Edwards v. Aguillard* that teaching creation science in the public schools was unconstitutional. Therefore, when a school board in Dover, Pennsylvania, required teachers to teach ID, lawyers for the plaintiffs in the subsequent 2005 federal district court trial *Kitzmiller v. Dover* sought to demonstrate historical links between creation science and ID. They were successful: such links were crucial in the judge's decision to declare ID a religious rather than a scientific view and that the teaching of ID therefore violated the Establishment Clause.

With ID's roots firmly in creation science, why have the two most prominent YEC organizations, the Institute for Creation Research and Answers in Genesis, attacked ID? Part of the ire of the YECs toward ID arises because of a strategy of ID leaders to omit biblical themes such as the flood of Noah, the special creation of Adam and Eve, and a young age of the earth. OECs tend to outnumber YECs in the leadership of ID and, in fact, ID was largely unknown other than to creationism watchers until 1991, when University of California, Berkeley, law professor Phillip Johnson's book *Darwin on Trial* was published. It is not unusual for antievolution tracts to emerge from creationist institutions or Bible colleges, but such tomes rarely emanate from faculty at major secular universities.

Johnson arguably put ID on the map, as far as the public was concerned, and Johnson's leadership in shaping the legal and philosophical approach of ID was substantial during the 1990s and early 2000s, until ill health required him to take a lower profile. Among other things, Johnson contended that all Christian creationists should unite to attack evolution, setting aside their young-earth versus old-earth squabbles and other differences until they had convinced the public of the scientific and religious shortcomings of evolution. Once evolution was defeated, all the creationists could have a polite discussion over their differences. This strategy may have been appealing to individual creationists, but established creationist organizations were resistant to stepping back from their cherished positions. Eventually they declared that ID—correct in its bashing of evolution—nonetheless was doomed to failure because it would not bring the public to Christianity unless it put the Bible at the center of its mission. And that, of course, was at the heart of the matter. Merely persuading the public that evolution was unsupported by science and inherently atheistic was inadequate: it was necessary to replace evolution with special creation.

4. WHAT DOES THE FUTURE HOLD?

Disagreements among leaders of the creationist movements are only part of the story. What is perhaps more significant is how these movements are viewed by the public. The differences between YECs and OECs are stark, and a choice must be made between an earth that is billions of years old or only a few thousand years old. ID, meanwhile, is commonly considered to be an adjunct to either the YEC or the OEC perspective. Rather than viewing ID as the sophisticated scientific argument dreamed of by its proponents, most members of the public who are familiar with it see it as a generalized form of creationism—which, in fact, it is.

Since the *Kitzmiller v. Dover* trial, the ID star has burned a bit more dimly. Leaders of the movement, affiliated with the Seattle-based Center for Science and Culture at the Discovery Institute, are now encouraging legislation and regulations that would encourage the teaching of "evidence against evolution." Sometimes, evolution is bundled with other "controversial issues" such as global warming and human cloning for special treatment in the curriculum. A common tactic calls for teachers to be given "academic freedom" to bring in "alternative views" to those expressed in the textbook or state standards. In the case of evolution, of course, "alternative views" is a euphemism for creationism. It appears to be a popular strategy for promoting creationism: more than 70 "academic freedom"–style bills were proposed in various state legislatures in the period 2003–2015, although opponents managed to defeat almost all of them in committee. When such bills reach the floors of their respective chambers, however, they are often difficult for elected officials to publicly oppose. Two bills have passed: one in Louisiana in 2008 and one in Tennessee in 2012.

In contrast, YEC is thriving, although explicit attempts to promote the teaching of creationism in the public schools are rare. Particularly prominent is the Answers in Genesis ministry, which since its founding in 1994 has been remarkably successful at capturing the market for creationism. This success is apparently due in part to its adopting a style heavier on evangelism and lighter on science than the Institute for Creation Research and in part to its use of the latest technology, including a well-crafted website. Answers in Genesis also opened a lavish "Creation Museum" in northern Kentucky in 2007, which will be joined in the future by a Noah's Ark theme park. The Institute for Creation Research has moved to an expanded new campus in Dallas, Texas, and is raising funds to rebuild its own museum in that city. Elsewhere in the country, several smaller creationism museums are in the planning stages or have already opened.

All in all, it appears that the creationism movement in the United States is prospering. And given its fragmented history, it is safe to say that even if some constituents fall out of favor, new varieties will emerge somewhere on the continuum.

FURTHER READING

Berkman, M., and E. Plutzer. 2010. Evolution, Creationism, and the Battle to Control America's Classrooms. New York: Cambridge University Press. *Invaluable for its extensive, thoughtful, and fruitful use of survey data, especially its rigorous national survey of high school biology teachers.*

Forrest, B., and P. R. Gross. 2007. Creationism's Trojan Horse: The Wedge of Intelligent Design. Rev. ed. New York: Oxford University Press. *The definitive exposé of intelligent design as a strategy of rebranding creationism, updated with a chapter on events after the Kitzmiller trial.*

Larson, E. 2003. Trial and Error. 3rd ed. Oxford: Oxford University Press. *The authoritative history of the legal struggles over the teaching of evolution in the United States, although it stops short of the Kitzmiller trial.*

Matzke, N. J. 2015. The evolution of antievolution policies after Kitzmiller v. Dover. Science 351(6268): 10–12. *A recent analysis of the "academic freedom" laws that have largely replaced legislation calling for equal time for creationism.*

McCalla, A. 2006. The Creationist Debate: The Encounter between the Bible and the

Historical Mind. New York: Continuum. *A synoptic history of the creationism/evolution controversy, focusing on the development of the historical sciences and how the ways of interpreting the Bible developed in response.*

National Science Board. 2016. Science and Engineering Indicators 2016. Arlington, VA: National Science Foundation. *A biannual review of statistical information on science and engineering, including a survey of adult Americans' knowledge of science content and science methodology.*

Numbers, R. L. 2006. The Creationists: From Scientific Creationism to Intelligent Design. Exp. ed. Cambridge, MA: Harvard University Press. *A monumental work on the history of the creationist movement, newly updated with a chapter on intelligent design.*

Pennock, R. T. 1999. Tower of Babel: The Evidence against the New Creationism. Cambridge, MA: MIT Press. *The first, and still a valuable, examination of the intelligent design movement, by a philosopher who testified at the Kitzmiller trial.*

Ruse, M. 2005. The Evolution-Creation Struggle. Cambridge, MA: Harvard University Press. *The distinguished philosopher and historian of science attempts to understand the roots of the creationism/evolution controversy.*

Scott, E. C. 2009. Evolution vs. Creationism: An Introduction. 2nd ed. Berkeley: University of California Press. *A comprehensive history, commentary, and sourcebook on the creationism/evolution controversy.*

CHAPTER 19

EVOLUTION AND THE MEDIA

Carl Zimmer

OUTLINE

1. Evolution and the birth of modern science communication
2. Evolution and creationism: The dangers of false balance
3. Evolution and the rise of new media
4. The *Darwinius* affair: A cautionary tale
5. Conclusion

On March 28, 1860, the *New York Times* ran a very long article on a newly published book called *On the Origin of Species*. The *Times* explained that the dominant explanation for life's staggering diversity at the time was the independent creation of every species on earth. "Meanwhile," the anonymous author wrote, "Mr. DARWIN, as the fruit of a quarter of a century of patient observation and experiment, throws out, in a book whose title has by this time become familiar to the reading public, a series of arguments and inferences so revolutionary as, if established, to necessitate a radical reconstruction of the fundamental doctrines of natural history."

Today, more than 150 years later, evolutionary biologists are continuing to reconstruct natural history, and journalists are still documenting that reconstruction. Each week brings a flood of reports on new research into evolution, ranging from fossil dinosaurs to the emergence of new strains of viruses to evolutionary clues embedded in the human genome. The *New York Times* continues to publish articles about evolution, as do many other newspapers and magazines. But reports on evolution can also take many new forms that were inconceivable in Darwin's day. They can be the subject of television shows, blogs, podcasts, and tweets. This chapter examines the ways in which media has treated evolution over the past four decades, and the rapid changes currently unfolding. (To learn about the fascinating relationship of evolution and the media in earlier periods of history see Browne 2001 and Larson 1998.)

1. EVOLUTION AND THE BIRTH OF MODERN SCIENCE COMMUNICATION

To understand the relationship between evolution and the media, it helps to take an evolutionary perspective. The journalistic coverage of evolution as we know it today began to take shape in the 1970s. Newspapers, especially in the United States, were growing rapidly at the time and developing new features to attract readers. Many newspapers hired reporters who specialized in science, and many science writers focused much of their attention on evolutionary biology. For example, Boyce Rensberger, a science writer for the *New York Times*, wrote a string of stories about evolution in the 1970s. In one typical Rensberger article (April 12, 1975), titled "East Africa Fossils Suggest That Man Is a Million Years Older Than He Thinks," he described the discovery of a 3-million-year-old fossil of a hitherto-unknown species of hominin, *Australopithecus afarensis*.

Four years later, the *Times* founded a weekly section dedicated to science. It was the first science section ever included in an American newspaper, but in the next few years, many other newspapers followed suit. A number of science magazines were also launched. Old standards like *Scientific American* were joined by start-ups such as *Discover* and *Omni*. All these new publications gave special attention to evolution.

One reason for this focus was that evolutionary biology itself had entered an exciting period of renewal, and so there were many stories for reporters to write about. New fossils like *A. afarensis* provided paleontologists with fresh insights into human evolution. Dinosaurs, which had long been considered sluggish and slumped, received a makeover. During the 1970s, the Yale paleontologist John Ostrom oversaw the reconstruction of dinosaurs as fast-running, warm-blooded creatures—an upgrade from *Godzilla* to *Jurassic Park*.

Geologists were also adding to evolution's cinematic appeal. In the late 1970s Walter Alvarez of the University of California at Berkeley and his colleagues discovered clues that an asteroid smashed into earth 65 million years ago. That collision happened to coincide with the end of the Cretaceous period, a time of mass extinctions that claimed the dinosaurs Ostrom was rehabilitating. Alvarez made a radical connection between the impact and the mass extinctions. Mass extinctions had long been thought to stretch across millions of years, caused by slow-moving processes such as gradual sea level change. Alvarez and his colleagues offered a vision of sudden disaster: the asteroid impact threw dust and rock high into the atmosphere, causing a global environmental catastrophe—darkness for months, acid rain, global warming, and more. In a geological flash, millions of species became extinct.

Alvarez was arguing for a catastrophic mode of evolution. To understand evolution 65 million years ago, we could not simply extrapolate back from the small, incremental changes natural selection produces today from one generation to the next. As a result, some scientists argued, the end-Cretaceous extinctions did not fit into the framework of the Modern Synthesis. The *Modern Synthesis*—an integration of genetics, paleontology, ecology, and other branches of biology—explained life predominantly as the result of natural selection operating on small differences among individuals over vast periods of time.

Challenges to the Modern Synthesis came from studies not just on mass extinctions but on more tranquil periods of the fossil record. Paleontologists Niles Eldredge and Stephen Jay Gould argued that the fossil record revealed a pattern of stasis and change, a pattern they dubbed *punctuated equilibria*: species remained stable for millions of years, and new species rapidly branched off in just thousands of years. Eldredge and Gould argued that this pattern of evolution allowed selection to take place not just between individuals but perhaps also between species.

Science writers chronicled these challenges to the Modern Synthesis, but they also reported on other scientists who were expanding its scope. In 1976 the British zoologist Richard Dawkins, building on the work in the 1960s of George Williams and William Hamilton, published *The Selfish Gene*. Dawkins argued that evolution was best understood from a gene-centered perspective. The Harvard biologist Edward O. Wilson undertook a similar project, interpreting a vast range of behaviors—from the selfless work of sterile worker bees to the bloodshed of human warfare—as strategies for genes to get themselves replicated. In 1975 he unveiled his synthesis in the book *Sociobiology*. Rensberger (May 28, 1975) reported its publication on the front page of the *New York Times* in his article "Sociobiology: Updating Darwin on Behavior."

As evolution was appearing on the front pages of newspapers, science programming was also emerging on television. In 1974, for example, the Public Broadcasting Service developed the *Nova* series. New research on evolution figured prominently in these shows as well. In his 1980 series *Cosmos*, Carl Sagan discussed the basic principles of evolution, along with new ideas about the role of comets and other impacts on the history of life. And in 1981, Walter Cronkite, having just retired from his nightly television news show, hosted a series of science shows called *Cronkite's Universe*. On one episode his guests were Donald Johanson— one of the discoverers of *A. afarensis*—and the paleoanthropologist Richard Leakey. Johanson and Leakey engaged in a heated debate about the place of *A. afarensis* in human evolution. Johanson believed it was on the line that led to *Homo sapiens*, while Leakey considered it a side branch. During the program, Johanson held up a chart showing his version of hominin phylogeny. Next to it was a blank space where he asked Leakey to draw his hypothesis. Instead, Leakey drew an X through Johanson's tree. In its place, he drew a large question mark (Wilford 2011).

Evolutionary biologists debated on television, and they also debated in print. As Dawkins and Wilson garnered attention for their expansion of the Modern Synthesis, Stephen Jay Gould and other scientists launched scathing criticisms, arguing that adaptationists ascribed far too much power to natural selection. They condemned sociobiology as "just-so stories"— plausible-sounding tales of adaptation rather than carefully constructed and tested hypotheses. Most of Gould's attacks took place not in the pages of scientific journals but in popular publications such as *Natural History* and the *New York Review of Books*. Dawkins, Wilson, and others responded in kind, and the debate gave rise to a number of hugely popular books, such as Gould's *Wonderful Life* (1998) and Dawkins's *The Blind Watchmaker* (1996).

2. EVOLUTION AND CREATIONISM: THE DANGERS OF FALSE BALANCE

In December 1981 a number of the top science journalists in the United States converged on Little Rock, Arkansas, to cover a story about evolution. The story did not concern a new fossil, or a new hypothesis about speciation, but a trial. Earlier that year, the Arkansas legislature had passed a law requiring that public school teachers present "creation science" alongside evolution in their biology classes. A group of teachers and religious figures filed a lawsuit challenging the law as an unconstitutional promotion of religion.

Of all the sciences, evolutionary biology attracts an unmatched amount of social controversy. Organized religious opposition to the teaching of evolution in the United States first emerged in the 1920s, leading to the famous Scopes "monkey trial" of 1925. Conflicts over the teaching of evolution have continued to break out in the decades since then. The 1981 case *McLean v. Arkansas* led to the banning of "creation science" from classrooms. But it did not stop the conflict over the teaching of evolution. Journalists have continued to report on the attempts of some state and local school board members to question the validity of evolution and to promote creationism in its various forms. (See chapter 18 for more on the history of creationism in the United States.) Much of the coverage of evolution found in newspapers, magazines, and television news programs addresses these social conflicts, rather than the science of evolution itself. This bias is an unfortunate result of the nature of modern journalism: editors and journalists seek easily explained conflicts between people. Another weakness in much modern journalism is a craving for false balance. If one side in court trial says that evolution is true, then a journalist may feel obligated to unquestioningly quote someone from the other side. This "he said, she said" form of journalism can be legitimate in political reporting, but it is unacceptable in science reporting. It implicitly gives equal credibility to opposing sides, even if one side has no science whatsoever to back up its case. False balance promotes the mistaken impression that evolution is controversial within the scientific community, rather than the foundation of modern biology.

3. EVOLUTION AND THE RISE OF NEW MEDIA

In some ways, the relationship between evolution and the media has changed little since the 1970s. Television networks periodically air shows dealing with paleontology and human origins. In 2014, astrophysicist Neil deGrasse Tyson presented a new version of *Cosmos* on the

Fox television network that featured an entire episode on the subject of evolution. Meanwhile, Richard Dawkins and Edward Wilson continue to write best-selling books, along with many talented younger evolutionary biologists, such as Steven Pinker, Jared Diamond, Olivia Judson, Sean Carroll, and Neil Shubin. Evolution still inspires abundant journalism in newspapers and magazines. And journalists continue to cover controversies over evolution, including the 2005 *Kitzmiller v. Dover* case and Louisiana's 2008 law protecting creationist science teachers in the name of "academic freedom."

Yet, tremendous changes are under way. People are rapidly moving to the Internet to learn about science, including evolution. Evolution first went online in the 1990s, when a few evolutionary biologists and evolution aficionados began to set up online discussion groups such as the one at talk.origins.org. They posted comments about new advances in evolutionary biology and the attempts of creationists to block the teaching of evolution. Later these sites also hosted lists of frequently asked questions about evolution, such as, "If we evolved from monkeys, why are there still monkeys?"

Talk.origins and other evolution discussion groups were founded at a time when few people outside universities had even heard of the Internet. As the number of Internet users grew exponentially, programmers invented more powerful ways to post information online. Blogs allowed people to self-publish their writing; they also made it possible to post podcasts, video, and other media. Today, thanks to the Internet, far more biologists are regularly writing about evolution than ever before (Goldstein 2009).

As blogs bloomed, the older venues for news on evolution struggled. A number of science magazines launched in the 1970s and 1980s, such as *Omni* and *Science 80*, eventually folded. Science coverage in newspapers suffered in the 1990s. In 1989, a total of 95 newspapers ran science sections. By 2013 that number had shrunk to just 19. Those shuttered science sections were the victims of an industry-wide blight. Newspapers were being squeezed for greater profits, even as their readerships were declining. They offered their senior staff buyouts to reduce labor costs. A number of the science writers who had been part of the field's first efflorescence left the business.

Many newspapers and magazines now see the Internet as an essential part of their future. As of 2014, the *New York Times*, for example, sells 1.1 million copies of the Sunday edition. But 41 million people access its website each month. The news on its site also radiates outward across the World Wide Web as people comment on it in blogs and forums.

These huge changes in readership are changing the way evolution and other branches of science are reported. The print edition of the *New York Times* still includes a science section every Tuesday, but it also offers many untraditional kinds of coverage of evolution. For example, the newspaper has published blog posts by evolutionary biologists about their work, and offers podcasts and even short videos about evolution. In 2009 it posted the *Origin* in an online form, with annotations from some of the world's leading scientists.

But the *New York Times* and other publications have to compete with scientists themselves to present evolution to the public. The University of California, Berkeley, has set up a major website called Understanding Evolution (evolution.berkeley.edu), which presents not only the basic concepts of evolution but also new scientific developments. In 2009, Casey Dunn, an evolutionary biologist at Brown University, established a blog called *Creaturecast* about animal evolution (creaturecast.org), where he and his co-bloggers regularly publish innovative videos. One episode explains how single-celled organisms made the evolutionary transition to multicellularity, for example. The film is a stop-action animation of purple modeling clay, which morphs into cells, which then join together into bodies. The video is at once charming and surprisingly enlightening. And most important, it was something no one would have imagined a few years earlier. In 2013, the *New York Times* began to regularly feature Creaturecast videos on their Science page.

4. THE *DARWINIUS* AFFAIR: A CAUTIONARY TALE

Creative efforts such as *Creaturecast* inspire hope for the future of evolution and the media, but they should not inspire a blind optimism. The Internet is also home to a great deal of misinformation about evolution, especially on creationist sites. Some of these sites are relatively obvious, such as Creation Safaris (creationsafaris.com). Other sites cloak their creationism. A site with the harmless-sounding name All About Science

(AllAboutScience.org) has a long page titled "Darwin's Theory of Evolution: A Theory in Crisis." It takes a bit of snooping around to discover that All About Science is produced by a group called AllAboutGod.com.

Too often, journalists for major media provide poor information about evolution online. In fact, the very nature of twenty-first-century media fosters bad reporting on evolution. One of the most instructive events took place in May 2009, when journalists reported on the unveiling of a new fossil of a primate dubbed *Darwinius masilae*.

The unveiling was unique in the annals of paleontology. At the American Museum of Natural History, New York City Mayor Michael Bloomberg and other luminaries gazed at the slab preserving a 47-million-year-old specimen (known as Ida, named after the daughter of one of the paleontologists who described the fossil). Minutes before the press conference was to commence, the electronic journal *PLoS ONE* published a paper about the fossil. Some of the paper's authors, speaking at the press conference, described the fossil as both the holy grail of paleontology and the lost ark of archaeology.

The scientists were not the only ones to speak that morning. Nancy Dubuc, an executive at the History Channel, said that the fossil "promised to change everything that we thought we understood about the origins of human life" (Pilkington 2009). Why was Dubuc there? Because the unveiling of *Darwinius* was actually a television phenomenon, years in the making.

Television producers had started putting together a big-budget show about *Darwinius* even as the scientists were analyzing the fossil and writing up their results. The documentary's main message was also the chief argument in the *PLoS ONE* paper: *Darwinius* belonged to the lineage that led to monkeys, apes, and humans. As a result, it illuminated how our ancestors diverged from more distantly related primates, such as lemurs. As the air date for the documentary approached, the History Channel cranked up a massive publicity machine. A trade book was rushed into print; ads appeared; YouTube videos spread like viruses. The History Channel set up an elaborate website called Revealing the Link (revealingthelink.com). It featured hyperbolic claims from the scientists, such as "When our results are published, it will be just like an asteroid hitting the Earth."

As press manipulation, the strategy worked well. Newspapers, magazines, and even television news programs ran stories about *Darwinius* on their websites on the day of its grand unveiling. Few of them would have ever considered covering the discovery of an Eocene primate, it's safe to say, without the elaborate publicity. Unfortunately, most reporters simply relayed hyperbolic quotes from their sources. They also demonstrated some deep misunderstandings about evolution. "Fossil is evolution's missing link," announced the *Sun*, falling prey to the common misbelief that paleontologists could ever determine our direct ancestors (Soodin 2009). (In fact, paleontologists compare related species to determine the pattern by which new traits emerged in different lineages.) Given the upheavals in the media these days, it's not surprising that the press was so swayed by the *Darwinius* publicity machine. The number of skilled science writers who can report a story like this one with the proper skepticism is dwindling. And all media organizations are racing to be the first to get news online.

A few veteran journalists tried to obtain the paper to show it to other experts on fossil primates to get their opinion on its importance, but they were thwarted by both *PLoS ONE* and the authors. Ann Gibbons, a correspondent for *Science*, finally got her hands on the paper the weekend before the press conference, but only after signing a nondisclosure agreement with the television company that produced the *Darwinius* documentary. Gibbons promised not to show the paper to anyone before the press conference (Zimmer 2009).

The first wave of articles about *Darwinius* was based entirely on the press conference and claims from the scientists who had published the paper. Days later, Gibbons and a handful of other science writers published articles that offered a broad look at *Darwinius*, rather than the breathless press conference coverage that dominated the news. Nearly all the other experts reporters contacted thought the fossil was impressive but that the claims of its kinship with humans unjustified. "This hypothesis now lies well outside the scientific mainstream, and the discovery and description of Ida have done little to rehabilitate it," wrote Christopher Beard of the Carnegie Museum of Natural History in the September 2009 issue of *American Scientist*.

In October 2009, five months after the *Darwinius* circus had folded its tents and moved on, paleontologists Erik Seiffert and colleagues published an important new paper on early primate evolution. They described another early primate fossil, called *Afradapis*, and compared its anatomy with that of *Darwinius* and a wide range of other primate fossils. Their analysis placed *Darwinius* on the branch that led to lemurs, not to us. The reaction from the press for this paper was a stark contrast to the pandemonium that greeted *Darwinius* in May. Very few newspapers and other publications even mentioned the new study. Perhaps if there had been a big-budget documentary on *Afradapis*, things might have been different.

5. CONCLUSION

Information about evolution is now available in a staggering range of forms. But readers, listeners, and viewers cannot simply assume that everything they encounter is accurate. People must learn to think critically about what they read, watch, and listen, and should also strive to develop a strong understanding of the basic principles of evolutionary theory. They can also tap into the collective wisdom of the blogosphere. And finally, they should resist the rapid-fire allure of the Internet. After all, the scientific process does not run on a 24-hour-a-day news cycle. It takes years for scientists to gather data and present hypotheses, and for other scientists to test them. Journalism, it is often said, is the first draft of history. In the history of evolutionary biology, that first draft is sometimes wrong.

FURTHER READING

Beard, C. 2009. The weakest link. American Scientist (September–October), american-scientist.org/bookshelf/pub/the-weakest-link.

Browne J. 2001. Darwin in caricature: A study in the popularisation and dissemination of evolution. Proceedings of the American Philosophical Society 145 (December): 4.

Dawkins, R. 1986.The Blind Watchmaker: Why the Evidence of Evolution Reveals a Universe without Design. New York: W. W. Norton.

Goldstein, A. 2009. Blogging evolution. Evolution: Education and Outreach 2: 548–559

Gould, S. J. 1989. Wonderful Life: The Burgess Shale and the Nature of History. New York: W. W. Norton.

Larson, E. 1998. Summer of the Gods: The Scopes Trial and America's Continuing Debate over Science and Religion. Cambridge, MA: Harvard University Press.

Pilkington, E. 2009. To get a glimpse of the Ida fossil, the media make monkeys of themselves. Guardian, May 19, 2009, guardian.co.uk/science/2009/may/19/ida-fossil-primate-media-us.

Seiffert, E. R., J.M.G. Perry, E. L. Simons, and D. M. Boyer. 2009. Convergent evolution of anthropoid-like adaptations in Eocene adapiform primates. Nature 461: 7267 (October 22): 1118–1121.

Soodin, R. 2009. Fossil is evolution's "missing link." Sun, May 19, 2009, thesun.co.uk/sol/homepage/news/article2437749.ece.

Wilford, J. 2011. Tracking human lineage through a bramble. New York Times, May 9, 2011.

Zimmer, C. 2009. Science held hostage. Loom, May 21, 2009, phenomena.nationalgeographic.com/2009/05/21/science-held-hostage/.

PART V

Nature and Nurture

CHAPTER 20

LINGUISTICS AND THE EVOLUTION OF HUMAN LANGUAGE

Mark Pagel

OUTLINE

1. What is language?
2. When did language evolve?
3. Why did language evolve?
4. The evolution of human languages
5. Languages adapt to speakers
6. The future of language evolution

This chapter discusses how human language differs from all other forms of animal communication, when and why it evolved, and how elements of language evolve over long periods of time. It closes with a brief account of how languages might evolve in an increasingly globalized world.

GLOSSARY

Cognate. The term in linguistics analogous to *homology* in evolution. Two words are cognate if they derive from a common ancestral word (e.g., the English *water* and the German *wasser* are cognate words, but neither is cognate to the French *eau*).

Homology. A term used in genetics and evolution to identify two genes or two traits thought to derive from a common ancestor. Human and chimpanzee hands and fingers are homologous, as are many human and chimpanzee genes. Bat wings, by comparison, are not homologous to bird wings, even though both are used for flying.

Language. A form of communication unique to humans consisting of discrete words formed into sentences composed of subjects, objects, and verbs.

Phylogeny. A tree diagram like a family tree but conventionally depicting the evolutionary relationships among a group of species. The diagram can be generalized to any evolving objects transmitted from one generation to the next, including languages.

Regular Sound Correspondences. A term used in linguistics to describe statistical regularities in the way sounds change in words over long periods of time. One of the best known of these is the regular replacement of a *p* sound at the beginning of a Latin word with an *f* sound in the Germanic languages, as in the change from Latin *pater* to *father*.

1. WHAT IS LANGUAGE?

We instinctively recognize that human language is unique among all forms of biological communication, but what do we mean by that? Most animals communicate, but humans are the only animals with *language*. Human language is distinct in having the property of being compositional: we alone communicate in sentences composed of discrete words that take the roles of subjects, objects, and verbs. This makes human language a digital form of communication as compared with the continuously varying signals that typify the grunts, whistles, barks, chest thumping, bleating, odors, colors, chemical signals, chirruping, or roars of the rest of life. Those familiar sights, sounds, and smells might signal an animal's status or intentions, or indicate its physical prowess; they might tell a predator it has been spotted, or send a message to nearby relatives of an imminent danger. But lacking subjects, verbs, and objects, these acts of communication do not combine and recombine to produce an endless variety of different messages. Thus, your pet dog can tell you it is angry, and even how angry it is, but it cannot recount its life story.

2. WHEN DID LANGUAGE EVOLVE?

No one knows when the capacity to communicate with language evolved, but we can narrow the range of possibilities. Our closest living relatives, the modern chimpanzees, cannot speak, but we can, and so this tells us that language evolved sometime in the 6–7 million years between the time of our common ancestor and the late arrival of modern humans, around 160,000 to 200,000 years ago. For about the last 2.5 million years of that time, the evolutionary lineage of *Homo* or "human" species that would eventually give rise to modern humans, or *Homo sapiens*, was evolving in Africa. One of the best known of these *Homo* ancestors is *Homo erectus*, which probably evolved about 2 million years ago and had a relatively large brain, although at around 700 to 800 cc, it was still only a little more than half the size of the modern human brain. *Homo erectus* skulls reveal the impressions of two slightly protruding regions of the brain—known as Broca's and Wernicke's areas after their discoverers—that neuroscientists have identified as being involved in speech, at least in humans. This finding had led some researchers to suggest that *Homo erectus* had the capacity for speech, even if perhaps a rudimentary one. But Broca's and Wernicke's areas are also enlarged in some apes, so their presence is not by itself a definitive indicator that a species had language. There is also no evidence to suggest *Homo erectus* had anything even remotely close to the complex societies, tools, or other artifacts that we recognize as fully human.

A question that excites great differences of opinion is whether the Neanderthals spoke. This sister species to modern humans last shared a common ancestor with us sometime around 500,000 to 700,000 years ago. Neanderthals had large brains, they could control fire, and they had managed to occupy much of Eurasia by perhaps 300,000 years ago. Speculation that they had language has been fanned by the recent discovery that the Neanderthals had the same variant of a segment of DNA known as FOXP2, possessed by modern humans, that has been implicated, among many other effects, in influencing the fine motor control of facial muscles that is required for the production of speech. FOXP2 is not a gene itself, even though it is often described as one, but rather a short segment of DNA that affects how other genes are expressed. In fact, FOXP2 affects the human brain by causing about 50 genes to be expressed more and another 50 to be expressed less.

But, just as was true of Broca's and Wernicke's areas in *Homo erectus*, having the same variant of FOXP2 as modern humans do doesn't tell us that the Neanderthals had language—they might have, but we cannot conclude with certainty that they did on the basis of this short segment of DNA. An analogy explains why. Most people's cars have engines and so do Ferraris. But this doesn't make every car a Ferrari. Closer to our own case, we know the modern human brain differs greatly from those of the Neanderthals in having a more fully developed and highly interconnected neocortex. This is the evolutionarily new and enlarged uppermost layer of the brain that is implicated in symbolic thinking and language. In simple terms, that thick layer of our cortex is why modern humans have foreheads and the Neanderthals did not. Given these differences, there is no compelling reason to conclude that FOXP2 affected Neanderthals' brains as it does ours. Indeed, recent evidence shows that small changes to the human version of FOXP2 mean that, despite having a primary sequence identical with the Neanderthal version, human FOXP2 is expressed differently in our brains.

Further evidence against the Neanderthals' having language comes from the archaeological record, which shows they did not produce the range of tools and symbolic objects that we associate with the cognitive complexity of *Homo sapiens*. For example, by perhaps 60,000 years ago modern humans were producing abstract and realistic art, and jewelry in the form of threaded shell beads, teeth, ivory, and ostrich shells; they used ochre and had tattoos; they had small stone tools in the form of blades and burins; they made artifacts of bone, antler, and ivory, as well as tools for grinding and pounding; and their improved hunting and trapping technology included spear throwers, nets, and possibly even bows. The Neanderthals did not produce any of these things—for instance, there is no evidence they even had needles that would have allowed them to produce sewn clothing. Even where there is some evidence they might have had jewelry or body decoration, it cannot be ruled out that they acquired these from modern humans who inhabited Europe contemporaneously. And whereas modern humans somehow had the creative capacity to spread out around the entire world, building the technologies that allowed them to inhabit territories from the Arctic to the scorched deserts of Saharan Africa, the Neanderthals never left Eurasia.

This line of reasoning leads to the conclusion that language evolved with the arrival of our own species. Our capacity for language was almost certainly present in our common ancestor, because today all human groups speak, and speak equally well. There are no languages that are superior to others, and no human groups that speak primitive as opposed to advanced languages. If this hypothesis is correct, then language is no more than around 160,000 to 200,000 years old, although some anthropologists think language arose even later, pointing to a sharp increase beginning around 70,000 to 100,000 ago in evidence of symbolic thinking and the complexity of human societies. Even though our species arose prior to 100,000 years ago, it is just possible that all modern humans alive today trace their ancestry to common ancestors who lived around that time, so language evolving at this later date, though unlikely, is possible.

3. WHY DID LANGUAGE EVOLVE?

Evolution does not produce complex adaptations like language in a single moment. Usually, something pulls the trait and its precursors along so that it pays its way by granting some advantage to its bearers. To most people the advantage of having language is obvious—it allows us to communicate. All animals could benefit from being able to communicate as humans do, however, so this does not explain why humans have language and no other species does. Instead, we need to search for ways that having language granted benefits to speakers in our species but not in others. But searching for an advantage to language leads immediately to two evolutionary predicaments. One is that much of what a speaker has to say might benefit someone else, and potentially at a cost to the speaker. Natural selection never promotes naive altruism such as this, and so language would quickly have died a silent death, as selfish people all too willing to listen would have profited but with no intention of returning the speaker's favor. Alternatively, perhaps language evolved to help humans mislead or trick others. It is in fact a fundamental tenet of communication that animal signals evolve to benefit the signalers. This solves the problem of altruism, because language might help speakers benefit at others' expense. But this poses the second predicament: if others know that speakers' acts of communication are designed to benefit themselves, surely this will favor people who don't listen.

To understand how language overcame these two predicaments and to grasp why language evolved, we need to appreciate something that is true of human social behavior but that is virtually absent in the rest of the animal kingdom. Members of the human species are the only ones who routinely cooperate with other *unrelated* members of their species.

Whereas most cooperation elsewhere in nature is limited almost entirely to helping kin or other relatives—that is, to acts of nepotism—humans routinely exchange favors, goods, and services with people other than those in their immediate families. We have an elaborate division of labor, we engage in task sharing, and we have learned to act in coordinated ways, such as when we go to war or simply combine or coordinate our energies to complete some task. But as powerful as this form of social behavior is, it is risky because in each of these situations we run the risk of being taken advantage of by other members of our group who might hold back, enjoying the spoils of cooperation but without having to pay the costs.

Elsewhere I have suggested that the role of language in this complex social behavior—and the reason it evolved—was to act as the conduit for carrying the information needed to make our form of cooperation work and for keeping people's selfish instincts in check. It is a highly specialized piece of social technology, and so great were the benefits deriving from the cooperation language made possible that it easily paid its way in evolutionary terms. Both evolutionary predicaments are solved because now speakers *and* listeners benefit from having language. We use it to negotiate exchanges, to make plans, to remember lessons from the past, and to coordinate actions. Persons who try to take advantage of this cooperation, perhaps by failing to return someone else's goodwill or by not contributing their fair share, can be exposed as "cheats" and their reputation tarnished. These are all complicated social acts that other animals don't perform, and they require more than grunts, chirrups, odors, and roars. This explains why we and we alone have language: our particular brand of sociality could not exist without it.

4. THE EVOLUTION OF HUMAN LANGUAGES

It is by now widely accepted that modern humans first arose in East or possibly southern Africa, and then later swept "out of Africa" in two waves, one perhaps 120,000 years ago and a second one sometime between 50,000 and 70,000 years ago. Opinions differ about the success of the first wave; some researchers suggest modern humans got no farther than what is known today as the Levant region of the Middle East, where they eventually went extinct. Others believe that people from this first wave survived and spread eastward, establishing the populations that would eventually move into Southeast Asia, Indonesia, Papua New Guinea, and Australia. Whichever scenario is true, by the time of the second wave, modern humans had gone on to establish their presence permanently outside Africa and in a matter of a few tens of thousands of years had occupied nearly every environment on earth.

Why this sudden change in the behavior of a single species? The answer is almost certainly "language." While all other species are largely confined to the environments to which their genes are adapted, having language and the social systems it made possible granted humans the capacity to adapt at the cultural level by producing new technologies, and it was via successive bouts of innovative cultural adaptation that our ancestors came to occupy the world. Language had granted humans the capacity to share ideas. This simple development meant that later generations could benefit from the accumulated wisdom of the past, and new ideas could quickly spread among the members of a population. Once a species has the ability to cooperate with unrelated individuals, a vast store of knowledge and wisdom is unleashed as the society can then draw on a wider range of talents and skills than would be available to a single individual or even to a family. We therefore expect this style of cooperation to be associated with an explosion of complexity, and this is precisely what the archaeological record reveals.

As these newly talkative people moved around the world they evolved different languages. They could do so because, unlike with genes, whose precise sequence of nucleotides determines the proteins they produce, there is no necessary connection—save perhaps for the onomatopoeic words—between a word or sound and its meaning. Currently there are approximately 7000 different languages spoken, or 7000 mutually incomprehensible systems of communication around the world. This makes humans unique—and somewhat bizarrely so—among animals in not being able to communicate with other members of their own species. As it happens, the 7000 different languages are not evenly distributed around the world. Instead, *language density*, or number of different languages found in a given geographic area, follows almost exactly the pattern of biological species diversity. Both show strong latitudinal gradients, such that both the number of different languages and number of

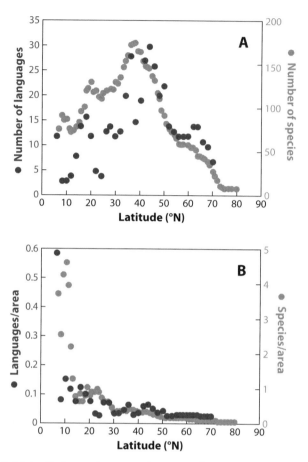

Figure 20-1. Relationships between languages and species. Numbers of North American human language–cultural groups (before European contact) and mammal species are distributed similarly across degrees of latitude. (A) Numbers of languages and numbers of mammal species at each degree of north latitude in North America. The trends reflect the shape of the continent, being narrow in the south regions and growing wider at higher latitudes. Both trends peak at approximately 40°N, where North America is ~4800 km wide. (B) Densities of languages and mammal species, calculated as the number of each found at the specific latitude divided by the area of the continent for a 1° latitudinal slice at each latitude. (Figure and data are reproduced, with permission, from Nature Reviews Genetics [Pagel 2009].)

different species in a given area are small in the northernmost regions of the world, and both become more tightly packed in tropical regions (figure 20-1). In some tropical regions, languages can be so densely packed that a different language can be spoken every few kilometers! For instance, the Vanuatu island of Gaua in the South Pacific Ocean covers 342 km², and like so many of the islands in this region, it is the roughly

Table 20-1. Some analogies between biological and linguistic evolution

Biological evolution	Language evolution
Discrete heritable units (e.g., nucleotides, amino acids, genes, morphology, behavior)	Discrete heritable units (e.g., words, syntax, and grammar)
Mechanisms of replication	Teaching, learning, imitation
Mutation—various mechanisms yielding genetic alterations	Innovation (e.g., formant variation, mistakes, sound changes, introduced sounds and words)
Concerted evolution	Regular sound changes
Homology (genes that are related by descent from a common ancestral gene and are now found in different species [e.g., ribosomal genes in nearly every species])	Cognates (words that are related by descent from a common ancestral word and are now found in different languages [e.g., father and pater])
Natural selection	Social selection, trends
Drift	Drift
Cladogenesis (e.g., allopatric speciation [geographic separation] and sympatric speciation [ecological/reproductive separation])	Lineage splits (e.g., geographic separation and social separation)
Anagenesis	Linguistic change without split
Horizontal gene transfer	Borrowing
Hybridization (e.g., horse, zebra, wheat, strawberry)	Language Creoles (e.g., Surinamese)
	Correlated genotypes/phenotypes (e.g., allometry, pleiotropy)
Correlated cultural terms (e.g., five and hand)	
Geographic clines	Dialects/dialect chains
Fossils	Ancient texts
Extinction	Language death

circular remnant plug of an ancient volcano. Gaua is just 19–21 km in diameter, but this speck of an island supports five distinct languages—*Lakon* or *Vurē*, *Olrat*, *Koro*, *Dorig*, and *Nume*.

It is as if human language groups act like distinct biological species, and this linguistic isolation is clearly evident in the ways that languages evolve. Most of us learn our language from our parents and those immediately around us. This dominant form of language transmission from parent to offspring or from older generation to younger generation means that language evolution shares a number of features with genetic evolution (table 20-1). Thus genes and words are both discrete elements that evolve by a process of descent with modification as they are passed over many generations: genes can acquire mutations, and words can acquire sound changes. For example, the Old English *brōthor* evolved into the modern English *brother*. Words can also be borrowed from a donor language and incorporated into a recipient language—the English word *beef* is borrowed from the French *boeuf*—just as genes can sometimes be transmitted from one species to another, such as when two closely related species form hybrids.

These and the other parallels between genetic and linguistic evolution mean that when people spread out over an area, and new languages evolve, those languages share many features and form what we recognize as families of languages. Indeed, in *The Descent of Man* (1871), Darwin noted that "the formation of different languages and of distinct species ... [is] curiously parallel...." This parallel had in fact been recognized in the late eighteenth century, nearly 100 years before Darwin, by an English judge, Sir William Jones, working in colonial India during the rule of the English King George III. To process court papers Sir William found it necessary to learn Sanskrit, and in doing so he became aware of curious similarities among Sanskrit, Latin, and Greek. Jones described these to a meeting of the Asiatic Society in Calcutta in 1786, saying:

> The Sanskrit language ... [bears] ... a stronger affinity ... [to Greek and Latin] ... both in the roots of verbs and in the forms of grammar, than could possibly have been produced by accident; so strong, indeed, that no philologer could examine them all three, without believing them to have sprung from some common source, which, perhaps, no longer exists.— Sir William Jones, Calcutta, February 2, 1786

Jones had identified what linguists would later recognize as the Indo-European language family, which arose around 8000 to 9000 years ago with the advent of farming in the Fertile Crescent, or what is roughly present-day Turkey and Iraq. Farmers and farming technology then spread out from that region, seeding the languages currently spoken all over Europe, parts of central Asia, and the Indian subcontinent. These include the Germanic languages German, Dutch, and English; Romance languages that derive from Latin including French, Spanish, and Italian; Slavic languages; and even Persian, Hindi, and Punjabi.

By comparing the similarities and differences among a group of languages such as those that form the Indo-European family, it is possible

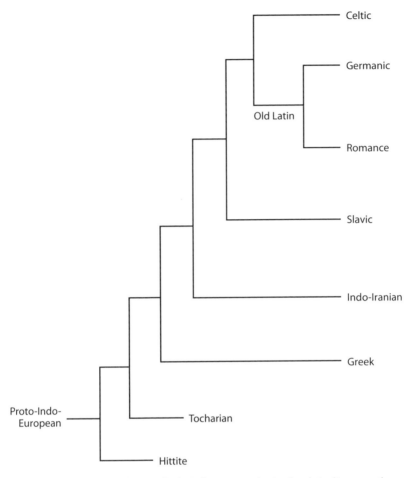

Figure 20-2. The Indo-European language family. Evolutionary tree showing the relationships among the major branches of this language family. Celtic languages include Irish, Breton, and Welsh; Germanic languages include German, Dutch, and English; Romance or Latinate languages include French, Italian, and Spanish; Slavic languages include Russian, Czech, and Polish; Indo-Iranian languages include Persian, Afghan, Hindi, and Punjabi. Hittite and Tocharian are two extinct languages. The base or root of the tree represents the proto-Indo-European language that might have existed 8000 to 9000 years ago.

to construct evolutionary or *phylogenetic* trees depicting the probable course of evolution of those languages from their common ancestral or protolanguage. Figure 20-2 displays just such a tree for a selection of Indo-European languages, including some of its main branches. The most popular and widely used approach to inferring linguistic phylogenies draws on the close parallels between linguistic and genetic evolution such as are listed in table 20-1. In particular, whereas evolutionary biologists infer phylogenetic trees of species by studying *homologous* genes, evolutionary linguists infer linguistic trees from sets of *cognate* words. Just as slight differences in the sequence of a gene can identify sets of closely related species, slight differences in words can identify sets of closely related languages.

For example, the word *madre* means *mother* in both Spanish and Italian, and we instinctively recognize that both derive from the Latin *mater*. Likewise, we recognize that the English *mother* is similar to the German *mutter*, and again both of these seem to derive from *mater*. In fact, these comparisons identify English as part of the Germanic branch of the Indo-European languages and identify Spanish and Italian as part of the Romance or Latinate branch of this same family (figure 20-2).

Not all such comparisons are this simple, but linguists make use of what they call *regular sound correspondences* (table 20-1) to help them identify whether two words are cognate and how closely related they are. Regular sound correspondences arise when a specific *phoneme*, say *p*, changes to the same other phoneme, say *b*, in many words in the lexicon—a phenomenon known as *regular sound change*. These changes are strikingly similar to what happens in the phenomenon known as *concerted evolution* in biology, in which the same genetic change can occur in parallel at many sites in a gene or genome. One of the best known of these regular sound changes is the replacement of a *p* sound at the beginning of a word with an *f* sound, as in the Latin *pes* or *ped*, which becomes *foot*, and *pater* becomes *father*. Other regular sound correspondences reveal to linguists that the English *five* is closely related to the German *fünf*, that the French *cinq* is related to the Spanish *cinco*, and less obviously that all four of these words derive from the Latin *quinque*.

Of course, not all words are cognate. The Spanish *agua* is cognate to the Italian *acqua*, but neither is cognate to the English *water* or to the German *wasser*. This lack of cognacy serves to reinforce that the Romance languages are a distinct group from the Germanic languages, even though both branches trace their ancestry to a pre-Latin language that might have been spoken 5000 years ago or more.

Once a set of comparisons is made among a group of languages, identifying cognate and noncognate words for a large number of different meanings (a *meaning* being what a word refers to), the resulting data can be used along with formal methods to infer the phylogenetic tree. A variety of methods are commonly used to infer trees, including parsimony, distance, and likelihood methods, and these are broadly similar whether applied to genetic or linguistic data. *Parsimony methods* seek a tree that minimizes the number of evolutionary events along its branches. Thus, parsimony methods would favor a tree that put English with German, and French with Italian or Spanish, over one, for example, that showed German as more closely related to Spanish than to English. To put German next to Spanish would suggest that the word *water* or *wasser* had somehow evolved twice. *Distance methods*, as their name implies, seek to define a distance between all pairs of languages, and those distances are then used to construct a tree. *Likelihood methods* use formal statistical models to estimate the probability of changes in words through time. A tree is constructed that makes the observed set of changes most probable, given the model of evolution.

Trees such as the one depicted in figure 20-2 have now also been produced for the Austronesian languages, the Bantu languages of Africa, the Arawak languages of South America, the Semitic languages, the Uralic languages of Northern Europe, and some Melanesian languages, and this is an important and growing area of the field of evolutionary linguistics. The existence of sets of languages that comprise families of related languages shows that at least some elements of language evolve slowly enough to preserve signals of their ancestry dating back thousands of years. For example, linguists recognize that the word for *two* of something is probably derived in *all* Indo-European languages from a shared ancestral sound that has been conserved for many thousands of years. Thus, in Spanish the word is *dos*, it is *twee* in Dutch, *deux* in French, *due* (doo-ay) in Italian, *dois* in Portuguese, *duo* (δύο) in Greek, *di* in Albanian, and *do* in Hindi and Punjabi; Julius Caesar would have said *duo*. This conservation leads to the proposal that the original or proto-Indo-European word that was spoken perhaps 9000 years ago was also "two"—as it sounds—and indeed, some scholars suggest it was *duwo* or *duoh*.

A handful of other words, including *three, five, who, I,* and *you,* are also highly conserved. For example, the English word *three* is *tre* in Swedish and Danish, *drei* in German, *tre* in Italian and *tres* in Spanish,

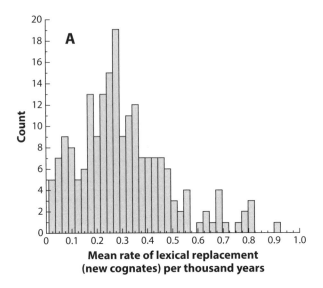

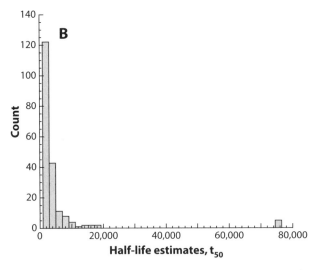

Figure 20-3. Rates of lexical replacement. (A) Counts of the rate at which a word comes to be replaced by a new unrelated word in the Indo-European languages for 200 common vocabulary words, measured in units of numbers of new words per 1000 years of evolution. Fastest to slowest rate represents more than a 100-fold difference. The average = 0.3 ± 0.18 new words per 1000 years (or roughly one every 3000 years), median = 0.27, range = 0.009 to 0.93. (B) Counts of the word half-life estimates as derived from the rates of lexical replacement for the same 200 words. The half-life measures the expected amount of time before a word has a 50 percent chance of being replaced by a new, unrelated word. The average half-life is 5300 years, with a median of 2500 years, and ranges from 750 to a theoretical value of 76,000. (Figure and data are reproduced, with permission, from Nature Reviews Genetics [Pagel 2009].)

tria in Greek, *teen* in Hindi and *tin* in Punjabi, and *tri* in Czech, leading to the suggestion that the proto-Indo-European word for *three* might have been *trei*. These conserved words are closely followed by pronouns such as *he* and *she*, and by the *what*, *where*, and *why* words, all of which often show a striking degree of similarity among many Indo-European languages. Other words, however, can vary considerably across these same languages, meaning that they have evolved or changed at far higher rates. The English word *bird*, for example, is *vogel* in German, *oiseau* in French, and *pajaro* in Spanish.

When long lists of words are studied for their frequency of change among the languages of a language family such as Indo-European, it is possible to derive estimates of their rates of change using statistical methods. It turns out that most words can be expected to last somewhere around 1500 to 2500 years, with some lasting far longer and others for shorter amounts of time (figure 20-3). It is possible from these rates of change to calculate a word's linguistic *half-life*, defined as the amount of time it takes for there to be a 50 percent chance a word will be replaced by some new, unrelated word (figure 20-3). This approach shows that some words might be expected to last more than 10,000 years. The linguist Merritt Ruhlen has even proposed a list of "global etymologies" that he thinks are signals leftover from the last common ancestor to all human languages, or our *mother tongue*. Ruhlen's list includes words for *who, what, two, water, finger,* and *one*. The mere existence of a mother tongue is controversial, and many linguists dispute the list, but Ruhlen cites evidence that traces of these words are found in many language families from all over the world.

5. LANGUAGES ADAPT TO SPEAKERS

But why should words last as long as they do? Given all the opportunities for neighboring language groups to borrow words from one another, and all the ways that words can change, language changes far more slowly than we might expect; that is, from the standpoint of effective communication, words do not need to last much more than around three generations—the time span covered in a typical population of speakers. If they changed faster than that, then we might not be able to talk to our grandparents. But most words last somewhere around 1500 to 2500 years, so why is there such fidelity in language?

One answer comes from thinking of elements of languages as "replicators" that must adapt to the environment of the human mind. Words are sounds that compete for attention in the mind of speakers. So, to be successful words must get themselves replicated by being transmitted to someone else's mind. It follows that those words that are easiest to remember and say will tend to be those retained. We can see the current competition for space in our mind in common words like *sofa* and *couch*, or *living room*, *sitting room*, *reception room*, and *parlor*. Which of these forms will win? There are no simple answers, but precisely because there is seldom any necessary connection between a sound and its meaning, the competition often focuses on characteristics of the sounds themselves, and thus we can expect the competition to be more intense the more often a word is used. There is a huge disparity in how often different words get used; in fact, so great is the disparity that about 25 percent of all human speech is made up from a mere 25 words. According to the *Oxford English Dictionary*, the English language's top 25 include *the, I, you, he, this, that, have, to be, for,* and *and* (while *they, we, say,* and *she* make it into the top 30). Not surprisingly, perhaps, the words used most frequently have common characteristics: they are short, often monosyllabic, distinct, and easy to pronounce. And this begins to explain why some words last longer than perhaps they need to. It is not that these frequently used words are somehow more important but that they might be stable over long periods because they have become so highly adapted to our minds that it becomes difficult for a new form to arise and outcompete or dislodge them.

6. THE FUTURE OF LANGUAGE EVOLUTION

Our native language is, perhaps, one of our most intimate traits, being the voice of the "I" or "me" that defines our conscious self. It is the language of our thinking, and it is the code in which many of our memories are stored. Thus it is not surprising that one of the greatest personal losses a people can suffer is the loss of their native language. And yet, currently somewhere about 15 to 30 languages go extinct every year as small traditional societies dwindle in numbers or get overwhelmed by larger neighbors, and younger generations choose to learn the languages of larger and politically dominant societies. Whatever the true numbers of languages going extinct, the loss of languages greatly exceeds the loss of biological species as a proportion of their respective totals.

Some projections say that only a handful of languages will see out this century. This raises the question of which language will win if ever a single language should succeed all others on earth. Currently, three languages are spoken by a far greater number of people than any of their competitors. About 1.2 billion people speak Mandarin, followed by around 400 million each for Spanish and English, and these are closely followed by Bengali and Hindi. It is not that these languages are better than their rivals; it is that they have had the fortune of being linked to demographically prosperous cultures. On these counts Mandarin might look like the leader in the race to be the world's language, but this ignores the fact that vastly more people learn English as a second language—including many people in China—than any other. Already it is apparent that if there is a worldwide *lingua franca* it is English.

Still, English itself might be transformed as it is bombarded by the influences of such large numbers of nonnative English speakers who bring along their accents, grammar, and words to English when they speak it. This ability of English to take in so-called foreign words has been the key to its adaptability for at least the millennium since the Norman conquest of the English in 1066 brought an influx of Norman French vocabulary. Just as words must adapt to be competitive in the struggle to gain access to our mind, languages have to adapt as a whole to remain useful to their speakers, and those that do so will be the survivors. Self-appointed human "minders" in the form of reactionary grammarians, sticklers for spelling, or those who deliberately try to exclude some words and phrases will succeed in controlling the rate at which their languages naturally change, but in doing so they might consign these languages to the backwaters of international communication. This might already be happening to French and German, as both governments have ministries devoted to language "purity." The alternative to this control is not the free-for-all that some might fear. If communication is important, languages will never change at rates that imperil the very reason for which they exist.

FURTHER READING

Cavalli-Sforza, L. L., A. Piazza, P. Menozzi, and J. Mountain. 1988. Reconstruction of human evolution: Bringing together genetic, archaeological, and linguistic data. Proceedings of the National Academy of Sciences USA 85: 6002–6006. *A widely cited early attempt to link genetic and linguistic diversity.*

Enard, W., M. Przeworski, S. E. Fisher, C. S. Lai, V. Wiebe, T. Kitano, A. P. Monaco, and S. Pääbo. 2002. Molecular evolution of FOXP2, a gene involved in speech and language. Nature 418: 869–872.

Fitch, W. T. 2010. The Evolution of Language. Cambridge: Cambridge University Press.

Gray, R. D., and Q. D. Atkinson. 2003. Language-tree divergence times support the Anatolian theory of Indo-European origin. Nature 426: 435–439. *This study used language phylogeny to test a historical hypothesis for the timing of the origin of Indo-European languages.*

Hruschka, D., et al. 2015. Detecting regular sound changes in linguistics as events of concerted evolution. Current Biology 25(1): 1–9. doi:10.1016/j.cub.2014.10.064.

Maricic, T., et al. 2013. A recent evolutionary change affects a regulatory element in the human *foxp2* gene. Molecular Biology and Evolution 30: 844.

Pagel, M. 2009. Human language as a culturally transmitted replicator. Nature Reviews Genetics 10: 405–415. *Provides a general overview of language evolution, including a description of methods of phylogenetic inference, and statistical studies of how languages evolve.*

Pagel, M. 2012. Wired for Culture: Origins of the Human Social Mind. New York: W. W. Norton. *One chapter of this book presents many of the arguments given here about why language evolved.*

Pagel, M., and R. Mace. 2004. The cultural wealth of nations. Nature 428: 275–278.

Pagel, M., Q. D. Atkinson, and A. Meade. 2007. Frequency of word use predicts rates of lexical evolution throughout Indo-European history. Nature 449: 717–719. *A general explanation for variation in rates of word evolution.*

Renfrew, C. 1987. Archaeology and Language: The Puzzle of Indo-European Origins. Cambridge: Cambridge University Press.

Wade, N. 2006. Before the Dawn: Recovering the Lost History of Our Ancestors. New York: Penguin.

CHAPTER 21

CULTURAL EVOLUTION

Elizabeth Hannon and Tim Lewens

OUTLINE

For most of the twentieth century, evolutionary theory focused on phenotypic variation underpinned by inherited genetic variation. Any comprehensive account of evolution must acknowledge that this is at best a simplification of the processes underlying change and stasis in organic lineages. In the human species, and also in various animal species, habits, skills, and technology—what we might consider cultural traits—can also contribute to survival and reproduction. Moreover, these traits are often maintained, in our own species at the very least, by learning from others; that is, they are inherited nongenetically. Further, these traits often show patterns of cumulative improvement as discoveries made in one generation are conserved in, and modified by, later generations. There also exist subgroups within many species with distinct traits, again often generated and maintained through learning. Theories of cultural evolution aim to build rigorous accounts of change in species over time, in ways that incorporate the important roles of learning from others. How exactly these accounts should be fashioned, what relationship cultural evolution should have with organic evolution, and how culture itself should be conceived of remain open questions. This chapter describes some recent attempts to address these questions.

GLOSSARY

Cultural Evolution. A process of change in the traits manifested within a population that is explained by various forms of social learning among species members.

Horizontal Transmission. Transmission within a generation, sometimes also used to refer to transmission from any nonparent.

Meme. A cultural entity, intended to be analogous to a gene, capable of being replicated and transmitted between individuals.

Replicator. An entity capable of being replicated and capable of influencing its own chances of being replicated through its effects on the world.

Vertical Transmission. Transmission from parent to offspring, usually of genetic material.

1. WHAT IS CULTURAL EVOLUTION?

Not all evolutionary approaches that seek to account for cultural phenomena are theories of cultural evolution. Some evolutionary psychologists, for example, tend to regard human behavior and culture as the output of species-typical cognitive adaptations, and they assume that the most important mechanism producing such cognitive adaptations is natural selection acting on genetically inherited variation (see chapter 6). These evolutionary psychologists acknowledge that changes in cultural environments can affect the behavioral outputs of cognitive adaptations, but they tend to downplay the role of nongenetic inheritance systems in evolution. As a consequence, they also tend to be skeptical of the thought that cognitive adaptations might sometimes be generated over time by the cumulative effects of learning.

The primary characteristic of full-blown theories of cultural evolution is that they aim, usually through the fashioning of formal mathematical models, to incorporate phenomena whereby individuals learn from each other—"social learning"—into mainstream understandings of evolution. Often, these efforts take the form of models of gene–culture coevolution. Cultural inheritance and genetic inheritance are processes that can affect each other in important ways. Cultural changes bring about alterations to species' environments, which in turn affect how genes act in development and what selection pressures act on genes. Dairy farming, for example, is a cultural innovation that is thought to have arisen sometime between 6000 and 8000 years ago. It appears to have created a modified selective environment that favored genes for lactose tolerance. Conversely, dairying itself seems to have been preferentially adopted in those communities where genes for lactose tolerance made it easier for individuals to use this potential new source of calories. Gene–culture coevolutionary theory suggests that, far from removing us from the evolutionary fray, our changing cultural environments are implicated in the ongoing evolution of our physiological natures.

If models that take account of social learning are the primary tools of cultural evolutionary theory, what are the theory's primary explanatory targets? Many of the questions the theory asks contrast humans with other species. Why, for example, is our species so much better able than others to sustain and augment increasingly refined bodies of knowledge? And why are humans conspicuously beneficent in their dealings with each other, even when interacting with nonkin whom they may encounter only once in a lifetime? These broad comparative questions lead to more fine grained ones about the specifics of human learning: Are we equally likely to learn from any individual we might be exposed to, or do we show selective forms of receptivity to certain types of individuals? How does the frequency at which an idea or practice appears in a population affect our chances of learning it? What reproductive advantage might alternative ways of learning confer on their bearers? Is it possible that social learning within subpopulations can maintain culturally distinctive groups, in ways that might give rise to group-level adaptations via group-level competition? And might such a process of "cultural group selection" explain our beneficence toward other members of our social groups?

The general tone of these questions reminds us not to exaggerate the extent to which cultural evolutionary theory marks a break with longer-standing approaches to evolutionary psychology. Cultural evolutionists ask how various tendencies to learn from others may have augmented reproductive fitness in our species' past, and they often assume that these learning dispositions are inherited genetically. But cultural evolutionists also argue that their questions cannot be answered unless we can find some way of understanding what happens at the level of populations when individuals learn from one another in characteristic ways. Hence the need, once again, to find models that can combine genetic evolution with social learning. Moreover, once cultural evolutionists have fashioned tools that can help us understand what happens in large groups of learners, they can then move away from evolutionary questions about why the human capacity for culture emerged in the first place and can instead begin to ask questions about why ongoing cultural phenomena—such as the spread of innovations of all kinds through modern populations—take the form they do.

2. MEMETICS

A theory of cultural evolution needs some systematic way of modeling the effects of cultural inheritance. One such approach is *memetics*. Memetics has attracted considerable attention in the popular scientific literature, but many of the most productive and influential theorists of cultural evolution have either explicitly rejected the *meme* concept or shown neutrality to it. Initially developed by Richard Dawkins, and following the "gene's-eye view" of natural selection he popularized, memetics takes the view that to explain the sort of transgenerational resemblance needed for cumulative evolutionary change, entities that have the ability to make faithful copies of themselves—so-called replicators—are required. In standard biological models of evolution, it is assumed that genes are the relevant replicators; genes make copies of themselves, and (so the story goes) this explains why offspring resemble their parents. For culture to evolve, memeticists argue that cultural replicators—called *memes*—are once again required. Dawkins lists some exemplary memes: "tunes, ideas, catch-phrases, clothes fashions, ways of making pots or of building arches." Note that while it is sometimes assumed that all memes are ideas (and vice versa), Dawkins's list includes other types of things, such as ways of making pots, which are techniques.

Memetics proposes that ideas, skills, practices, and so on are entities that can be understood to hop from mind to mind, making copies of themselves as they go. For example, you hear a song on the radio as you leave your house in the morning and you sing it at work that day. Your colleague later whistles it as she prepares dinner that evening. Her child hums the song in school the next day and passes it on to his classmates. The meme—in this case, the tune—spreads through its being "catchy." Like genes, memes have differential success in replication: some songs are catchier than others. The rate at which a meme may replicate itself is thought to be dependent on the same factors that determine the rate at which a gene may replicate itself—namely, its effects on the organism it inhabits and on the local environment (partly constituted by the downstream effects of other memes) in which the organism finds itself.

Critics of memetics put pressure on the claim that cultural inheritance is analogous to genetic inheritance. Such criticisms are as likely to originate from those sympathetic to the broad project of developing evolutionary approaches to culture as they are from those outside this project and doubtful of its merits. The remainder of this section outlines some of the key criticisms leveled against memetics.

First, cultural items rarely behave like replicators, and imitation is often very error prone. If you see us dance the tango, and this inspires in you the desire to also dance the tango, we will almost certainly not dance exactly the same steps. Our dances will have been influenced by a wide variety of factors such as who taught us the dance and our own particular physiologies. If this is copying, it is very bad copying indeed.

A second and closely related criticism of memetics draws on the fact that while genetic replication allows us to trace a token copy of a gene back to a single parent, ideas are rarely copied from a single source in a way that allows us to trace clear lineages. Perhaps you learned the tango from several teachers, and your style has been influenced by watching expert dancers. There is no clear single origin for your "tango" meme. Within the realm of biological evolution, an understanding of Mendel's laws has been important in explaining some aspects of evolutionary dynamics. Mendel's laws rely on an understanding of genes as discrete, transmitted units. But if token ideas can appear in an individual by virtue of that individual's exposure to several sources, then it is unlikely that anything close to Mendel's laws will be discovered within cultural evolution. Such an objection need not be fatal for theories of cultural evolution in general, as we shall see, but it does threaten the tight analogy drawn by memetics between ideas and genes.

Third, memetics seems to demand that we be able to divide culture into discrete units. But it is not clear how this should be done. Ideas stand in logical relation to one another. It is impossible to believe in the theory of relativity without understanding it, and one cannot understand it without holding many additional beliefs relating to physics. It is not at all clear that it makes sense to think that the theory of relativity can be isolated from the rest of physics as an individual meme. One might respond that we have delineated the meme incorrectly here, and those ideas to which the theory of relativity stands in logical relation all form a part of a single more inclusive meme. The worry, now, is that even if we "step back" and consider some broader group of theories, we cannot understand even them without further basic mathematical training, understanding of the operation of measuring apparatus, and so forth. A form of holism looms, according to which single memes will correspond to massive complexes of belief.

These criticisms focus on whether memes exist.

A final worry stems from asking whether, even if memes do exist, the meme concept is of any use. The charge here is that memetics is not particularly enlightening: it only dresses up familiar explanations in a slightly different guise. So, for instance, we might allow that clothes fashions are memes, but even if that is the case, memetics does not explain why fashion memes differentially replicate. To explain why one such meme propagates throughout the population while another one perishes in obscurity still requires reference to local conditions, consumer psychology, and so on. Any value memetics can bring to the explanation of why one meme is fitter than another is parasitic on conventional work done in psychology. And if individual preferences are subject to change over time, then there may be no general and informative theory of cultural evolution to be had; rather, we will have to settle for local explanations that look to shifting preferences. The upshot of this, the argument goes, is that memetics never gets beyond conventional narrative cultural history and cannot provide us with a new scientific framework for understanding culture.

3. MODELING CULTURAL EVOLUTION

Another line of investigation, pioneered by Luigi Luca Cavalli-Sforza, Marcus Feldman, Peter Richerson, and Robert Boyd, concerns the ways in which cultural inheritance can affect evolutionary processes. These models do not assume that cultural inheritance works in the same way as genetic inheritance and thus they differ significantly from memetic approaches. Indeed, they model cultural inheritance in ways that depart quite markedly from genetic inheritance. So it is that many of Robert Boyd and Peter Richerson's models explicitly assume that an individual's cultural makeup is an error-laden blend—synthesized from, and influenced by, many cultural "parents"—rather than a collection of discretely transmitted self-replicating gene-like particles. Their work then focuses on the population-level evolutionary consequences of such an inheritance system. Moreover, they tend to concentrate on this form of modeling while remaining uncommitted regarding the precise way in which cultural variants are physically realized.

Such a move can be defended by an appeal to history. Darwin's theory of evolution by natural selection lacked a plausible material theory of inheritance for some time, but this did not prevent Darwin's theory from being useful in the interim. Even without an account of what exactly is inherited in cultural inheritance, work can be done to explain the changes in (cultural) trait frequencies in a population by focusing on the population-level consequences of (cultural) inheritance, selection, mutation, and other forces. So although cultural evolutionary theorists may deny that cultural change should be understood in just the same way as biological change is understood, their approach remains recognizably evolutionary in style.

All the same, one might think that even if cultural change does not require cultural replicators such as memes, cultural replicators are necessary if cultural change is to produce adaptations. Beneficial traits must somehow be preserved over reasonably long stretches of time for cumulative selection processes to operate and for complex adaptations to arise. One obvious concern in this context stems from the fact that learning is often very error prone. If an individual hits on a fitness-enhancing behavior, that trait may be lost to future generations either because it is miscopied or because it is combined with other, less adaptive traits to produce an averaged, "blended" behavior (recall that a particular version of the tango may be an amalgamation of the influences of several teachers and famous dancers). Again, we might fear that cultural traits will not persist long enough for selection to act on them.

Richerson and Boyd argue that these problems are not fatal, and they have developed a number of valuable models to demonstrate why this is the case. These models assume that individuals will pick up cultural traits from a variety of sources and will frequently make mistakes. They also assume that we possess certain kinds of cognitive biases, and they show how these biases can dampen the spread of error in the population. So even if errors are occasionally made, these isolated errors will tend not to be imitated by others if we possess a *conformist bias*, such that we are disproportionately likely to imitate or learn the traits we most commonly encounter. A *prestige bias*, whereby we are more likely to imitate the traits of members of the population who are deemed to be successful, is also thought to keep error in check. It is likely that at least some of the traits possessed by a successful individual will have been instrumental in causing that individual's success. A bias toward imitating successful individuals therefore increases the chances that those success-generating traits will be imitated by others, to their advantage. But if the trait that has in fact generated success is not correctly identified by the imitator, the result is that the imitator is unlikely to become successful and so will not become a target for future imitators. Overall, the thought goes, prestige bias helps preserve, at the population level, traits that cause success in individuals. Cognitive biases such as these, it is argued, allow for mistakes to be made at an individual level but protect against those mistakes being repeated so widely that they undermine the inheritance of cultural traits in a population.

While the existence of such biases can dampen the spread of error, it is far from obvious that they will keep error in check to the extent that cultural inheritance will be robust enough for selection processes to operate. To show that these properties of individual psychology combine to yield population-level inheritance requires some abstract mathematical modeling, and much of the novel explanatory payoff of recent work in cultural evolutionary theory comes from the insights gained from this sort of modeling. The establishment of such population-level consequences is important, for it enables the investigator to revise the constraints one might naively think must bear on cultural inheritance if cumulative cultural evolution is to occur.

This approach allows cultural evolutionists to agree that cumulative evolution requires that fitness-enhancing cultural traits be preserved in the offspring generation as a whole while denying that this requires faithful transmission between individuals. This reasoning also answers one of our earlier criticisms of meme theory: in taking a population-level perspective, cultural evolution offers genuinely novel explanatory resources that go beyond cosmetic redescriptions of what we already know.

One may ask why it should be the case that we are able to learn from nonparents at all, given that *horizontal transmission* enables the spread of maladaptive traits. Cultural evolutionists have defended the thought that the overall adaptive benefits of learning from nonparents outweigh the overall adaptive costs. Determining how best to live in an environment can be difficult, even dangerous, if one attempts to do so without guidance; one may not be able to tell until it is too late which foodstuffs are nutritious and which are poisonous, for example. Similarly, a prestige bias is an achievable solution to a tricky problem: "determining who is a success is much easier than determining how to be a success" (Richerson and Boyd 2005, 124). The contention here is that cognitive biases that incline us toward imitating or learning from certain individuals may not rule out all maladaptive traits spreading through the population, but these biases are nonetheless more adaptive, overall, than available alternatives.

4. NONHUMAN ANIMAL CULTURAL EVOLUTION

There is widespread acceptance of the existence of some degree of culture (sometimes referred to as "tradition") in nonhuman species. For example, distinct dialects exist within the songs of certain species of birds, and tool use in some primates can vary from group to group within a single species. Japanese macaques are a particularly well studied population, and differences in everything from their grooming behavior to diet have been documented. The macaques of Koshima Island have developed some remarkable behaviors. To attract the macaques to open land so they could be observed, primatologists left sweet potatoes on a beach. This technique was effective, but it did leave the potatoes covered in sand. One member of the troop solved this problem by washing the potatoes in a nearby stream. Soon, her peers followed suit. After a while, the group began to use the sea instead of the stream for washing, preferring the taste that the saltwater imparted, and their young took to playing in the sea for the first time. The group also discovered fish discarded by local fishermen and added this item to their diet. These changes in their behavioral repertoires have taken place over the course of 50 years and identify them as distinct from other troops of macaques.

This example appears to have some key ingredients for cultural evolution: an individual hits on an innovation, and the innovation spreads throughout the population, which creates behaviorally distinct populations within the species. However, critics argue that this process alone is unlikely to secure cumulative evolution. Not all cultural inheritance involves "observational learning" or imitation, and the worry is that only these forms of social learning will allow the appearance of complex adaptations. We can see why observational learning might be considered crucial with the following example.

Certain populations of blue tits learned to remove the foil tops from milk bottles to gain access to the milk inside. The birds' attention was drawn to the milk bottles by the activity of their conspecifics. But it was only through trial-and-error exploration of the milk bottles that each tit worked out how to get to the milk inside. If an individual bird happened on a particularly efficient means of removing the foil top, this technique could not be transmitted to any other bird. Social learning of this sort, which does not rely on observational learning or imitation, means that innovations cannot be combined and built on, and cumulative evolution is unlikely to get off the ground.

Although these sorts of considerations have left some pessimistic about the possibility of significant cultural evolution in nonhuman animals, at worst they merely make complex adaptations as the result of cumulative evolution unlikely. Cultural variations may still play an important role in any evolutionary story of a given species. In the same way that dairy farming led to selection pressure for the ability to digest lactose, cultural changes in nonhuman species may alter the selective environment of those species and instigate a sequence of evolutionary changes. Further, as the macaques of Koshima Island demonstrated, although novel behaviors may not be built on and made more complex, one new innovation can open up previously unexplored parts of the environment and inspire further innovation. At the very least, this sort of example leaves room for the kind of gene–culture coevolutionary models briefly discussed in section 2.

Cultural evolution in nonhumans is an exciting area of research that is beginning to uncover the importance of forms of social learning, not only in primates but in birds, fish, and even insects. Further work will help us establish the significance of culture on the overall evolutionary trajectory of a species, as well as the extent to which we may speak of distinct cultural evolutionary processes in nonhuman species.

5. CULTURAL PHYLOGENIES

We have seen that much of the most influential work in cultural evolutionary theory offers adaptive rationales for the emergence of our capacities to learn from each other. This is not the only way to approach culture from the perspective of evolution. Darwin himself famously argued that all of life could be represented as a great tree, with later diversified forms understood to be the modified descendants of a few common ancestors. Darwin also knew that cultural items—he used the example of the world's languages (see chapter 20)—could potentially be represented in the same way. Linguists have long depicted changes in languages over time using genealogical trees that are strongly reminiscent of the phylogenetic trees we now use to show evolutionary relationships between biological species.

This raises the question of whether we can use biological methods of phylogenetic analysis to reconstruct plausible genealogies in the realm of culture. It is reasonable to infer that badgers, bats, and whales all share a common ancestor because this gives a good explanation for structural similarities in their forelimbs, in spite of those limbs having the very different functions of digging, flying, and swimming, respectively. More generally, just as shared ancestry can help explain similarities in the features of distinct biological species, so shared origins can explain the appearance of similar word forms in different languages, similar patterns of typographical errors in texts, or similar decorative patterns on pots and textiles.

Proponents of cultural phylogenetics argue that sophisticated tools developed for understanding relations between biological species can be valuable for reconstructing the history of cultural items such as languages, texts, and textiles. The image of biological evolution we have inherited from Darwin shows the branches of evolutionary trees expanding ever outward, never joining with one another. But it is evident in the cultural realm that the language of one population, for example, can be influenced by the languages of several of its neighbors, or that innovations in automobile design can borrow from developments in hi-fi systems and spacecraft engineering. In all these cases, one cultural "lineage" can acquire valuable material directly from many other distinct "lineages," with the result that a tree of culture—as opposed to a traditionally drawn tree of life—would have branches entangled in a web-like formation.

Historically, these concerns about the "reticulation" of cultural evolution compared with organic evolution have made some commentators—most prominently Stephen Gould—skeptical of cultural phylogenetics. Most recent proponents of these methods acknowledge that cultural "lineages" are highly reticulated. Even so, the recognition in recent years that organic lineages—especially bacterial lineages—also show considerable cross-lineage borrowing means that phylogenetic tools are now sophisticated enough to enable decent reconstructions of reticulated patterns of cultural influence.

6. DEFINING CULTURE

According to Richerson and Boyd (2005), "culture is (mostly) information in brains." Cultural inheritance is then understood as the transmission of this information from one person to the next. So even though Richerson and Boyd deny any strong similarity between genes and cultural variants, they maintain that cultural variants "must be gene-like to the extent that they carry cultural information." If we are to understand what is meant by culture and cultural evolution, then we must understand what is meant by *information*. There is no consensus on the meaning of this term, and the definitions that are offered—where they are offered at all—are problematic. For example, in their earlier work, Boyd and Richerson (1985) offered a definition of information as "something which has the property that energetically minor causes have energetically major effects." This is a curious definition; presumably it is meant to evoke intuitive examples whereby small informational "switches" (whether they are literally switches in a designed control system or metaphoric "genetic switches" in developmental pathways) have magnified downstream effects on the systems they influence. However, there are plenty of cases of information-bearing relations in which the energetic inequality is reversed. An instrument's display screen can carry information about solar flares; here, an energetically major cause has an energetically minor effect. Perhaps because of these oddities, their more recent work describes information as "any kind of mental state, conscious or not, that is acquired or modified by social learning and affects behavior" (Richerson and Boyd 2005). They later qualify this description with the concession that in some cases, "cultural information may be stored in artefacts." Alex Mesoudi (2011), however, does not offer a definition of information in his cultural evolutionary work but states that it is "intended as a broad term to refer to what social scientists and lay people might call knowledge, beliefs, attitudes, norms, preferences, and skills," while also insisting that culture is "information rather than behaviour." But because skills involve practiced, embodied behaviors, it is unclear how Mesoudi can count them as a form of cultural information.

Eva Jablonka and Marion Lamb defend the use of "information" on the grounds that it provides us with a term that can free us from worrying about the specifics of modes of transmission. It is taken to cover what is transmitted in genetic material, epigenetic material, environmental structures such as nests, behaviors learned from conspecifics, and the kind of knowledge stored in books. Although this kind of abstraction allows us to formulate hypotheses and theories that bear on all these cases of transmission, grouping them together will highlight what differences exist among them, too, which may encourage us to attend to features of certain types of information and its transmission that we might otherwise overlook. For example, repositories of symbols, most obviously in the form of libraries and computer databases, are vital inheritance systems for humans, allowing the preservation and accumulation of knowledge across generations. Nonsymbolic transmission occurs when some birds inherit their song from adult birds around them. Jablonka and Lamb use the characteristic differences among typical modes of social inheritance in animals and humans to illuminate the impact our own symbolic transmission systems have on human cultural evolution.

In sum, there is some confusion here over what is meant by information and thus how we define culture. The worry is that the term *information* masks some serious issues that any theory of cultural evolution ought to be addressing; we really ought to be clear about what it is we are trying to explain with our theories.

Developing a more fine-grained analysis of cultural inheritance, as Jablonka and Lamb suggest the concept of "information" may allow, can only add to the explanatory power of theories of cultural inheritance, but more work is needed first to clarify some conceptual confusion. While more research exists on human cultural evolution than on nonhuman cultural evolution, both areas are in their infancy. Thus, we should not be surprised to find that we are faced with a paucity of data and concepts in need of some untangling. But although the precise details have yet to be ironed out, the research so far has at least demonstrated that cultural evolution is both possible and plausible.

FURTHER READING

Boyd R., and P. J. Richerson. 1985. Culture and the Evolutionary Process. Chicago: University of Chicago Press.

Boyd, R., and P. J. Richerson. 1996. Why culture is common but cultural evolution is rare. Proceedings of the British Academy 88: 73–93. *An overview of nonhuman culture and difficulties for theories of cultural evolution in nonhuman animals.*

Cavalli-Sforza, L., and M. Feldman. 1981. Cultural Transmission and Evolution: A Quantitative Approach. Princeton, NJ: Princeton University Press. *One of the first attempts to construct a theory of cultural evolution.*

Dawkins, R. 1976. The Selfish Gene. Oxford: Oxford University Press. *Chapter 11 contains Dawkins's first discussion of his meme concept.*

Gray, R., S. Greenhill, and R. M. Ross. 2007. The pleasures and perils of Darwinizing culture (with phylogenies). Biological Theory 2: 360–375. *An overview of the application of phylogenetic methods to culture, including a very useful summary of the diverse uses to which these methods can be put.*

Jablonka, E., and M. Lamb. 2005. Evolution in Four Dimensions: Genetic, Epigenetic, Behavioral, and Symbolic Variation in the History of Life. Cambridge, MA: MIT Press. *Among other things, deals with research into animal culture and defends a coevolutionary model for nonhuman animals. Also includes a discussion of the concept of "information" and its place in evolutionary theory.*

Lewens, T. 2015. Cultural Evolution: Conceptual Challenges. Oxford: Oxford University Press. *An assessment of recent disputes over cultural evolutionary theory, with specific focus on criticisms from social scientists.*

Mesoudi, A. 2011. Cultural Evolution: How Darwinian Theory Can Explain Human Culture and Synthesize the Social Sciences. Chicago: University of Chicago Press. *A detailed and sympathetic overview of the growing field of work in cultural evolution, including valuable discussion of work on cultural phylogenetics and experimental approaches to cultural evolution.*

Odling-Smee, F. J., K. Laland, and M. W. Feldman. 2003. Niche Construction: The Neglected Process in Evolution. Princeton, NJ: Princeton University Press. *One of the most developed accounts of gene–culture coevolution.*

Richerson, P., et al. 2016. Cultural group selection plays an essential role in explaining human cooperation: A sketch of the evidence. Behavioral and Brain Sciences 39: e30. *A thorough summary of evidence for the role of selection at the level of cultural groups in explaining human cooperative tendencies.*

Richerson, P., and R. Boyd. 2005. Not by Genes Alone: How Culture Transformed Human Evolution. Chicago: University of Chicago Press. *An accessible introduction to cultural evolution by two of the founders of the field.*

Sterelny, K. 2012. The Evolved Apprentice: How Evolution Made Humans Unique. Cambridge, MA: MIT Press. *An important recent effort to understand why humans became so good at building and maintaining storehouses of socially transmitted information.*

CHAPTER 22

EVOLUTION AND NOTIONS OF HUMAN RACE

Alan R. Templeton

OUTLINE

1. The biological meaning of race
2. Do biological races exist in chimpanzees?
3. Do biological races exist in humans?
4. Do adaptive traits define human races?
5. Do human races exist: The answer

Races exist in humans in a cultural sense, but it is essential to use biological concepts of race that are applied to other species to see whether human races exist in a manner that avoids cultural biases and anthropocentric thinking. Modern concepts of race can be implemented objectively with molecular genetic data, and genetic data sets are used to see whether biological races exist in humans and in our closest evolutionary relative, the chimpanzee.

GLOSSARY

Admixture. Reproduction between members of two populations that previously had little to no reproductive contact.

Alleles. Alternative forms of homologous genes within a species that constitute the most basic type of genetic diversity.

Evolutionary Lineage. A population that maintains genetic continuity and identity over many generations because of little to no reproductive interchange with other populations.

Evolutionary Tree. A depiction of the ancestral relationships that interconnect a group of biological entities through a diagram in which ances-

tral nodes can split into two or more descendant types but that does
not allow fusion of previously split types.

Gene Flow. Movement of individuals or gametes from the local population of birth to a different local population followed by successful
reproduction.

Genetic Differentiation. Differences among populations based on particular alleles they possess, the frequencies of shared alleles, or both.

Haplotype. A specific nucleotide sequence existing among the homologous copies of a defined DNA region, whether a gene or not.

Isolation by Distance. A model of gene flow in which most genetic interchange is between neighboring populations but in which genes can
spread to distant populations over many generations because there
are no absolute barriers to movement between neighboring populations.

Local Population. A collection of interbreeding individuals of the same
species that live in sufficient proximity that most mates are drawn
from this collection of individuals.

Race. A subpopulation within a species, also called a *subspecies*, that has
sharp geographic boundaries separating it from the remainder of the
species, with the boundaries characterized by a high degree of genetic differentiation defined either through a quantitative threshold
or qualitatively as a separate evolutionary lineage.

1. THE BIOLOGICAL MEANING OF RACE

Do human races exist? Many people would answer yes because they
have a strong sense of their own racial identity and feel they can classify
other people into racial categories. However, the ability to classify oneself and others into races does not mean that races actually exist as a
culture-free, biological category. For example, Lao and coworkers (2010)
assessed the geographic ancestry of self-declared "whites" and "blacks"
in the United States by the use of a panel of genetic markers. It is well
known that the frequencies of alleles (different forms of a gene) vary
over geographic space in humans. The differences in allele frequencies
are generally so modest that any one gene yields only a little information
about the geographic origins of one's ancestors. However, with modern

DNA technology, it is possible to infer the geographic ancestry of individuals by scoring large numbers of genes. Self-identified "whites" from the United States are primarily of European ancestry, whereas US "blacks" are primarily of African ancestry, with little to no overlap in the amount of African ancestry between US "whites" and "blacks." In contrast, Santos and coworkers (2009) did a similar genetic assessment of Brazilians who self-identified as "whites," "browns," and "blacks" and found extensive overlap in the amount of African ancestry among all these "races." Indeed, many Brazilian "whites" are surprised to learn that they are considered to be "blacks" when they visit the United States, and similarly, some US "blacks" are considered to be "whites" by Brazilians. Obviously, the culturally defined racial categories of "white" and "black" do not have the same genetic meanings in the United States and Brazil. It is clear that an objective, culture-free definition of race is required before the question about the existence of biological races can be answered.

One way of ensuring a culture-free definition of race is to use a definition that is applied to species other than humans. The word *race* is not commonly used in the nonhuman literature; instead, the word *subspecies* is used to indicate the major types or subdivisions within a species. There is no consensus on what constitutes a species, much less a subspecies. Because the US Endangered Species Act mandates the protection of endangered subspecies of vertebrates as well as endangered species, conservation biologists have developed operational definitions of race or subspecies applicable to all vertebrates. We will apply these culture-free definitions to humans to avoid an anthropocentric definition of race.

Biologically, *races* are geographically circumscribed populations within a species that have sharp boundaries separating them from the remainder of the species. In traditional taxonomic studies, the boundaries were defined by morphological differences, but increasingly these boundaries are defined in terms of genetic differences that can be scored in an objective fashion in all species. Most *local populations* within a species show some degree of *genetic differentiation* from other local populations by having either some unique *alleles* or different frequencies of alleles. If every genetically distinguishable population were elevated to the status of race, then most species would have hundreds to tens of thousands of races. This would make the concept of race nothing more than a synonym for a local population. There is a consensus that race or subspecies should refer to a degree or type of genetic differentiation that is well above the level of genetic differences that exist among local populations. Both quantitative and qualitative criteria are used to define these racial genetic boundaries.

Quantitatively, one commonly used threshold is that two populations with sharp boundaries are considered to be different races if 25 percent or more of the genetic variability that they collectively share is found as between-population differences. One of the oldest measures used to quantify these differences is a statistic known as f_{st}. Consider drawing two homologous genes at random from all the genetic variation collectively shared by both subpopulations. The frequency with which these two randomly drawn genes from the total population are different alleles is designated by H_t, the expected heterozygosity of the total population. Now consider drawing two genes at random from just a single subpopulation. Let H_s be the average frequency with which these randomly drawn genes from the same subpopulation are different alleles. Then, $f_{st} = (H_t - H_s)/H_t$. In many modern genetic studies, the degree of DNA sequence differences between the randomly drawn genes is measured, often with the use of a model of mutation, instead of just determining whether the two genes are the same or different alleles. When this is done, the analysis is called an *analysis of molecular variance* (AMOVA). Regardless of the specific measure, the degree of genetic differentiation can be quantified in an objective manner in any species. Hence, human "races" can indeed be studied with exactly the same criteria applied to nonhuman species. The main disadvantage of this definition is the arbitrariness of the threshold value of 25 percent.

A second definition of race defines the genetic differences qualitatively. Sharp boundaries exist in this case, because the species is subdivided into two or more *evolutionary lineages*. An evolutionary lineage is created within a species when an ancestral population is split into two or more subpopulations, often by some sort of geographic barrier, such that there is no or extremely limited genetic interchange after the split. This means that the subpopulations tend to evolve mostly independently of one another, causing the lineages to accumulate genetic differences with increasing time since the split. Immediately after the split, the subpopulations would share most ancestral polymorphisms (gene loci with more than one allele) and would therefore be difficult to diagnose as separate lineages. With increasing time after the split, genetic divergence accumulates, and diagnosing the separate lineages becomes easier. Unlike the f_{st} definition of race, no arbitrary threshold of differentiation is set *a priori*. A split into separate lineages also means that the genetic differences among the races would define an *evolutionary tree* analogous to an evolutionary tree of species. Statistical methods exist for testing the null hypothesis that the genetic variation within a species has a treelike structure, and other statistics test the null hypothesis that the entire sample defines a single evolutionary lineage. Therefore, just as with the f_{st} definition, the lineage definition of race can be implemented for all species in an objective fashion using uniform criteria, thereby avoiding a human-specific or cultural definition of race.

2. DO BIOLOGICAL RACES EXIST IN CHIMPANZEES?

Before addressing the existence of human races, we first apply these definitions of race to our closest evolutionary relative, the chimpanzee. In this manner, the definitions can be applied in a context that avoids the emotion and cultural biases that inevitably creep into discussions of human race.

Based on morphological differences, the common chimpanzee (*Pan troglodytes*) has been subdivided into five races or subspecies: *P. t. verus* in the Upper Guinea region of western Africa, *P. t. ellioti* in the Gulf of Guinea region (southern Nigeria and western Cameroon), *P. t. troglodytes* in central Africa, *P. t. schweinfurthii* in the western part of equatorial Africa (mostly southern Cameroon), and *P. t. marungensis* in central and eastern equatorial Africa. Gonder and coworkers (2011) genetically surveyed chimpanzees throughout their range. They discovered sharp genetic differences separating the Upper Guinea and Gulf of Guinea populations from all other populations, but with less sharp genetic boundaries between the equatorial African populations. Table 22-1 shows the pairwise AMOVA results for these populations. The Upper Guinea and Gulf of Guinea populations are above the 25 percent threshold for contrasts with each other and with all other chimpanzee populations. However, the three regions sampled in equatorial Africa are all well below the 25 percent threshold. Hence, three races or subspecies of common chimpanzees are delineated by the threshold criterion: *P. t. verus* in the Upper Guinea region, *P. t. ellioti* in the Gulf of Guinea region, and the chimpanzee populations from equatorial Africa.

If chimpanzees are subdivided into separate evolutionary lineages, the genetic differences among lineages should define a treelike structure characterized by splits and isolation. There are genetic differences between different geographic areas (table 22-1), but such genetic differentiation can also arise when *gene flow* (genetic interchange associated with individuals who disperse from their birth population) occurs but is restricted by geography. For example, gene flow can be restricted when most dispersal is limited to nearby local populations. Because genes are passed on from generation to generation, a new allele can still spread throughout a species' range over multiple generations by using local populations as "stepping-stones" to reach more distant local populations. Such stepping-stone models yield a pattern of *isolation by distance* in which the degree of genetic differentiation between two populations increases with increasing geographic distance between them.

Genetic differentiation structured by isolation by distance can be distinguished from genetic differentiation due to lineage splits by testing for constraints on genetic distances. Consider three hypothetical populations (A, B, and C) such that A is closer to B than to C, and B and C are the closest geographic pair. Under isolation by distance, the genetic distance (measured, say, by the f_{st} value between a pair of populations) should increase with increasing geographic distance; that is, the f_{st} between A and B should be less than the f_{st} between A and C. In contrast, suppose populations A, B, and C represent separate evolutionary lineages (races) such that A split from the common ancestral population of B and C in the past, followed by a more recent split between populations B and C. This results in an evolutionary tree of populations such that genetic distance between populations increases with the time since their split from a common ancestral population. In this hypothetical case, the genetic distances between populations A and B and between populations A and C should be the same, since they both involve a split from the same ancestral population. Hence, the expected pattern of genetic distances differs for trees versus isolation by distance, and formal statis-

Table 22-1. Genetic differentiation among populations of chimpanzees as measured by R_{st}

	Upper Guinea	Gulf of Guinea	Southern Cameroon	Central Africa
Gulf of Guinea	0.41			
Southern Cameroon	0.43	0.25		
Central Africa	0.46	0.27	0.07	
Eastern Africa	0.44	0.28	0.05	0.03

Source: Modified from Gonder et al. 2011.
Note: R_{st} is related to f_{st} but incorporates a mutational model for microsatellites.

tical tests exist to determine whether the pattern of genetic differentiation is consistent with the special constraints imposed by an evolutionary tree.

Another method for testing for a treelike structure is based on finer geographic sampling. As more sites are sampled under an isolation-by-distance model, the geographically intermediate populations should also have intermediate genetic distances. In contrast, when the populations are grouped into a smaller number of evolutionary lineages, genetic distances among populations within a lineage should be relatively small, although they may show an isolation-by-distance pattern within the geographic range occupied by a particular lineage. However, the genetic distances are expected to show a large, sudden increase when the geographic boundary between two lineages is crossed.

When the chimpanzee genetic data are used to estimate an evolutionary tree of populations, the resulting tree has the Upper Guinea population splitting off first, followed by the Gulf of Guinea population, and then splits among the equatorial Africa populations. This tree predicts that the Upper Guinea population should be equally distant from all the other populations, and table 22-1 shows that this prediction is supported when the error in estimating the distances is taken into account. This tree also predicts that the Gulf of Guinea population should be equally distant from all the equatorial African populations but that this distance should be smaller (less time since the split) than the distances involving the Upper Guinea population. Table 22-1 shows that this prediction is also supported. However, the genetic distances among the three equatorial African populations show the isolation-by-distance pattern on an east-west axis. These three populations are therefore collapsed into a single lineage. Hence, chimpanzees do show a treelike structure of genetic differentiation with three lineages: Upper Guinea, Gulf of Guinea, and the combined equatorial African populations. Hence, races do exist in chimpanzees under the lineage definition, and they correspond exactly to the same three races defined by the quantitative threshold definition of race.

3. DO BIOLOGICAL RACES EXIST IN HUMANS?

Do human races exist according to the same criteria applied to chimpanzees? In 2002, Rosenberg and others performed a genetic survey of 52 human populations. They used a computer program to sort individuals or portions of their genomes into five groups and discovered that the genetic ancestry of most individuals was inferred to come mostly from just a single group. Moreover, the groups corresponded to five major geographic populations: (1) sub-Saharan Africans; (2) Europeans, Near and Middle Easterners, and Central Asians; (3) East Asians; (4) Pacific populations; and (5) Amerindians. This paper was the most widely cited article from the journal *Science* in 2002, and many of these citations claimed that this paper supported the idea that races were biologically real in humans. However, Rosenberg and coauthors were more cautious in their interpretation. When they increased the number of groups beyond five, they also obtained an excellent classification into smaller, more regional groups. Hence, they showed that with enough genetic markers, it is possible to discriminate most local populations from one another. Recall that genetic differentiation is *necessary but not sufficient* to define races, so even if there is a consensus that five groups is the right number, genetic discrimination alone does not necessarily mean that these five groups are races.

Assuming for now that the five groups are the meaningful populations, do these groups satisfy the quantitative threshold definition of race? Table 22-2 shows the AMOVA results for these five groups, along with a comparable analysis of the three races of chimpanzees that satisfy both the threshold and lineage definitions of race. Table 22-2 shows how the genetic variation is hierarchically partitioned into differences among individuals within the same local population, differences among local

Table 22-2. AMOVA of genetic variation in chimpanzees and in humans

			Genetic variance components		
Species	Number of "races"	Number of populations	Among individuals within populations	Among populations within races	Among races
Chimpanzees	3	5	64.2%	5.7%	30.1%
Humans	5	52	93.2%	2.5%	4.3%

Sources: Chimpanzees—data from Gonder et al. 2011; humans—data from Rosenberg et al. 2002.

populations within the same "race," and among "races." Table 22-2 confirms the reality of race in chimpanzees using the threshold definition, as 30.1 percent of the genetic variation is found in the among-race component, a result expected from the pairwise analysis shown in table 22-1. In contrast with chimpanzee races, the five major "races" of humans account for only 4.3 percent of the genetic variation—well below the 25 percent threshold. The genetic variation in our species is overwhelmingly variation among individuals (93.2 percent). According to the threshold definition, there are no races in humans.

As for the lineage definition, a treelike structure of genetic differentiation has been strongly rejected for *every* human data set subjected to testing for the constraints expected from an evolutionary tree of populations. Increased geographic sampling further undermines the idea of separate lineages. When Rosenberg and coworkers published their results in 2002, their geographic sampling was coarse. It is now known that the computer program used in these studies generates well-differentiated populations as an artifact of coarse sampling from species characterized by isolation by distance. Figure 22-1 shows a plot of the pairwise f_{st} values of humans as a function of geographic distance. The results fit well with the predictions of an isolation-by-distance model. Consequently, it is not surprising that when Behar and coworkers (2010) sampled Old World populations more finely and used a similar computer program to the one used in the 2002 study, most individuals showed significant genetic inputs from two or more populations, indicating that most human individuals have mixed ancestries. The "races" so apparent to many who cited Rosenberg and coworkers simply disappeared with better sampling. These results and figure 22-1

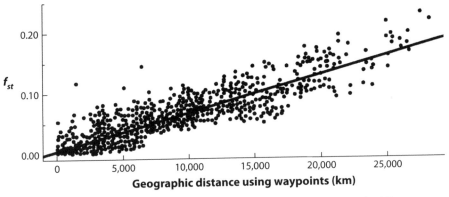

Figure 22-1. Isolation by distance in human populations. (Modified from Ramachandran et al. 2005. Copyright 2005 National Academy of Sciences, USA.)

falsify the hypothesis that humans are subdivided into evolutionary lineages.

Another way of testing for distinct lineages is through a technique known as *multilocus nested-clade phylogeographic analysis* (ML-NCPA). Many regions of the human genome experience little to no recombination. The distinct genetic states that exist in such regions (called *haplotypes*) reflect the accumulation of various mutations during evolutionary history. This evolutionary history, called a *haplotype tree*, is the history of the genetic variation in that genomic region and is *not* necessarily the history of the populations that bear this variation. Indeed, if a species has sufficient gene flow, there can be no evolutionary tree of populations, because there are no population splits; however, there will still be haplotype trees for each nonrecombining region of the genome. Haplotype distributions can be influenced by population-level history, but the population-level information embedded in a haplotype tree must be extracted carefully. It is never justified to equate a haplotype tree directly to an evolutionary history of populations. ML-NCPA provides a statistically rigorous method for making inferences about population history from haplotype trees. No other technique of phylogeographic inference has been so extensively validated as ML-NCPA by both positive controls (data sets for which outside information exists that indicates a known historical event or process) and computer simulation. These validations show that ML-NCPA is not prone to making false-positive inferences about past splits and is very powerful in detecting separate lineages, even when the split is relatively recent and results in haplotype trees affected by retention and sorting of ancestral haplotypes. Moreover, ML-NCPA can detect lineages even when there is some, but very limited, genetic interchange. ML-NCPA does not require an *a priori* model of the evolutionary history of a species; rather, the history is inferred directly from the haplotype trees using explicit criteria applicable to all species. Finally, each inference made with ML-NCPA is subject to a statistical test for significance.

Figure 22-2 shows the inferences from ML-NCPA about human evolution. The oldest inferred event is an out-of-Africa range expansion into Eurasia genetically dated to about 1.9 million years ago—the same time that the fossil evidence indicates that *Homo erectus* spread out of Africa into Eurasia during a major wet period in the Sahara. The paleoclimatic data indicate that the Sahara region experienced repeated minor wet periods, such that the Sahara is unlikely to have been a dispersal barrier on a timescale of tens of thousands of years. Consistent with these paleoclimatic data, ML-NCPA infers limited genetic interchange with isolation by distance between sub-Saharan Africa and Eurasia starting no later than 700,000 years ago in the Pleistocene. The null hypothesis of complete genetic isolation during the Pleistocene is decisively rejected. Consequently, even during the Pleistocene, Old World human populations were not subdivided into isolated and independently evolving lineages.

The next major event shown in figure 22-2 is a second population expansion out of Africa into Eurasia around 700,000 years ago, corresponding to the spread of the Acheulean tool culture out of Africa into Eurasia during the second major Saharan wet period of the Pleistocene. The null hypothesis of no *admixture* between the expanding population and the Eurasian populations is rejected. Hence, the Acheulean expansion was marked by additional genetic interchange between African and Eurasian populations, further weakening the hypothesis of isolated Pleistocene lineages of humans. Gene flow then continued until a third major expansion of humans out of Africa into Eurasia occurred around 130,000 years ago, the time of the last major Saharan wet period. The fossil record indicates that modern humans began expanding out of sub-Saharan Africa at 130,000 years ago and reached China no later

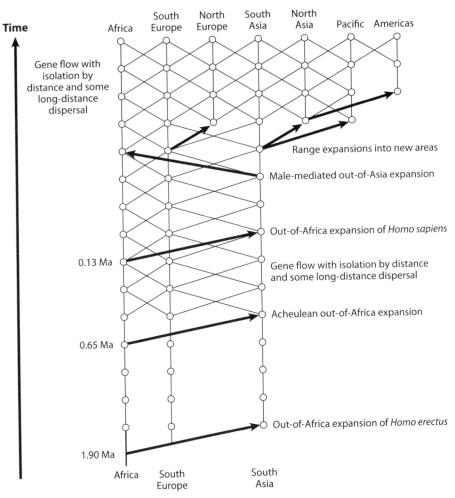

Figure 22-2. Significant inferences about human evolution from multilocus, nested-clade phylogeographic analysis. Geographic location is indicated on the x-axis, and time on the y-axis, with the bottom of the figure corresponding to 2 million years ago. Vertical lines indicate genetic descent over time, and diagonal lines indicate gene flow across space and time. Thick arrows indicate statistically significant population range expansions, with the base of the arrow indicating the geographic origin of the expanding population. Lines of descent are not broken, because the population range expansion events were accompanied by statistically significant admixture when they involved expansion into a previously inhabited area. (Modified from Templeton 2005.)

than 110,000 years ago. The null hypothesis of no admixture is over-whelmingly rejected for this expansion event. This strong rejection of strict replacement without admixing was the most controversial ML-NCPA inference, because the dominant model of human evolution at the time was the out-of-Africa replacement model (see chapter 3). The inference of admixture has since been supported by studies on fossil DNA of archaic Eurasian populations.

Following the expansion with admixture of modern humans from Africa, there have been additional expansions, mostly into areas not formerly occupied by humans (figure 22-2). Wherever humans lived, gene flow was established, mostly limited by isolation by distance but with some long-distance dispersal as well. On a timescale of tens of thousands of years, there is not one statistically significant inference of splitting or isolation during the last 700,000 years. Fossil DNA studies have also shown that ancient human populations were linked by limited, although possibly persistent, gene flow. Because of gene flow and admixture, humans are a single evolutionary lineage. Hence, there are no races in humans under the lineage definition.

4. DO ADAPTIVE TRAITS DEFINE HUMAN RACES?

Races or subspecies, when they exist, always occupy a subset of the geographic range of their species. Sometimes, environmental factors vary over the geographic range of the species, and some of these environmental factors can induce natural selection that results in local adaptation. Hence, when races exist, they sometimes display local adaptations to the environment associated with their geographic subrange that are not adaptive in the remainder of the species' geographic range. This reasoning leads to the idea that local adaptations can sometimes be biological markers of racial status.

Variation in environmental factors can still induce natural selection that results in local adaptations in species with sufficient gene flow and admixture to prevent race formation. However, in this case, the geographic distributions of the local adaptations reflect the geography of the environmental factors and not racial boundaries. Frequently, different adaptive traits display discordant geographic distributions, thereby indicating that these are simply adaptations of local populations and not markers of higher groupings such as race.

Because humans are not subdivided into races by any of the definitions applied to other species, the locally adaptive traits of humans are not "racial" traits. Skin color is historically the locally adaptive trait most commonly considered a "racial" trait in humans. Skin color is an adaptation to the amount of ultraviolet (UV) radiation in the environment: dark skins are adaptive in high-UV environments to protect cells from radiation damage, and light skins are adaptive in low-UV environments to make sufficient vitamin D, which requires UV radiation. Skin color varies continuously among humans and does not fall into a few discrete "racial" types. Moreover, the geographic distribution of skin color follows the environmental factor of UV incidence and does not reflect overall genetic divergence. For example, the native peoples with the darkest skins live in tropical Africa and Melanesia. The dark skins of Africans and Melanesians are adaptive to the high UV found in these areas. Because Africans and Melanesians live on opposite sides of the world, they are more highly genetically differentiated than many other human populations (figure 22-1). Europeans, who are geographically close to Africa, are more similar at the molecular genetic level to Africans than Melanesians are to Africans, despite the fact that Europeans have light skins that are adaptive to the low-UV environment of Europe. Hence, skin color is not an indicator of the degree of genetic differentiation, as a true racial trait would be.

Adaptive traits in humans do not define coherent populations. For example, the adaptive trait of dark skin is widespread in sub-Saharan Africa. Another adaptive trait found in Africa is resistance to sleeping sickness, and the responsible gene is found at frequencies up to 80 percent in parts of western Africa where the parasite that causes sleeping sickness is common. However, this adaptive trait is virtually absent in East African populations. Hence, the distribution of sleeping sickness resistance is only a subset of the geographic distribution of dark skin in Africa. Another adaptive trait is resistance to malaria, which is widespread in African populations. However, malaria is also common in some areas outside Africa, and malarial resistance is found in many European and Asian populations as well. Indeed, one of the alleles underlying malarial resistance, the sickle-cell allele, is most frequent in certain populations on the Arabian Peninsula despite often being regarded as a disease of "blacks." Each adaptive trait in humans has its own geographic distribution that reflects the distribution of the underlying environmental factor for which it is adaptive. The discordance in the distributions of adaptive traits in humans makes them useless in defining races.

5. DO HUMAN RACES EXIST: THE ANSWER

Using culture-free, objective definitions of race, the answer to the question whether races exist in humans is clear and unambiguous: no. Human evolutionary history has been dominated by gene flow and admixture that unifies humanity into a single evolutionary lineage. This finding does not mean that all human populations are genetically identical. Isolation by distance ensures that human populations are genetically differentiated from one another, and local adaptation ensures that some of these differences reflect adaptive evolution to the environmental heterogeneity that our globally distributed species experiences. However, most of our genetic variation exists as differences among individuals, with between-population differences being very small. There are no biological races in humans; indeed, despite our global distribution, we are one of the most genetically homogeneous species on this planet.

FURTHER READING

Gonder, M. K., S. Locatelli, L. Ghobrial, M. W. Mitchell, J. T. Kujawski, F. J. Lankester, C.-B. Stewart, and S. A. Tishkoff. 2011. Evidence from Cameroon reveals differences in the genetic structure and histories of chimpanzee populations. Proceedings of the National Academy of Sciences USA 108: 4766–4771.

Lao, O., P. M. Vallone, M. D. Coble, T. M. Diegoli, M. van Oven, K. J. van der Gaag, J. Pijpe, P. de Knijff, and M. Kayser. 2010. Evaluating self-declared ancestry of U.S. Americans with autosomal, Y-chromosomal and mitochondrial DNA. Human Mutation 31: E1875–E1893.

Long, J. C., and R. A. Kittles. 2009. Human genetic diversity and the nonexistence of biological races. Human Biology 81: 777–798. *Formal statistics tests of the hypothesis that human genetic variation is structured in a treelike fashion. They show that the hypothesis of a tree of populations is strongly rejected for humans.*

Pääbo, S. 2015. The diverse origins of the human gene pool. Nature Reviews Genetics 16: 313–314. *This review article focuses on the ancient DNA studies from both Europe and*

Asia that indicate intermittent and perhaps persistent gene flow among archaic human populations.

Ramachandran, S., O. Deshpande, C. C. Roseman, N. A. Rosenberg, M. W. Feldman, and L. L. Cavalli-Sforza. 2005. Support from the relationship of genetic and geographic distance in human populations for a serial founder effect originating in Africa. Proceedings of the National Academy of Sciences USA 102: 15942–15947.

Relethford, J. H. 2009. Race and global patterns of phenotypic variation. American Journal of Physical Anthropology 139: 16–22. *This paper shows that skin color variation and other morphological traits are clinal and are not well described by discrete racial categories.*

Rosenberg, N. A., J. K. Pritchard, J. L. Weber, H. M. Cann, K. K. Kidd, L. A. Zhivotovsky, and M. W. Feldman. 2002. Genetic structure of human populations. Science 298: 2381–2385.

Santos, R. V., P. H. Fry, S. Monteiro, M. C. Maio, J. C. Rodrigues, L. Bastos-Rodrigues, and S.D.J. Pena. 2009. Color, race, and genomic ancestry in Brazil: Dialogues between anthropology and genetics. Current Anthropology 50: 787–819.

Templeton, A. R. 2003. Human races in the context of recent human evolution: A molecular genetic perspective. In A. H. Goodman, D. Heath, and M. S. Lindee, eds., Genetic Nature/Culture, 234–257. Berkeley: University of California Press. *This paper covers many of the same issues as this chapter, but with older data sets. These older data sets yield the same conclusions found in this chapter.*

Templeton, A. R. 2005. Haplotype trees and modern human origins. Yearbook of Physical Anthropology 48: 33–59. *The results of multilocus nested-clade phylogeographic analysis of humans, showing that gene flow and admixture have been such common features of recent human evolution that there is only one evolutionary lineage of humanity.*

Wolpoff, M., and R. Caspari. 1997. Race and Human Evolution. New York: Simon & Schuster. *The authors strongly argue against typological thinking in anthropology that explains human variation in terms of a few types or "races" rather than dealing with the entire range of variation found in living humans. Although the book ignores the nonhuman literature and tends to hyperbole, it remains an excellent introduction to the intertwined topics of racism and models of modern human origins.*

CHAPTER 23

THE FUTURE OF HUMAN EVOLUTION

Alan R. Templeton

OUTLINE

1. Can we predict how humans will evolve?
2. Has human evolution stopped?
3. Future nonadaptive evolution
4. Future adaptive evolution
5. Eugenics and genetic engineering

How humans will evolve in the future is highly speculative because the process of evolution depends critically on random processes such as mutation, recombination, and genetic drift, and because adaptive evolution is strongly influenced by changing environments. Because the human environment includes culture, which can change quickly, it is difficult to predict future environments and hence future adaptive evolution. Nevertheless, some predictions can be made based on a basic understanding of evolutionary mechanisms.

GLOSSARY

Eugenics. Programs designed to direct evolutionary changes in the human population by controlled breeding, selective abortions, and sterility operations.

Gene Flow. Movement of individuals or gametes from the local population of birth to a different local population followed by successful reproduction.

Gene Pool. The set of genetic variants collectively shared by a reproducing population.

Genetic Drift. The evolutionary force associated with random sampling events that alters the frequencies of genetic variants in the gene pool.

Genetic Engineering. The deliberate modification of characteristics of an organism by manipulating its genetic material.

Heterozygosity. The condition in which the two homologous segments of genetic material inherited from the parents are of a different state.

Mutation. A variety of molecular-level processes by which the genetic material of an organism (usually DNA) undergoes a change.

Neutral Allele. An allele that is functionally equivalent to its ancestral allele in terms of its chances of being replicated and passed on to the next generation.

Recombination. The generation of new combinations of DNA segments that are unlike those that existed in the parents.

1. CAN WE PREDICT HOW HUMANS WILL EVOLVE?

The current biological state of humans is the product of our past evolution (see chapters 3 and 22). But what of our evolutionary future? We can infer to some extent what our ancestors were like 100,000 or a million years ago, but what will our descendants be like 100,000 or a million years into the future? This question is difficult to answer because the evolutionary process itself is strongly influenced by random factors. There can be no evolution of any sort without genetic variation, and this genetic variation is created by mutation and recombination. *Mutation* and *recombination* are molecular-level processes that create new genetic variants before these variants are expressed phenotypically in individuals living and reproducing in an environment. In this sense, the genetic variation that is the raw material for all evolutionary change is random with respect to environmental needs—a fundamental premise of Darwinian evolution. Moreover, once genetic variation is randomly created, sampling error further accentuates the randomness of evolution. For example, suppose a new autosomal mutation occurs such that its bearer would be expected to have 10 percent more offspring than the average for the population. Assuming the population was stable, the average number of offspring per individual would be two for the population as a whole, but for this mutant bearer, the average would be 2.2. This would be regarded as extremely intense natural selection for a highly favored mutation. However, the actual number of offspring of the individual bearing the new mutant is not the same as the expected number. Perhaps this individual died from an accident or a disease unrelated to the mutant effect before he or she could reproduce; perhaps this individual failed to find a mate; perhaps this individual had fewer or more than 2.2 children. It is impossible to predict the exact number of children any particular individual will have even if the average number of children is known. At best what is known is the probability distribution of the offspring number. Suppose this offspring probability distribution is a Poisson distribution (a standard, commonly used probability distribution) with an average of 2.2 offspring. Then, despite the large fitness value this mutant individual has, there is still an 11 percent chance that he or she will have no offspring at all, and hence the mutant will be lost. Moreover, there is the randomness of meiosis. Suppose the mutant individual had three children. Each child would have a 50:50 chance of receiving the mutant, so with a chance of $(1/2)^3 = 1/8$, none of the children would receive the mutant, and the mutant would be lost even though the bearer had more children than the population average. Using a more thorough mathematical analysis, the probability that this strongly favored mutant gene is ultimately lost by chance despite being strongly favored by natural selection is 82 percent! Thus, even strong natural selection cannot completely overcome the randomness inherent in the evolutionary process. These random events affecting the fate of a particular mutant gene are occurring for all genes in the population, and can strongly influence the course of evolution. The evolutionary force associated with these random sampling events that affect existing genetic variation is known as *genetic drift*. Genetic drift, mutation, and recombination ensure that evolution can never be completely predictable.

Another factor that makes future evolution difficult to predict is the environment. Adaptive evolution is always with respect to an environment, and it is difficult to predict what our environment will be like in the distant future. This is especially true for us, because we define much of our environment through our culture (see chapter 21). Culture in turn can change rapidly in unforeseen ways. Therefore, even the evolutionary trajectory of the nonrandom evolutionary force of natural selection is difficult to predict because the raw material on which natural selection operates has a large random component and because our environment, particularly our cultural environment, can change in a manner that is difficult to predict. Nevertheless, some predictions are possible using general evolutionary principles, but specific details must always be regarded as speculative.

2. HAS HUMAN EVOLUTION STOPPED?

One of the more popular predictions about future human evolution is that there will be none; that is, human evolution has already stopped. For example, the distinguished evolutionary biologist Stephen Jay Gould (2000) stated:

> There's been no biological change in humans in 40,000 or 50,000 years. Everything we call culture and civilization we've built with the same body and brain.

The basic rationale behind the conclusion that human evolution has stopped is that once the human lineage had achieved a sufficiently large brain and had developed a sufficiently sophisticated culture (sometime around 40,000–50,000 years ago according to Gould, but more commonly placed at 10,000 years ago with the development of agriculture), cultural evolution supplanted biological evolution through natural selection; that is, humans no longer adapt to their environment through natural selection but rather alter the environment to suit human needs through cultural innovations. However, other evolutionary biologists have come to exactly the opposite conclusion. For example, Cochran and Harpending (2009) concluded that "human evolution has accelerated in the past 10,000 years, rather than slowing or stopping, and is now happening about 100 times faster than its long-term average over the 6 million years of our existence."

There are two fundamental flaws in the proposal that human evolution has stopped. First, it ignores the fact that evolution can occur owing to factors other than natural selection. This flaw is discussed in the next section. The second flaw is the premise that cultural evolution eliminates adaptive evolution via natural selection. All organisms adapt to their environment, and we define much of our environment by our culture. Hence, cultural change can actually spur adaptive evolution in humans. Since the development of agriculture the human population has grown in a roughly superexponential fashion. Agriculture also induced a more sedentary lifestyle. As a result, even early agricultural systems resulted in large increases in local human densities. This in turn created a new demographic environment that was ideal for the spread of infectious diseases. For example, the Malaysian agricultural system, first developed in Southeast Asia, makes extensive use of root and tree crops that are adapted to wet, tropical environments. This tropical agricultural system was introduced to the African mainland about 2500 years ago, and malaria has become a common disease in these new agricultural areas. Because of agriculture, malaria became a major selective agent in African, and other, human populations. The result was that human populations began to adapt to malaria via natural selection. In sub-Saharan Africa, natural selection favored an increase in frequency of the sickle-cell allele at the hemoglobin β-chain locus, which confers resistance to malaria in individuals heterozygous for the sickle-cell allele. Similar selective forces were introduced wherever agriculture created the conditions to allow malaria to become a sustained, epidemic disease, and human populations in turn adapted to malaria by increasing the frequency of a number of alleles at many different loci in addition to the sickle-cell allele. In terms of the numbers of people affected, these antimalarial adaptations alone constitute the vast bulk of the classical Mendelian genetic diseases that afflict current humanity.

Agriculture produced a selective environment that also favored genes associated with risk for common systemic diseases in current human populations, one of the more common of which is type 2 diabetes. Much of the increased incidence of diabetes is due to environmental changes in diet and lifestyle. However, phenotypes arise from the interaction of genes with environment, so a strong environmental component to type 2 diabetes, and many other systemic diseases, does not preclude a genetic component due to adaptive evolution in recent human history. The idea that genes predisposing an individual to type 2 diabetes could represent recent adaptive evolution was first proposed by James V. Neel in 1962 as the *thrifty genotype hypothesis*. This hypothesis postulates that the same genetic states that predispose one to diabetes also are advantageous when individuals suffer periodically from famines. When food is more plentiful, selection against these genotypes would be mild because the age of onset of the diabetic phenotype is typically after most reproduction has occurred and because the high-sugar, high-calorie diets found in modern societies that help trigger the diabetic phenotype are very recent in human evolutionary history. There is now much evidence for the thrifty genotype hypothesis, including the genomic signatures of strong natural selection at genes shown to increase risk for diabetes in populations with a recent history of exposure to famines or calorie-restricted diets. The thrifty genotype hypothesis has often been portrayed as an example of past adaptation to a Paleolithic lifestyle despite the fact that the populations used to test this hypothesis all suffered from famines in historic times. Hence, the thrifty genotypes present in current human populations are an adaptation to recent events in agricultural systems prone to periodic failures and are *not* a legacy of human evolution having stopped in the Paleolithic.

Agriculture also induced positive selection for humans to adapt to the products of agriculture. For example, with the domestication of cattle and goats, milk and its derivatives became not only a source of nutrition but also a dietary component that protects against nutritional rickets, a common disease associated with high-cereal diets, another by-product of an agricultural environment. The phenotype of adult lactase persistence is determined by a single gene that allows the digestion of milk sugar. This specific allele shows one of the more powerful signatures of strong recent natural selection in the human genome.

As the preceding examples demonstrate, agriculture—and culture in general—did not stop human evolution via natural selection but rather induced it through its direct and indirect effects on the human environment. Cultural innovations indeed shield some traits from natural selection, but cultural evolution will likely induce further adaptive evolution of many other traits in humans.

3. FUTURE NONADAPTIVE EVOLUTION

Not all evolution is adaptive. Evolution within a species is a change in the type or frequencies of genes or gene combinations in the gene pool over time, with the *gene pool* being the set of genes collectively shared by a reproducing population. Natural selection is a powerful mechanism for altering the frequencies of genes in the gene pool, but developmental constraints, patterns of dispersal, system of mating, population size, mutation, recombination, and other factors can also cause alterations in the gene pool. Evolutionary change is determined not by one evolutionary mechanism operating in isolation but rather by several mechanisms operating in concert.

Because evolution emerges from the interaction of multiple evolutionary forces, even a relaxation of natural selection induces further evolution. Many traits are developmentally correlated, so if one trait is made selectively neutral by a cultural innovation, that in turn will alter the evolutionary balance at other, correlated traits, which in turn can induce further nonadaptive evolution via developmental correlations for the neutral trait. For example, most animals adapt to their diet in part through their teeth and jaws, but humans increasingly used tools and fire to prepare their food. These cultural innovations reduced the importance of jaw and tooth evolution as a means of adapting to the dietary environment. Rebecca Ackerman and James Cheverud (2004) tested the hypotheses of selected versus neutral evolution of human teeth and jaws by comparing various hominin fossil measurements with the expected developmental correlations among relative brain size, tooth size, and jaw size as inferred from modern-day humans, chimpanzees, and gorillas. Their analysis indicated the intensity of selection on the face diminished with time in the human lineage, and by 1.5 million years ago there was no longer any detectable selection on human teeth and jaws. This conclusion supports the hypothesis that cultural evolution in the human lineage had indeed eliminated natural selection on these traits. However, this does *not* mean that human teeth and jaws have not evolved over the last 1.5 million years. During the last 1.5 million years, there was a large increase in brain size in the human lineage driven by natural selection, and given the developmental constraints common to humans, chimpanzees, and gorillas, human jaws and teeth continued to evolve as a correlated effect of brain size evolution. In particular, jaws and teeth became relatively smaller for overall human head size as a correlated response to increased brain size, with the jaw becoming relatively smaller more rapidly than the teeth. The result of this correlated evolution is that humans have a small, flat face compared with those of chimpanzees and gorillas, and humans have jaws that tend to be too small for their teeth, leading to tooth crowding in the jaws. This nonadaptive evolution in turn favored the cultural evolution of the profession of orthodontics. One major past selective constraint on brain size has been the difficulty of passing a large-brained baby through the mother's birth canal. With the widespread use of cesarean sections, this selective constraint is being reduced in intensity. If this trend continues and if there is still selection for increased brain size, human jaws will become even smaller relative to the teeth. Therefore, the profession of orthodontics has a secure evolutionary future. As this example shows, the release of traits by culture from natural selection leads to further nonadaptive evolution of these traits—not evolutionary stasis.

As culture makes more mutant alleles effectively neutral with respect to natural selection, then genetic drift and mutation become the evolutionary forces that influence the evolutionary fate of these *neutral alleles*, and the rate of neutral evolution equals the mutation rate to neutral alleles. Hence, to the extent that cultural evolution reduces selective forces in humans, the mutation rate to neutral alleles will increase, which in turn will result in an increase in the rate of neutral evolution in humanity. At first this may seem to be a trivial factor in future human evolution, since by definition this accelerated evolution involves only alleles that have no adaptive significance. However, when a gene has many potential selectively neutral mutations, it is possible for that gene to accumulate many functionally equivalent alleles differing by a series of neutral mutations. In this manner, new forms of the gene can evolve via neutrality that are several mutational steps away from the ancestral gene form. The phenotypic effects of a mutation often depend on other mutations that have occurred previously, so that a mutation that would have been deleterious or neutral on the ancestral allelic background may be selectively favored on the new, derived allelic background. In this manner, neutral evolution can actually increase the adaptive potential of a population and allow for adaptive transitions that would otherwise be unlikely. Hence, cultural evolution that reduces natural selection can increase the long-term adaptive potential of the human species.

Another consequence of cultural evolution is that humans have experienced superexponential growth for the last 10,000 years. The resulting large population size interacts strongly with the random forces of mutation and genetic drift. A small population will have very few new mutations at any given time. For example, suppose a specific nucleotide mutation has a probability of 10^{-9} of occurring per gene per generation at an autosomal locus. In a diploid population of 500, there are 1000 copies of an autosomal gene, so the expected number of new mutations to this specific form in any given generation is 10^{-6}; that is, there is only one chance in a million of this mutation occurring in any given generation. Hence, the randomness of mutation plays a large role in the evolution of this population. The human population size is now at 6.8 billion, so for an autosomal locus we would expect 13.6 occurrences of this specific mutation every generation. The large human population size is causing humans to enter an evolutionary zone that few eukaryotic organisms have ever reached—the zone in which virtually every single-step mutational change occurs in every generation. This in turn greatly reduces the randomness of evolution induced by the mutational process. Recall that the sickle-cell allele became selectively favored in sub-Saharan Africa after the introduction of the Malaysian agricultural complex 2500 years ago. What is more remarkable is that this specific sickle-cell mutation went to high frequency in sub-Saharan African populations from at least four independent mutations of this specific nucleotide. The ability of large populations to produce a huge reservoir of mutational variants means that human populations are more evolutionarily responsive than ever to changes in the environment. As long as the human population size remains large, it will remain in this rare evolutionary zone that increases its adaptive potential.

An expanding population also increases the probability of long-term survival of a new mutant, thereby enhancing the reservoir of mutational variants beyond that of a population of fixed size. For example, consider a mutant with a 10 percent advantage in a stable population in which an individual had an average of two offspring with a Poisson offspring distribution. As indicated earlier, the chances of this highly favorable mutation being lost by chance alone is 82 percent. Now suppose this mutant occurs in a growing population in which the average number of children is three. Then, the probability of loss of the favorable mutant is reduced to 33 percent. However, there is an evolutionary price to be paid for this enhanced survival of favorable mutations. Consider a deleterious mutant that reduces the fitness of its heterozygous bearers by 10 percent. In a constant-size large population, such a deleterious mutant is eliminated by natural selection with a probability of 1, but in the growing population its chance of elimination is reduced to 53 percent. Hence, beneficial, neutral, and deleterious mutations all accumulate in the human gene pool owing to our unique demographic history. Indeed, recent studies in which the entire DNA sequence of some genes was determined in a sample of nearly 15,000 individuals reveal a large excess of rare variants due to recent mutations in the human gene pool, and many of these recent variants appear to be deleterious.

Exponential population growth cannot be sustained indefinitely in any world with finite resources, so it is inevitable that this phase of human demographic history will end in the future. Indeed, the rate of growth is already dropping. The only question is whether human population will continue to grow to a larger stable size, decrease in size, or fluctuate up and down. The change in demographic environment associated with population size stability or decline will end the era of enhanced survival of mutants, particularly deleterious ones. Indeed, natural selection will in the future start acting to eliminate the reservoir of deleterious variants that have accumulated in the gene pool during the last 10,000 years of population growth. However, as long as our population stabilizes at a large number, the reservoir of genetic variation will remain high, conferring a high degree of adaptability to the human species.

The changing demographic environment will also alter the balance of local genetic drift with gene flow. Although genetic drift causes random fluctuations in allele frequencies, it has some very predictable properties. First, genetic drift causes genetic variation to be lost, and the smaller the population size, the more rapidly genetic variation is eroded. Second, when a species is split into multiple local populations with little genetic interchange between them, genetic drift causes random changes in allele frequencies in all of them. Because the changes are random, they are unlikely to be in the same direction in every local population. Hence, genetic drift leads to genetic differences among local populations, and the smaller the local population sizes, the greater the expected differences among them. *Gene flow* occurs when either individuals or gametes disperse from one local population to another through reproduction. Gene flow can introduce a mutation that arose in one local population into the gene pool of another local population. Hence, gene flow tends to increase the amount of genetic diversity found within local populations. The genetic interchange associated with gene flow also reduces the genetic differences among local populations. Note that genetic drift and gene flow have exactly opposite effects on genetic variation *within* local populations (decreased by drift, increased by gene flow) and genetic differences *among* local populations (increased by drift, decreased by gene flow). As a result, the balance of genetic drift to gene flow is the primary determinant of how a species' genetic variation is distributed within and among its local populations.

There is no doubt that the balance of genetic drift to gene flow has been greatly altered in recent human evolution and continues to change at a rapid pace. The increased human population size associated with the development of agriculture weakens the evolutionary force of genetic drift, and a wide variety of cultural innovations have greatly increased the ability of people to move across the globe and thereby augment gene flow. In addition, our system of mating is changing in response to cultural changes. Currently, about 10.8 percent of human couples on a global basis are related as second cousins or closer, and this subset of human couples is associated with an increased incidence of genetic disease and systemic diseases with a genetic component, as well as increased susceptibility to infectious diseases. Preference for mating with a relative decreases the amount of gene flow, but this preference is rapidly declining with increased urbanization, improved female education, and smaller family sizes. If these cultural trends continue, gene flow and outbreeding will become even stronger in future human populations. All these alterations are increasing the level of genetic variation within local human populations and decreasing the genetic differences among human populations. As long as the ability to disperse over the globe remains high and the trend toward outbreeding continues, much of future human evolution over the next tens of thousands of years will be dominated by decreased local genetic drift and increased gene flow. The result will be increased levels of individual *heterozygosity* (that is, the two copies of an autosomal gene borne by an individual are increasingly likely to be of different allelic states). This rapid and ongoing shift to increased levels of heterozygosity in humans is already having discernible health effects. For example, in studies that control for diet, socioeconomic status, and other factors, several clinical traits have significant beneficial changes with increasing heterozygosity. Similarly, areas of the human genome that lack heterozygosity are associated with diseases with genetic components, such as schizophrenia and late-onset Alzheimer's disease. As heterozygosity levels continue to increase in humans owing to vastly increased abilities to disperse, these beneficial effects are expected to increase even more. This increased heterozygosity will also reduce the deleterious consequences of the many rare deleterious variants the species has accumulated during its phase of superexponential population growth.

The second effect of this new balance between drift and gene flow will be the eventual fusion of all human local gene pools into a single species-wide gene pool. As described in chapter 22, humans already are one of the most spatially homogeneous species on this planet in a genetic sense. The modest genetic differences observed today among different human populations will be further eroded, and with continual gene flow and large population sizes will eventually be eliminated. The only genetic differences that will be biologically meaningful in the human species will be the differences among individuals, which will be high because of the high levels of genetic variation in the common human gene pool.

One nonadaptive consequence of this genetic fusion of human populations will be the loss of local adaptations. For example, skin color in humans is an adaptation to the local level of ultraviolet radiation and is not a good indicator of "race" or overall genetic differentiation among populations (see chapter 22). The degree of local adaptation reflects the balance of local selective forces favoring genetic differentiation versus gene flow favoring homogenization. If gene flow and outbreeding continue to increase, human populations will display less local adaptation and more genetic homogeneity across the globe.

4. FUTURE ADAPTIVE EVOLUTION

Adaptive evolution is always with respect to an environment, and it is difficult to predict the details of the future human environment. However, much of the past evolution induced by cultural changes has been associated with the alteration of the human demographic environment, and some predictions can be made there. Continued exponential growth is ultimately unsustainable. Two extreme scenarios are possible. The optimistic scenario is that human population size will stabilize, perhaps at a level smaller than today but still quite large, without any major collapse of human civilization. Under this scenario, it is likely that the current trends toward increased dispersal and outbreeding will continue. The level of heterozygosity will increase, improving the overall genetic health of the human species. This demographic environment will also yield a large human population with an immense reservoir of genetic variation of neutral and beneficial mutations but fewer deleterious mutations than at present. The genetic differences among human populations, already small, will become even less significant, and there will be far less local adaptation. However, because of the large reservoir of new mutations and because culture-induced neutrality will allow greater exploration of the mutational state space, the adaptive potential of the human species as a whole will be enhanced. This may be important in adapting to global climate change.

There is a caveat about this greater adaptive potential of future human populations. Although the randomness of genetic drift has been emphasized until now, Sewall Wright, the man most responsible for the development of the theory of genetic drift, emphasized its significance for adaptive evolution. Evolution, including adaptive evolution, arises from the interaction of multiple evolutionary forces, including genetic drift and natural selection. Just as a series of mutationally linked neutral alleles can augment adaptability by allowing a more thorough exploration of the mutational state space, genetic drift can allow a more thorough exploration of the adaptive gene pool state space when there are multiple ways of adapting to the same environment. Multiple adaptive solutions are particularly common when adaptive traits emerge from interactions among multiple genes. Selection in large populations where genetic drift is weak therefore tends to fine-tune a single adaptive solution, whereas populations with stronger genetic drift are more likely to undergo major adaptive innovations. Hence, future humans under this optimistic demographic scenario will have greatly enhanced potential for fine-tuning human adaptations but are unlikely to make major or radical adaptive breakthroughs unless there are also major environmental changes affecting humans at the global level.

The pessimistic demographic scenario is that human population size and civilization will both crash. This will reverse the trends to increasing dispersal and outbreeding, leading to much population subdivision. Because of 10,000 years of population growth, the current human gene pool has a disproportionate number of recent deleterious mutations. With population fragmentation, some of these globally rare deleterious variants will become locally common, causing a major decline in the overall genetic health of the human species and inducing a period of strong natural selection against deleterious variants after the population crash. Balancing this negative selection, the enhanced reservoir of neutral and beneficial mutations that were also accumulated during the period of population growth when coupled with increased genetic drift makes it likely that some human populations will undergo major adaptive breakthroughs. The nature of these breakthroughs is difficult to predict because of the strong random role that genetic drift will play in this process.

5. EUGENICS AND GENETIC ENGINEERING

The success of agriculture in sustaining 10,000 years of population growth was possible because humans became strong and effective selective agents on crop and livestock species. More recently, the ability to manipulate agricultural species has been augmented with *genetic engineering*, in which humans directly manipulate the genetic material of domesticated species. One possibility for future human evolution is that humans will choose to direct their own evolution by selective breeding (*eugenics*) and/or genetic engineering.

Eugenic proposals and programs have a long history in human societies. However, this history does not engender much confidence in such an approach to controlling human evolution. For example, the "genetics" used by the American eugenics movement is ludicrous in light of modern genetics, yet this pseudogenetics led to forced sterilizations and major changes in immigration laws, and served as a model for the eugenic excesses of the Nazi regime. When people turn principles of selective breeding and genetic manipulation on themselves, scientific objectivity is frequently lost, and nonscientific social theories and prejudices dominate in shaping eugenic proposals. Moreover, current knowledge of human genetics indicates that the successes attained in plant and animal breeding for agricultural purposes are not likely to be replicated in humans. Phenotypes arise from genes interacting with one another and with the environment. Agricultural breeding is almost always done in stocks or lines that are far more homogeneous genetically than humans. Thus, the effects of any one gene are far more predictable in agricultural breeding and engineering than they would be in humans. The same gene could have dramatically different phenotypic effects on different human genetic backgrounds. Second, phenotypes emerge from genotype by environment interactions. In agriculture, humans select and engineer crop and livestock strains specifically for how their genes interact with simple, homogeneous environments. Human environments are not simple or homogeneous, so once again the impact of a single gene can vary tremendously. For example, the single gene locus most predictive of risk for coronary artery disease, the number one killer in the developed world, is the *Apoprotein-E* locus (*ApoE*), which has three common alleles in most human populations: $\varepsilon 2$, $\varepsilon 3$, and $\varepsilon 4$. A retrospective study indicated that individuals bearing the $\varepsilon 4$ allele had the highest incidence of coronary artery disease on average. The same study revealed that individuals in the highest tertile for total serum cholesterol level had the highest incidence of coronary artery disease compared with the middle and lower tertiles for cholesterol level. Cholesterol level in turn is affected by many interacting genes (including *ApoE*) and environmental variables such as smoking, diet, and exercise. When genotype and cholesterol levels were combined, the group of people with the highest incidence of coronary artery disease by far were people with high cholesterol levels and the $\varepsilon 2/\varepsilon 3$ genotype. Note that the genotype with the highest absolute incidence of coronary artery disease has the "good" alleles at the *ApoE* locus. This is the main problem with eugenic and genetic engineering programs for humans: genetic background and environment are highly heterogeneous in humans, so the consequences of manipulations can never be accurately predicted. Moreover, environments change very rapidly for humans, making eugenic predictions even more prone to error. Unless it is decided to create separate castes of relatively genetically homogeneous human strains and keep them in highly restricted environments, eugenics and genetic engineering is unlikely to play a significant role in future human evolution.

The development of powerful new systems for editing DNA has raised much interest in a new eugenics that would treat genetic diseases and defects by repairing the specific defective mutations or genes. Recent studies on human embryos show that such genome-editing technology is still not precise enough for clinical applications but is rapidly improving. Quite aside from the ethical implications, the validity of the new eugenics depends upon two premises. The first premise is that the genes and mutations that actually cause the disease or defect of interest are known. Current genomic analyses can identify associations between specific genes and disease, but association is not causation. In actuality, for most "disease" genes, causation is not truly known, only association. Moreover, it is now known that many diseases are most likely caused by changes in gene regulation, and the regulatory elements for the gene are often located closer to other genes within the genome, making causation extremely difficult to infer from genome association. However, in some cases there is sufficient evidence that a specific mutation or gene is indeed responsible for the disease. Then, the second premise is that the causative disease mutation or gene is indeed a "defective" gene. The path from genotype to phenotype is complex and often involves *pleiotropy*; that is, the same genotype can influence many different traits, with environmental interactions often playing a role. For example, nondiabetic end-stage renal failure, a costly and typically fatal disease, is strongly associated with specific mutations in the *APOL1* gene, and other evidence supports their being causative. However, most people with these "defective" mutations do not get end-stage renal failure, indicating that additional factors are needed. One strong candidate is certain viral infections. Moreover, these same mutations in *APOL1* also appear to cause resistance to trypanosomes, so in areas with sleeping sickness, these "defective" mutations increase, not decrease, the average health of their bearers.

Studies on diseases generally focus on individuals who are ill with the disease of interest and do not consider the other pleiotropic roles of the gene nor the variety of environmental contexts in which the gene is expressed. Even if it is decided to implement these new gene-editing technologies on human embryos, the difficulties outlined here indicate that the new eugenics will have at most only an extremely limited role in future human evolution.

FURTHER READING

Ackermann R. R., and J. M. Cheverud. 2004. Detecting genetic drift versus selection in human evolution. Proceedings of the National Academy of Sciences USA 101: 17947–17951.

Allen, G. E. 1983. The misuse of biological hierarchies: The American eugenics movement, 1900–1940. History and Philosophy of the Life Sciences. Section II of Pubblicazioni della Stazione Zoologica di Napoli 5: 105–128. *A brief history of the American eugenics movement and the impact it had on laws and policy in the United States and other countries.*

Bittles, A. H., and M. L. Black. 2010. Consanguinity, human evolution, and complex diseases. Proceedings of the National Academy of Sciences 107: 1779–1786.

Campbell, H., A. D. Carothers, I. Rudan, C. Hayward, Z. Biloglav, L. Barac, M. Pericic, et al. 2007. Effects of genome-wide heterozygosity on a range of biomedically relevant human quantitative traits. Human Molecular Genetics 16: 233–241. *Heterozygosity levels were measured in four different Croatian populations that differed greatly in their degree of gene flow among local populations but that had similar diets, socioeconomic status, and other factors. Several clinical traits were then regressed against relative heterozygosity, and all significant results indicated beneficial effects with increasing heterozygosity.*

Cochran, G., and H. Harpending. 2009. The 10,000 Year Explosion: How Civilization Accelerated Human Evolution. New York: Basic Books. *Debunks the idea that human evolution has stopped and instead argues that it has accelerated.*

Coventry, A., L. M. Bull-Otterson, X. Liu, A. G. Clark, T. J. Maxwell, J. Crosby, J. E. Hixson, et al. 2010. Deep resequencing reveals excess rare recent variants consistent with explosive population growth. Nature Communications 1: 131. *One of the first studies to do extensive DNA resequencing, revealing a plethora of recent, rare variants in the human gene pool. Many of these variants are predicted to have deleterious consequences.*

Gould, S. J. 2000. The spice of life. Leader to Leader 15: 19–28.

Kaiser, J., and D. Normile. 2015. Embryo engineering study splits scientific community. Science 348: 486–487. *Discusses the technical and ethical problems encountered when gene editing was applied to human embryos for the first time.*

Ku, C. S., N. Naidoo, S. M. Teo, and Y. Pawitan. 2011. Regions of homozygosity and their impact on complex diseases and traits. Human Genetics 129: 1–15. *This paper shows that areas of the human genome that lack heterozygosity are associated with increased risk for diseases with a genetic component.*

Neel, J. V. 1962. Diabetes mellitus: A "thrifty genotype" rendered detrimental by "progress." American Journal of Human Genetics 14: 353–362. *A classic paper that developed the idea that the genes underlying risk for diabetes could have been adaptive during famine conditions. Much recent work has supported this hypothesis, and variants of the thrifty genotype hypothesis have been proposed for other common diseases in humans.*

Peter, B. M., E. Huerta-Sanchez, and R. Nielsen. 2012. Distinguishing between selective sweeps from standing variation and from a *de novo* mutation. PLoS Genetics 8: e1003011. *Provides evidence for several mutations being favored in recent human evolution, including that for lactase persistence.*

Templeton, A. R. 1998. The complexity of the genotype-phenotype relationship and the limitations of using genetic "markers" at the individual level. Science in Context 11: 373–389. *Discusses why eugenics and genetic engineering should be ineffective in human populations owing to the complex interactions among genes and between genes and environments.*

Templeton, A. R. 2006. Population Genetics and Microevolutionary Theory. Hoboken, NJ: John Wiley. *This textbook gives the details of many of the examples used in this chapter and also shows how multiple evolutionary forces interact to influence the trajectory of evolution.*

Templeton, A. R. 2010. Has human evolution stopped? Rambam Maimonides Medical Journal 1(1): e0006. doi:10.5041/RMMJ. 10006. *Gives additional details on some of the examples used in this chapter. It also argues against the idea that human evolution has stopped, using arguments not found in the book by Cochran and Harpending.*

Tzur, S., S. Rosset, R. Shemer, G. Yudkovsky, S. Selig, A. Tarekegn, E. Bekele, et al. 2010. Missense mutations in the *APOL1* gene are highly associated with end-stage kidney disease risk previously attributed to the *MYH9* gene. Human Genetics 128: 345–350. *This paper illustrates how a strong association with one gene was probably not causative, and that more likely causative mutations in another gene influence a variety of clinical traits and not just the disease that was the focus of the original studies.*

INDEX

Permian, 239; the end-Permian extinction, 16, 29, 239
Pettay, J. E., 72
phage therapy, 98–99; drawbacks of, 98–99
phenology, 239, 241
phenotypes, 376
phenotypic plasticity, 21–22, 29, 239, 242
Philosophiae naturalis principia mathematica (Newton), 267
Philosophie zoologique (Lamarck), 265
phylogenetic diversity (PD), 221, 226–227
phylogenetic relationships, 20
phylogenetic trees, 20, 325; conservation of, 226; in cultural phylogenetics, 341–342; and the evolution of languages, 323–324, 323 (figure); methods used to infer phylogenetic trees, 325
phylogenetics, 221, 223. *See also* phylogenetic trees
phylogeny, 314
phylogeography, 223–224
Pinker, Steven, 305
Plasmodium falciparum, 43, 99–100
pleiotropy, 377
Pleistocene, 62, 82, 356; environment of as highly spatiotemporally variable, 88
Pleistocene-Holocene transition, 248
Plio-Pleistocene boundary, 36
Pluckthun, Andreas, 199
Poecilia reticulatus (guppy), 27
Pogona vitticeps (bearded dragon), increase in females of with male chromosomes due to rising temperatures, 247
policing, 46, 47, 57
polio, 96; as depicted on an Egyptian bas relief (ca. 1500 BCE), 95, 96 (figure)
Pollan, Michael, 186
polymerase chain reaction (PCR), 192, 194, 196
Pope John Paul II, 270, 272
Pope Pius XII, 270
population: local population, 347, 348; population bottlenecks, population isolation, and effective population size, 232–234
population genetics, 7, 19, 20, 224–225, 232–234; and enzyme electrophoresis techniques, 19; and the "Modern Synthesis" of the 1930s and 1940s, 19
population records (historical), 62, 70–71, 73–74
population rescue: demographic rescue, 229;

evolutionary rescue, 229–231; genetic rescue, 229
population size, measurement of: census population size (N_c), 234; effective population size (N_e), 234
posture: orthograde posture, 32; pronograde posture, 32, 33
power, 47, 55–56, 57
primates, living: adaptations for suspending their body below branches in, 33–34; ancestors of shared with humans, 33; and convergent locomotor evolution, 34; knuckle-walking in, 34. *See also specific primates*
primatology, 71
Principles of Geology (Lyell), 266
Principles of Moral and Political Philosophy, The (Paley), 260
programmed death hypothesis, 115, 121
protease, 192
Provine, William, 277
psychology, 63. *See also* evolutionary psychology
pulsed-field gel electrophoresis (PFGE), 165
Puma concolor coryi (Florida panther), genetic rescue of, 229, 231; and *Puma concolor stanleyana*, 229

Qbeta, 193; and the Qbeta replicase, 193–194
quasispecies, 192, 194

rabbit, and the *Myxoma* virus, 135, 138, 138–139, 140
race, 10, 346; the absence of biological races in humans, 28–29, 353–358, 350; adaptive traits as not defining coherent populations of humans, 358–360; biological meaning of, 347–350; biological races in chimpanzees, 350–353; culturally defined racial categories, 347–348; definition of, 347; and evolutionary lineages, 349–350; and the f_{st} statistic, 349; the use of "subspecies" instead of "race" in the nonhuman literature, 348
Ray, John, 258
Rechenberg, Ingo, on evolutionary strategies, 211
reciprocity, 47–48, 53–54, 58, 59; indirect reciprocity, 47, 50, 54, 58; in nonhumans, 50
Redi, Francesco, 265
relatedness, 48
religion, 255, 256; and ancestors, 56. *See also specific religions*; argument from design;